Overlapping Cosmologies In Asia

Crossroads - History of Interactions across the Silk Routes

Edited by

Angela Schottenhammer
(*Catholic University of Leuven, Belgium*)

VOLUME 4

The titles published in this series are listed at *brill.com/cros*

Overlapping Cosmologies In Asia

Transcultural and Interdisciplinary Approaches

Edited by

Bill M. Mak
Eric Huntington

BRILL

LEIDEN | BOSTON

Cover illustration: Kanchū, *Model of Mount Sumeru Inscribed*, 1848. Woodblock print with hand coloring, 118 × 54 cm. Ryūkoku University Library.

The Library of Congress Cataloging-in-Publication Data is available online at http://catalog.loc.gov
LC record available at http://lccn.loc.gov/2021062602

Typeface for the Latin, Greek, and Cyrillic scripts: "Brill". See and download: brill.com/brill-typeface.

ISSN 2589-885X
ISBN 978-90-04-51141-5 (hardback)
ISBN 978-90-04-51167-5 (e-book)

This book is printed on acid-free paper and produced in a sustainable manner.

Contents

Acknowledgments

The inspiration for this volume arose from a series of conferences that took place around the world between 2015 and 2017, starting with the "International Workshop on Traditional Sciences in Asia 2015: An Interdisciplinary Investigation into Overlapping Cosmologies" (June 17–19, 2015) and culminating in the "International Conference on Traditional Sciences in Asia 2017: East-West Encounter in the Science of Heaven and Earth" (October 25–28, 2017), both hosted by Kyoto University under the auspices of the Hakubi Center for Advanced Research and the Institute for Research in the Humanities, supported by the Grants-in-Aid for Scientific Research of the Japan Society for the Promotion of Science (Project ID #15K01118, #15KK0050). During these many meetings, researchers from different fields and backgrounds intensely debated and discussed ideas of overlapping cosmologies, proving the value of bringing together diverse interdisciplinary and transcultural approaches like the ones in this volume. Readers interested in additional outcomes of these conferences may wish to consult the proceedings of a 2017 conference published in Bill M. Mak and Tokimasa Takeda, eds., *East-West Encounter in the Science of Heaven and Earth* 天と地の科学 (Kyoto: Institute for Research in Humanities, Kyoto University, 2019).

Figures and Tables

Figures

Tables

Abbreviations

Language Glosses

The following abbreviations indicate the languages in which glosses are given. They generally appear when there is some potential for ambiguity or when glosses from multiple languages appear together.

Av.	Avestan
Ch.	Chinese
Cpt.	Coptic
Gk.	Greek
Heb.	Hebrew
Jp.	Japanese
Lat.	Latin
MC	Middle Chinese
MP	Middle Persian
Pth.	Parthian
Sgd.	Sogdian
Skt.	Sanskrit
Tib.	Tibetan

Texts, Manuscripts, and Collections

CMD	Chen Meidong 陈美东. *Zhongguo kexue jishu shi—Tianwen juan* 中国科学技术史・天文卷. Beijing: Kexue Chubanshe, 2003.
IF	Nīsābūrī, Niẓām al-Dīn. *Kashf al-ḥaqāʾiq-i zīj-i īlkhānī.* MS 3421, Fatih Millet Kütüphanesi, Istanbul.
KB	*Kenkon bensetsu* 乾坤辨説 (A discussion on heaven and earth with critical commentaries)
NU	*Nanban unkiron* 南蠻運氣論 (*Yunqi* theory of the southern barbarians)
RR	Nīsābūrī, Niẓām al-Dīn. *Kashf al-ḥaqāʾiq-i zīj-i īlkhānī.* MS Persian 1203, Raza Library, Rampur.
S	Stein collection of the British Museum. [See Lionel Giles, *Descriptive Catalogue of the Chinese Manuscripts from Tunhuang in the British Museum* (London: British Museum, 1957); "The Silk Road Online," *International Dunhuang Project*, http://idp.bl.uk.]

Suwen *Huangdi neijing suwen* 黃帝内經素問 (Huangdi's inner classic, basic questions)

T *Taishō shinshū daizōkyō* 大正新脩大藏經 [The Taishō Tripiṭaka]. Edited by Takakusu Junjirō 高楠順次郎, Watanabe Kaigyoku 渡辺海旭, and Ono Genmyo 小野玄妙. vols. 1–85. Tokyo: Issaikyō Kankōkai, 1924–1934.

ZKJDT Guo Shuchun 郭書春, ed., *Zhongguo kexue jishu dianji tonghui. Shuxue juan* 中國科學技術典籍通彙。數學卷. 5 vols. Zhengzhou: Henan jiaoyu chubanshe, 1993.

Notes on Contributors

Ryuji Hiraoka
received his PhD from Kyushu University, Japan, in 2010 with a dissertation on Jesuit cosmology in early modern Japan. He is currently an associate professor in the Institute for Research in Humanities, Kyoto University, Japan. His research interests include the history of astronomy and cosmology and Euro-Asian scientific exchange. He has published extensively on the history of science and technology in early modern Japan, including "Jesuits, Cosmology and Creation in Japan's 'Christian Century' (1549–1650)" (2018) and, with Christopher Cullen, "The Geneva Sphere: An Astronomical Model from Seventeenth-Century Japan" (2019).

Satomi Hiyama
is an art historian with a research focus on Buddhist wall paintings along the Silk Road from the fifth to seventh centuries CE. She has a BA from Keio University (Aesthetics and Art History), MA from Tokyo University (Comparative Literature and Culture), and PhD from Freie Universität Berlin (Art History). She has engaged in research on the collection of Central Asian Buddhist wall paintings at the Asian Art Museum of the State Museums of Berlin with several fellowships from 2010 to 2016. Since 2018, she has been an assistant professor at the Hakubi Center for Advanced Research / Institute for Research in Humanities, Kyoto University.

Eric Huntington
studies relationships between visual culture, ritual, and philosophy in Buddhism. He is currently a T.T. and W.F. Chao Assistant Professor of Transnational Asian Studies at Rice University, and he previously held fellowships at Rice, Stanford, Princeton, and Washington Universities and the Saint Louis Art Museum after receiving his PhD from the University of Chicago. More of his work on cosmology can be found in his monograph, *Creating the Universe: Depictions of the Cosmos in Himalayan Buddhism* (University of Washington Press, 2018).

Yoichi Isahaya
(PhD, The University of Tokyo, 2015) is currently an assistant professor of Eurasian studies at the Slavic-Eurasian Research Center, Hokkaido University, Japan. His research interest is in the exchange of knowledge in pre-modern Eurasia, especially for the period of the Mongol empire (1206–1368), in the field

of astral sciences. His publications include "The *Tārīkh-i Qitā* in the *Zīj-i Īlkhānī*: the Chinese Calendar in Persian" (*SCIAMVS* 14 (2013): 149–258). He is also the second author of *Thābit ibn Qurra's Restoration of Euclid's Data: Text, Translation, Commentary* (with Nathan Sidoli, Cham: Springer, 2018).

Catherine Jami
is a Senior Researcher at the French CNRS. A historian of science, she has published extensively on the mathematical sciences in seventeenth- and eighteenth-century China, on the circulation of scientific knowledge between China and Europe during the same period, and on the role of Jesuit missionaries in the construction of the Qing state. She also maintains an active interest in studies of science and empire and in the spatial dynamics of knowledge, as well as in digital humanities. Since 2017, she has been the editor of the journal *East Asian Science, Technology, and Medicine.*

Bill Mak
is a research fellow at the Needham Research Institute, Bye Fellow of Robinson College, Cambridge University, and Principal Researcher at the Tsz Shan Monastery Buddhist Art Museum, Hong Kong. He specializes in Buddhist philology, history of science in India and China, and historical Sino-Indian scientific exchange. He is currently completing a book project titled *Foreign Astronomy in China, from Six Dynasties to Northern Song*, to be published by Routledge under the Needham Research Institute Series.

D. Max Moerman
is a professor in the Department of Asian and Middle Eastern Cultures, Barnard College, Columbia University. His research interests lie in the visual and material culture of Japanese Buddhism. His publications have examined such topics as the representation of pilgrimage landscapes in painting, literature, and ritual; the burial of sutras in twelfth-century Japan; the death of the Buddha in Edo-period print culture; islands of women in the history of Japanese cartography; narrative and iconographic traditions of lepers and hot springs; talismans and oaths among warriors, aristocrats, and prostitutes; and the history of the Japanese Buddhist world map.

Adrian C. Pirtea
(PhD, 2017) is a postdoctoral researcher at the Freie Universität Berlin (chair of Byzantine Studies) and at the Berlin-Brandenburg Academy of Sciences (long-term project Corpus Coranicum). His research focuses on various aspects of late antique and early medieval intellectual and religious history in the East-

ern Mediterranean, Mesopotamia, and Iran. One of his current projects studies the transfer and transformation of scientific and religious ideas about eclipses in Syriac, Persian, Arabic, and Byzantine texts on astrology and cosmology. He currently holds a Marie-Skłodowska-Curie Fellowship at the Department of Byzantine and Modern Greek Studies of the University of Vienna (2020–2022).

John Steele
is Professor of the History of the Exact Sciences in Antiquity in the Department of Egyptology and Assyriology at Brown University. His research focusses on the history of the astral sciences in Babylonia, the circulation of astronomical knowledge both within Babylonia and between cultures, and the reception of ancient astronomy from the seventeenth century to today. He is the author of several books, including *The Babylonian Astronomical Compendium* MUL.APIN (co-authored with Hermann Hunger, 2019), *Rising Time Schemes in Babylonian Astronomy* (2018), and *A Brief Introduction to Astronomy in the Middle East* (2008), which has recently been translated into Chinese and Persian.

Dror Weil
is currently University Lecturer (Assistant Professor) in History of Early Modern Asia at Cambridge University. After receiving his PhD from Princeton University in 2016, he held postdoctoral fellowships at the Max Planck Institute for the History of Science, the Berlin Center for the History of Knowledge, and Tel Aviv University and served as a University Lecturer at King's College London 2019–2021. In his studies, Weil explores the intersection of religious texts and scientific theories through the prisms of scholarly practices and translation of Arabic and Persian texts in China.

Introduction

Bill M. Mak and Eric Huntington

1 What is Cosmology?

Ask a selection of modern scholars of cosmology about the subjects of their research and you will receive surprisingly diverse responses. Contemporary cosmologists *as such* generally study the origins and workings of the universe as we have come to know it only recently, originating with the Big Bang and filled with distant objects and energies that are imperceptible without the most advanced instruments. Historians, in turn, may focus on older forms of astral science that interpret the visible planets and stars, scholars of religion may read texts that describe mythological mountains at the center of the universe, and art historians may characterize architectural structures that mimic cosmogonic processes.[1] The objects that belong to the modern study of cosmology and its history seemingly range from the grandest features of the knowable universe to the most personal human creations.

Complicating the definition of cosmology further, not all languages and cultures divide realms of knowledge in the same way. Ancient texts sometimes combine into a single discourse topics that would now be strictly divided into separate fields of cosmology, astronomy, geography, biology, physics, and philosophy. For example, Vasubandhu's *Abhidharmakośa* describes the characteristics of living beings as intimately connected to the natures of different cosmological realms through dynamic physical and causal processes.[2] Conversely, traditional systems may also distinguish topics that modern scientists might consider related. For example, some ancient Chinese authors considered the general observation of celestial phenomena (*tianwen*, heavenly pattern) to be a different discipline than the precise computation of celestial motion (*li*, astronomical system/ephemeris/calendar).[3]

In seeking to understand cosmology across periods of history, regions of the world, and different academic disciplines, then, one must begin by looking broadly enough to be inclusive but also precisely enough to support productive analyses. At a very basic level, the English word "cosmology" ultimately

1 See, for example, Umasvati, *That which Is*, 75; Hardy, "Hindu Temples."
2 La Vallée Poussin, *Abhidharmakośabhāṣyam*, 2:365–550.
3 On the way mathematical astronomy was conceived by the Chinese and the four definitions of *li*, see Sivin, *Granting the Seasons*, 38–40.

 | DOI:10.1163/9789004511675_002

originates in the Greek "κόσμος" (world) and "-λογια" (discourse), referring to topics of understanding the world that attracted diverse opinions not only among the ancient Greeks but also elsewhere in the ancient world.[4] Indeed, impulses to understand the world, and especially humanity's place within it, seem to occur in some of the earliest surviving documents from ancient cultures. The three-thousand-year-old Mesopotamian *Epic of Gilgamesh* tells the story of a man from civilization and a man from the wilderness, with the character of each determined by their place of origin.[5] The similarly ancient Indic *Ṛg Veda* divides the world into four cardinal directions and three horizontal layers, with the creation of the world sometimes characterized as an act of measurement and compared to the process of building a house.[6] Perhaps as an indication of the centrality of cosmology to human experience, similar writings on humanity, nature, and the world seem to remain relevant even in the present day. One need only compare these millennia-old texts to more recent works, such as Henry David Thoreau's *Walden* or Stephen Hawking's *A Brief History of Time*, to see that such investigations continue to be compelling to large and diverse audiences. The desire to understand our world is early, longstanding, and widespread.

Given the centrality of such thinking to human experience, it is perhaps not surprising that some key features of cosmology appear almost universally in traditions across the globe. The sun is a focal point of theorization and worship in regions from ancient Greece and Egypt to India, Japan, and South America, despite the vast separations and differences in histories between these various places. Other elements that are widely recognized as important in human existence, like the moon, water, and earth, also frequently become crucial aspects of cosmological thinking and narratives across cultures. Among these, celestial phenomena have often been a focus of special attention not just because they are experienced so widely but also because some so remarkably align to patterns of human activity while others remain provocatively inscrutable. On the one hand, phenomena like solstices and equinoxes clearly connect to the changes of the seasons and therefore agricultural cycles and other aspects of human life. In Egypt, the reappearance of the star Sirius in late summer was identified with the all-important annual flooding of the Nile as early as 3000 BCE.[7] The ability to mark and predict such events provided unquestion-

4 On the pluralism of Greek cosmologies, see Lloyd, "Greek Cosmologies," 204–208.

5 Kovacs, *The Epic of Gilgamesh*, xvii.

6 While these ideas are implied by the text, the details are actually more complicated. See Jamison and Brereton, *The Rigveda*, 35; MacDonell, *Vedic Mythology*, 11.

7 Campion, *Astrology and Cosmology*, 87.

able utility in the ancient world, and cultures that tended to deify important natural phenomena often saw gods in the features of the sky (in Egypt, Sirius was identified with the goddess Isis). On the other hand, celestial events like eclipses and the appearance of comets have less obvious relationships with human activity but seem no less worthy of explanation. In some Indian traditions, eclipses were understood as the devouring of the sun or moon by the demon Rāhu, with the particular direction at which the shadow enters the disc of the sun or moon portending disaster for specific groups of people.[8] Because the features of the sky can be both mechanically predictable and unaccountably mysterious, the study of astral science has understandably become a focal point of many kinds of cosmological inquiry across human history.

At the same time, one also quickly sees that different people from diverse regions have developed wildly varying explanations of the world, often for reasons specific to their geography or culture. Several traditions of South Asia describe the inhabited world as essentially a mass of land surrounded by great oceans, just as the Indian subcontinent is a peninsula, while the people of ancient Egypt saw their cosmos as divided into two banks across the fertile Nile river and surrounded mostly by desert and mountains.[9] Because each model purports to capture the structure of the known or knowable universe, they are influenced by local geography, agriculture, and trade. Other differences in cosmology relate more to philosophical or religious factors. To the ancient Chinese, correlative thinking across different aspects of the cosmos prompted explanations in terms of abstract structures of political hierarchy, theories of *yin* and *yang*, and the five phases (*wuxing*).[10] An alternative Buddhist view divides the world into three vertically stacked realms that correlate to specific psychological, ethical, and meditative states, modeling a path to enlightenment.[11]

2 Historical Approaches to Cosmology and Cosmologies

Tackling such breadth and complexity across cultures, modern historians of cosmology have adopted a variety of approaches. On the one hand, many have

8 Bailey, "The Demon Seer," 35–36.

9 Sircar, *Cosmography and Geography*, 33–59; Campion, *Astrology and Cosmology*, 86; Plumley, "Cosmology of Ancient Egypt."

10 Needham, *Science and Civilisation*. On the School of Naturalists and *wuxing*, see 2:232–265; on *yinyang*, see 2:273–303.

11 Gethin, "Cosmology and Meditation."

favored description, working to thoroughly explain single traditions by becoming immersed in single cultural or linguistic domains, mathematical systems, or textual or artistic sources. This research is unquestionably foundational, but because it requires such high degrees of specialization, it sometimes problematically creates artificial divisions of topics based on differences in the languages, cultural knowledge, or techniques necessary to understand these different expressions. In the field of Buddhist studies, for example, it has often been assumed that certain Chinese and Indian traditions must diverge simply because of their distinct cultural and linguistic contexts, despite the fact that they have common forebears and features, and without critically examining how this supposed divergence affects contemporary theorization or relates to historical realities.[12] Following such an essentialist division by broad cultural and linguistic groups, a typical global survey of cosmology in history divides topics into supposedly distinct regional cosmologies, with separate chapters on Babylon, Egypt, India, Scandinavia, Mesoamerica, and so on, even though it might also instructively be organized by types of calendrical cycles, notions of sacred mountains and waters, or specific applications of technology.[13]

On the other hand, focusing on the clear commonalities and differences between such cosmologies, other historians have adopted more comparative approaches. One frequent goal of comparison has been to trace the origins and dissemination of persistent or influential ideas, especially with a teleological eye toward the development of modern science from ancient traditions. Such approaches were especially popular among the encyclopedists of the nineteenth and early twentieth centuries, including Louis-Pierre-Eugène Sédillot, Moritz Cantor, and George Sarton, whose *Introduction to the History of Science* was among the most ambitious works in its field. Revising the Hellenocentrism of earlier scholars, Otto Neugebauer later convincingly demonstrated the indebtedness of Greek astronomers to their Egyptian and Babylonian predecessors.[14] Indeed, many Egyptian and Babylonian astronomical conventions remain with us even to the present day, such as the twenty-four hour clock and sexagesimal units of measurement, revealing a continuous dissemination and development of ideas over the past three millennia and across significant linguistic and cultural boundaries.

The comparativist tendency to see cosmologies in terms of origins, transmissions, and the eventual development of modern science, while clearly produc-

---

12 Gimello, "Random Reflections."

13 For examples, see Blacker and Loewe, *Ancient Cosmologies*; Campion, *Astrology and Cosmology*; Kelley and Milone, *Exploring Ancient Skies*.

14 Neugebauer, *Exact Sciences in Antiquity*; see also Rowe, "Otto Neugebauer's Vision."

tive, also tends to further encourage the view that cosmologies are essentially independent systems associated with particular cultural groups or individuals. When contact between these putative monoliths occurs, it is often assumed that one must take priority over the other, either in terms of chronology (being earlier) or in terms of scientific accuracy (being more effective), thus allowing scholars to propose decisive conflicts and linear models of transmission between supposedly disparate traditions. This almost Kuhnian analysis often precludes other ways of thinking, despite the fact that the history of cosmology is not limited to the history of science. In recent decades, historians of science have noted this problem well, especially Nathan Sivin, who proposed the concept of cultural manifolds as part of a multidimensional approach to research on ancient science.[15] As he argued, the development of a body of knowledge such as cosmology is often driven by factors extrinsic to the knowledge itself, including political, bureaucratic, and personal ones. Consequently, statements asserting that modern astronomy is fundamentally Babylonian in origin or that religious cosmologies have been rejected in favor of more convincing observational science are deeply problematic, since the real histories are usually not teleological and depend on a variety of complex factors.[16]

The traditional teleological interest in the development of modern science has also led to emphasis on Mediterranean and European traditions, such as Hellenic-Judaic-Christian cosmologies and Arabic astronomies, even though Asian cosmologies have been equally impactful on cultures both in the past and in the present, not to mention their diversity and richness in their own right. When Asian cosmologies do take primary focus, analyses often rely on Western categories and models of progress. David Pingree and Joseph Needham, for example, used comparative approaches to detail cosmological and astral traditions in India and China, respectively.[17] Particularly in the case of Needham, these efforts were often driven by a positivist agenda, giving disproportionate importance to matters that appealed to modern scientific sentiments even while affirming the value of non-Western traditions.[18]

Such focus on origins and accuracy also poses other difficulties, since both characteristics imply judgements of value and therefore incite nationalistic controversy. Some Indian, Chinese, and Japanese scholars, for example, have

15 Lloyd and Sivin, *Way and the Word*, xi; Sivin, "Multi-dimensional Approach."

16 See, for some examples, Lopez, *Buddhism & Science*, 39–72.

17 The most notable examples are Pingree, *Jyotiḥśāstra*; Needham, *Science and Civilisation*.

18 This is not to undermine the important contribution of Needham, but note the critical remarks in Nakayama, "Joseph Needham, Organic Philosopher" and Cullen, "Joseph Needham."

vigorously debated the possibilities of foreign-origin versus native-origin of their knowledge, and foreign scholars also weigh in on these issues.[19] Of course, the tendency to view history in relation to modern political boundaries is inherently problematic and tends to efface many different kinds of social identity, cultural interaction, and historical change. Recognition of this complexity has been one factor in helping scholars turn away from approaches to history built around nation-based cultural identities.[20]

As attention to questions of primacy have similarly subsided, scholars have more widely recognized that premodern cosmologies were not just abstract systems that occasionally met in brief moments of competition. In fact, many ancient cultures, especially those across Asia, were in continuous contact throughout the *longue durée* of history. It has become clear that, at any given moment in any particular region, there likely existed multiple bodies of cosmological knowledge and practice interacting with each other in myriad ways. Foreign ideas or new indigenous interpretations were regularly introduced and engaged in ongoing processes of negotiation with tradition, resulting in widely varying patterns of appropriation, conservation, and transformation. Many apparently stable cornerstones of cosmology actually never stopped evolving, as they were continually reengaged and reinterpreted by endless networks of individuals over time. Other cosmological forms seem almost to have fossilized, becoming markers of tradition or regional difference maintained without critique as novel models and practices developed around them.[21]

With such diversity in mind, one alternative to viewing cosmologies as monolithic systems becomes to focus especially on particulars: individual thinkers, texts, artworks, ideas, transmissions, and other forms of evidence that provide snapshots of cosmological thinking or historical processes in their own contexts. With this approach, a cosmology is not a coherent and broadly applicable system of ideas; rather it may be an individualized expression of a singular agenda in one specific text, artwork, or other form. In this way, one may productively contrast the cosmologies of Vasubandhu and Buddhaghosa,

---

19 For some representative scholarly works on the subject by Indian and Chinese scholars, see Subbarayappa, *Tradition of Astronomy*, 1–68, 279–342; Jiang, *Tianxue zhenyuan*. For the debate among Japanese scholars, most notably Shinzō and Eijima, see Yampolsky, "Twenty-eight Lunar Mansions." See also Needham, *Science and Civilisation*, 2:354, 3:175, 3:252–259, 3:273, *et passim*; Sivin, "Copernicus in China"; *Science in Ancient China*; and the chapter by Steele in this volume.

20 See, for example, Alpers, *The Indian Ocean*.

21 See, for example, the interaction between traditional cosmologies and more progressive theories in seventeenth-century India described by Pingree ("The Purāṇas and Jyotiḥśāstra").

two contemporary Buddhist thinkers who describe nearly identical models of the world, but do so for entirely disparate purposes in different contexts.[22] The absolutist position that there is some single abstract or rational system that unites such diverse artifacts becomes untenable.

At the same time as one looks to particulars instead of absolutes, however, the opposite problem arises: relativism. If every instance of cosmology is its own unique and fascinating expression, how can one weigh them against each other to characterize larger historical processes of innovation, stabilization, transmission, or change? Not every expression of cosmology is equal, and one must be able to explain as a matter of history not only why Einstein's theories advance on Newton's but also why traditional cosmologies of Buddhism or Daoism remain central to ethics and ritual even as religious leaders and practitioners agree that they are disproven by modern science.[23]

In the twentieth and twenty-first centuries, different academic disciplines have grappled with these problems of absolutism, relativism, and finding the in-between in their own ways. Historians of science articulated the need for a middle ground decades ago, and historians of religion and art have similarly moved through phases of Structuralism, Post-modernism, and beyond.[24] This book seeks, in part, to find new ways of thinking about this balance between absolutism and relativism by extending the discussion beyond any single field to look more generally across disciplines. By identifying values besides rationality and progress (in the history of science) or ritualism and symbolism (in the history of religions), new ways of thinking about the nature of cosmologies and their positions in history can emerge. As an edited volume, this book emphasizes finding diverse views from experts in different fields who may define cosmologies and their interactions quite differently. A cosmology may be changed simply by being expressed in a different form (see the chapter by Isahaya), or it may be defended against an opposing cosmology using the techniques of its competitor (Moerman). New cosmological knowledge may be adopted not as part of a coherent system but only slowly and piecemeal (Mak and Jami), or it may be accepted only to be interpreted in a radically different way (Pirtea). Seemingly contradictory systems may even exist unproblematically alongside each other and be used to express each other (Hiraoka and Huntington).

22 Huntington, *Creating the Universe*, 27–31.

23 See, for example, Gyatsho, *Universe*, 3–4; *Science and Philosophy*, 289.

24 Gimello, "Random Reflections," 69–70.

3 Overlapping Cosmologies

The picture that begins to emerge is one of multiple kinds of cosmologies, from abstract systems of ideas to singular expressions or practices, that relate to each other in widely varying ways, from contradiction and competition to adaptation, assimilation, translation, conflation, reinterpretation, and conservation. Even in a single time and place or a single text by a single author, one does not necessarily find the monolithic systems of isolated traditions, representing either historical culs-de-sac or teleological connections to more progressive ideas. Rather, there are widely varied bodies of cosmological knowledge and practice that have been continually reinterpreted and renewed by each person or community who has brought their own assumptions and skills to bear. As traders traveled between distant regions, intellectuals interpreted texts, or itinerant artists moved from community to community, cosmological ideas, expressions, and practices continually overlapped and transformed in myriad ways.

With this new picture of Asian cosmologies as complex and overlapping, a whole variety of new questions about the history of knowledge becomes possible. Instead of simply asking which group knew something first, more accurately, or more influentially, one can more broadly and deeply interrogate the variety of interactions that can occur when different ideas and practices meet in particular historical moments. What kinds of factors other than accuracy were considered by particular historical agents in deciding whether to adopt or reject a novel or foreign idea? Who were the mediators of knowledge, and did these people see themselves as such? Were changes in practice necessarily accompanied by changes in paradigm? How did the bodies of knowledge and practice of different kinds of people (scholars, artists, bureaucrats) interact with others? What were the roles of particular social groups, especially minorities, in preserving, destroying, developing, or communicating cosmologies?

Recent scholarship has started to address some of these issues, most notably in relation to the introduction of Western scientific cosmology to different Asian traditions, a process that often took decades or centuries and involved vigorous debates between various individuals readily identifiable in the historical record.[25] The wealth of available archival materials has provided a highly nuanced picture of cultural interactions in this relatively recent turn of events, but because the end result of these exchanges has often been the eventual acceptance of Western science, even this relatively refined story may some-

25 See, for example, Hammerstrom, *Science of Chinese Buddhism*, 50–79.

times amount to little more than a slower kind of scientific revolution, one measured not in terms of paradigm shifts but rather in terms of specific actors and intermediate processes of hybridization, adaptation, and interpretation.

In reality, even more varied encounters of cosmologies have occurred in numerous ways throughout history. Asian traditions were richly complicated, circulating and developing in dynamic, multilayered, and multidirectional interactions over time. Persian astronomers sought and read texts of their counterparts thousands of miles away (Isahaya), artists in Central Asia intermixed the cosmic mythologies of groups that met along transcontinental trade routes (Hiyama), minority Muslim communities influenced cosmological knowledge at the Ming and Qing courts (Weil), and the interpretation of such histories by modern scholars has been influenced by factors completely outside of Asia (Steele). As more and more complex interplay between traditions comes to light, it becomes increasingly difficult even to provisionally apply the traditional cultural, linguistic, and geographic names (such as "Chinese," "Persian," "Indian," and "Japanese"), let alone reify such distinctions as historical realities. For the sake of clarity, however, this volume does allow the use of such terms at each author's discretion and definition, whether it be to capture a specific geographic region, language of discourse, or cultural complex. Such labels will continue to remain useful in the service of individual arguments and histories, even as their varied usage across this volume begins to deconstruct them and show the possibility of other forms of conceptualization.

Indeed, the convergence of these discussions highlights another form of cosmological overlap that is key to this book, the coincidence and multiplication of ideas and practices not just in historical contexts but also across modern academic disciplines. In the historical sense, one may speak of overlapping cosmologies in terms ranging from the competition of scientific paradigms to the uncritical incorporation of mathematical techniques or artistic motifs. In the modern academic sense, one may consider how the slow adoption of foreign astral sciences in China (Mak) might provide insights into the hybridization of mythic and artistic cosmologies at Mogao (Hiyama). While these two examples have few cosmological concepts or practices in common, such comparisons can lead to insights into historical processes of adoption, adaptation, and hybridization that extend beyond the histories of either science or art.

It is hoped that the case studies selected for this volume, each expressing a different perspective on overlapping cosmologies, will help lay a groundwork for more significant theorization and study in the future. While subtler than the radical changes of scientific revolution, the processes described in the following pages are no less profound for understanding the history of cosmological thinking. The bodies of knowledge of pre-modern Asia were not static and conservative but active sites of interpretation by specific individuals and groups. In other words, cosmology has been a vigorous and dynamic force in Asian intellectual history.

4 Structure of the Book

In order to highlight cosmology as a dynamic intellectual tradition rather than static bodies of knowledge, this book eschews the normal organization by supposedly monolithic cultural groups or division by historical periods. Rather, given the paradigm of overlapping cosmologies, we ask the questions of what overlapped and how, who the agents were, and what resulted. The chapters of this volume are thus organized thematically to explore: i) varying dynamics of cultural interactions; ii) the importance of agents of interpretation; iii) intersections of mathematical technique; and iv) intersections of religious thought and visual imagery. Instead of presenting an historical overview of cosmology in Asia, then, these articles reveal specific subjects, methodologies, and insights into context that may help the field of cosmological studies grow in new directions. Further, beyond the interdisciplinary approach of the volume itself, several of the authors also use notably interdisciplinary methods, helping to bring the study of cosmology beyond any single field to broadly connect the history of science, religious studies, art history, and more. The result of all this is greater power to explain what a cosmology or cosmologies might be in a given context, the precise ways that cosmologies can overlap and interact, and how the theories and methods of seemingly disparate disciplines may combine to form a picture of history greater than the sum of the parts.

The first section of the book, *Transmissions of Knowledge*, examines some of the ways that ideas can travel and transform across vast cultural and linguistic boundaries, as well as the importance of perspective in understanding these cultural histories. Its chapters establish a key premise of this volume, that transmissions and overlaps of cosmologies are complex and multivalent, may happen in multiple or unpredictable stages, and must constantly be reexamined from new viewpoints.

Steele's chapter begins the volume with an example of why the continual reinterpretation of history is necessary, not just because of new sources of evidence but also because of changes in the disciplines in which historical questions are posed and answered. Analyzing not only transmissions of knowledge in Asia but also lineages of interpretation in Western scholarship, it focuses on the bases for a mistaken understanding of relationships between ancient

Babylonian and Chinese systems. Presenting three case studies of claims of Babylonian influence on Chinese astronomy and astrology, Steele reveals not only the errors in these arguments but more importantly the intellectual milieux that supported these interpretations, demonstrating the necessity of understanding transmissions of knowledge not only in the past but also in modern scholarship.

Mak's chapter continues the story by confronting the paradigm-shift model of cosmological history and showing that a number of Hellenistic astral concepts were repeatedly introduced to East Asia in multiple waves by Syriac Christians, Islamic astronomers, and the Jesuits. Bringing concepts of sexagesimal measurement, planetary weekdays, zodiac signs, and horoscopy in translated texts, these three different versions of Hellenistic astral science all carried unusual features of their own and met very different fates as they were encountered by the Chinese, who ultimately developed their own strategies to accommodate these new concepts. Foreign ideas were rejected in some cases and acculturated in others, thus challenging the idea of Hellenism as a monolithic cultural force that dispersed throughout Eurasia.

While the first section of the book highlights complex relationships between the larger cultural groups that are typically associated with cosmological bodies of knowledge, the second section, *Interpretive Communities*, draws closer attention to the importance of small groups and individuals as agents of knowledge and change, including not only mainstream scientists and scholars but also religious leaders and members of cultural minorities. Cosmologies can be understood to overlap not only in terms of vast cultures but also in terms of smaller groups with varied statuses and roles inside larger communities.

In a chapter that shows the value of combining histories of religion and science, Pirtea relates mythological stories about eclipses to astronomical concepts of lunar nodes (points in the sky where the paths of the sun and moon intersect, allowing eclipses). Examining accounts of a dragon that swallows the sun and moon to create eclipses, often treated as a kind of pseudo-planet, Pirtea reveals a fascinatingly complex history of the transmission and interpretation of ideas by particular groups for specific goals. Especially focusing on the work of Mani and his followers, Pirtea suggests a highly skilled adaptation of foreign traditions into a particular theological and soteriological system.

In the subsequent chapter on a Jesuit Aristotelean-Ptolemaic cosmology, Hiraoka argues that an author employed Chinese *yunqi* theory not to obscure the Christian origins of the knowledge but rather to make sense of novel ideas using well-established Chinese intellectual tools, an approach that was inherited and strengthened by subsequent scholars who came to form a community of interpretation. Neither wholesale adoption of foreign ideas nor a compromising syncretism, this account of a group of intellectuals interpreting and commenting on diverse theories reveals a subtler form of intellectual engagement.

Weil's contribution expands beyond a singular community to examine the complex interactions of several different groups in a single historical milieu, focusing on minority Muslims in the Ming and Qing empires. Between the fourteenth to eighteenth centuries, several different kinds of communities were responsible for transmitting and interpreting aspects of Arabo-Persian astral knowledge to various Chinese groups, including the imperial court and literati circles. Here, a special emphasis is placed on a cultural subgroup not only as a source of knowledge but also as part of a complex process of mediation.

The third section of the book, *Mathematical Techniques*, departs from these historical accounts to take a closer look at technical aspects of cosmology, revealing how the translation of foreign concepts, terms, and practices may itself be seen as a process of negotiation and overlap. Just as entire systems are not adopted in toto, so too must individual ideas and methods be interpreted and adapted to unique local circumstances, transforming in each case.

Isahaya's contribution identifies a rare example of Ptolemaic geometry being used to understand Chinese astronomical concepts in a fourteenth century Persian almanac. While some Western astronomical traditions are traditionally understood in terms of geometric representations of the heavens, Chinese systems are traditionally framed in terms of numerical values and tables. Here, the author al-Nīsābūrī inventively uses the mathematical tools of one system to explain the logic of another. The rarity of this kind of example, rather than minimizing its historical importance, gives some indication of the wide variety of interpretive strategies open to commentators on diverse traditions.

In the following chapter on the Chinese reception of the Euclidean concept of the circle in the seventeenth century, Jami identifies three separate periods of interpretation in which this foreign knowledge is adapted not only with different levels of sophistication but also in relation to different scientific instruments and technical applications. Focusing particularly on cartography and surveying, Jami shows that cosmological enterprises of ordering the world did not only focus on the heavens but also on the earth, a consideration that importantly reappears in the chapters by Huntington and Moerman. Once again, the adoption of foreign ideas is revealed not to be a wholesale and monovalent conversion to a new knowledge system. Rather, each new context and application of ideas is a new kind of overlap, providing further opportunities for adaptation and reformulation.

The final section of the book, *Religious Images*, reverses the technical focus of the third section to look at the impact of cosmological knowledge on religious and artistic traditions. While some scholars of cosmology focus almost exclusively on a history-of-science perspective, in reality, cosmology has been intimately tied to numerous other aspects of traditional life, everything from the procedures of ritual performances to the architecture of sacred temples. In focusing specifically on visual representations, this section also highlights another important medium for the transmission of knowledge beyond the expected texts and oral communications. Indeed, artistic examples often complement or contradict what is known from texts, providing unique and compelling narratives of cosmological history. This section also reveals how cosmological concepts can carry important meaning beyond providing tools for calculation or astral observation, such as establishing the basis for ritual programs at a specific site.

Moerman's chapter examines conflicts over the introduction of Jesuit cosmology to Edo Japan, focusing especially on the reactions of Buddhists who sought to justify their traditional model of the world that was centered around an enormous mountain called Sumeru. Entsū and his followers created various kinds of hybrid maps and devices that combined European cartography and science with Buddhist cosmology, exemplifying diverse kinds of accommodation, adaptation, and rejection of the new foreign worldview.

Contrasted with this account of conflict between traditional and foreign cosmologies in Japan, Huntington's chapter examines a negotiation between two mutually contradictory systems of cosmology that exist simultaneously in a purely Buddhist context. Recent mural paintings in Bhutan depict side-by-side two opposed cosmological systems, one of *abhidharma* philosophy and the other of Kālacakra tantrism. Close examination reveals that these images borrow characteristics from each other in unique kinds of conflation and comingling that could not be represented in textual sources, providing a new way of understanding how cosmologies might be understood to overlap.

Hiyama's contribution similarly examines a combination of multiple cosmological systems, this time at the Silk Road site of Mogao Cave 285 at Dunhuang. In this single place, elements of prototypically Indian and Chinese cosmological imagery were blended in an eclectic but coherent set of murals that represents another unique overlapping of cosmologies in history. While Dunhuang is already well-recognized as a site of significant multi-cultural interaction, specific relationships between cosmological bodies of knowledge must be investigated through rare examples such as this complex and extraordinary cave.

While the chapters of this volume cannot hope to capture the full complexity of interacting cosmological traditions across Asian history, they provide a thematic overview of many of the key issues, methodologies, and sources that are propelling the field into a new generation of scholarship. Looking at visual and religious sources, understanding the shifting interpretations of individual scholars, and identifying influential communities and intra-cultural interactions are all crucial to developing a more complete and nuanced picture of the intellectual history of the many ever-changing and overlapping cosmologies of history. Instead of being defined systematically by monolithic and isolated cultures or even elite individuals, cosmologies were constantly in flux. New or foreign ideas were transmitted, often without the context of an entire cosmological system, and many different kinds of people interpreted and adapted these ideas using their own bases of knowledge and as suited to diverse purposes. The notion of overlapping cosmologies provides a groundwork for ever-growing sophistication in future explorations of the cosmologies of traditional civilizations.

Bibliography

Alpers, Edward A. *The Indian Ocean in World History*. Oxford: Oxford University Press, 2013.

Bailey, Cameron. "The Demon Seer: Rāhula and the Inverted Mythology of Indo-Tibetan Buddhism." *Journal of the International Association of Buddhist Studies* 38 (2015): 33–72.

Blacker, Carmen, and Michael Loewe, eds. *Ancient Cosmologies*. London: Allen and Unwin, 1975.

Campion, Nicholas. *Astrology and Cosmology in the World's Religions*. New York: New York University Press, 2012.

Cullen, Christopher. "Joseph Needham on Chinese Astronomy." *Past & Present* 87, no. 1 (1980): 39–53.

Gethin, Rupert. "Cosmology and Meditation: From the Aggañña Sutta to the Mahāyāna." *History of Religions* 36, no. 3 (1997): 183–217.

Gimello, Robert M. "Random Reflections on the 'Sinicization' of Buddhism." *Society for the Study of Chinese Religions Bulletin* 5 (1978): 52–89.

Gyatsho, Tenzin. *Science and Philosophy in the Indian Buddhist Classics, Vol. 1: The Physical World.* Edited by Thupten Jinpa. Translated by Ian Coghlan. Somerville, MA: Wisdom Publications, 2017.

Gyatsho, Tenzin. *The Universe in a Single Atom: The Convergence of Science and Spirituality*. New York: Morgan Road Books, 2005.

Hammerstrom, Erik J. *The Science of Chinese Buddhism: Early Twentieth-Century Engagements*. New York: Columbia University Press, 2015.

Hardy, Adam. "Hindu Temples and the Emanating Cosmos." *Religion and the Arts* 20, no. 1–2 (2016): 112–134.

Hawking, Stephen. *A Brief History of Time*. New York: Bantam Books, 1998.

Huntington, Eric. *Creating the Universe: Depictions of the Cosmos in Himalayan Buddhism*. Seattle: University of Washington Press, 2018.

Jamison, Stephanie W., and Joel P. Brereton. *The Rigveda: The Earliest Religious Poetry of India*. Oxford: Oxford University Press, 2014.

Jiang Xiaoyuan 江曉原. *Tianxue zhenyuan* 天學眞原. Shenyang: Shenyang jiaoyu chubanshe, 1991.

Kelley, David H., and Eugene F. Milone. *Exploring Ancient Skies: A Survey of Ancient and Cultural Astronomy*. Second edition. New York: Springer, 2011.

Kovacs, Maureen Gallery, trans. *The Epic of Gilgamesh*. Stanford: Stanford University Press, 1989.

La Vallée Poussin, Louis de. *Abhidharmakośabhāṣyam*. Translated by Leo M. Pruden. Berkeley: Asian Humanities Press, 1990.

Lloyd, G.E.R. "Greek Cosmologies." In *Ancient Cosmologies*, edited by Carmen Blacker and Michael Loewe, 198–224. London: George Allen & Unwin, 1975.

Lloyd, G.E.R., and Nathan Sivin. *The Way and the Word: Science and Medicine in Early China and Greece*. New Haven: Yale University Press, 2002.

Lopez, Donald S. *Buddhism & Science: A Guide for the Perplexed*. Chicago: University of Chicago Press, 2008.

MacDonell, A.A. *Vedic Mythology*. Strassburg: Karl J. Trübner, 1897.

Nakayama, Shigeru. "Joseph Needham, Organic Philosopher." In *Chinese Science: Exploration of an Ancient Tradition*, edited by Shigeru Nakayama and Nathan Sivin, 23–43. Cambridge: MIT Press, 1973.

Needham, Joseph. *Science and Civilisation in China*. Cambridge: Cambridge University Press, 1954–.

Neugebauer, Otto. *The Exact Sciences in Antiquity*. Second unabridged and corrected edition. New York: Dover, 1969.

Pingree, David. *Jyotiḥśāstra: Astral and Mathematical Literature*. Wiesbaden: Harrassowitz, 1981.

Pingree, David. "The Purāṇas and Jyotiḥśāstra: Astronomy." *Journal of the American Oriental Society* 110, no. 2 (1990): 274–280.

Plumley, J.M. "The Cosmology of Ancient Egypt." In *Ancient Cosmologies*, edited by Carmen Blacker and Michael Loewe, 17–41. London: Allen and Unwin, 1975.

Rowe, David E. "Otto Neugebauer's Vision for Rewriting the History of Ancient Mathematics." *Anabases* 18 (2013): 175–196.

Sarton, George. *Introduction to the History of Science*. Baltimore: Williams & Wilkins Company, 1927–1948.

Sircar, D.C. *Cosmography and Geography in Early Indian Literature*. Calcutta: Indian Studies: Past & Present, 1967.

Sivin, Nathan. "Copernicus in China." *Studia Copernicana* (Warsaw) 6 (1973): 63–122.

Sivin, Nathan. *Granting the Seasons: The Chinese Astronomical Reform of 1280, with a Study of Its Many Dimensions and a Translation of Its Records*. New York: Springer, 2009.

Sivin, Nathan. "A Multi-Dimensional Approach to Research on Ancient Science." *East Asian Science, Technology, and Medicine*, no. 23 (2005): 10–25.

Sivin, Nathan. *Science in Ancient China*. Brookfield, Vermont: Variorum, 1995.

Subbarayappa, B.V. *The Tradition of Astronomy in India: Jyotiḥśāstra*. New Delhi: Centre for Studies in Civilization, 2008.

Thoreau, Henry David, and Jeffrey S. Cramer. *Walden: A Fully Annotated Edition*. New Haven: Yale University Press, 2004.

Umasvati. *That Which Is: Tattvartha Sutra*. Translated by Nathmal Tatia. New Haven: Yale University Press, 2010.

Yampolsky, Philip. "The Origin of the Twenty-eight Lunar Mansions." *Osiris* 9 (1950): 62–83.

CHAPTER 1

The Influence of Assyriology on the Study of Chinese Astronomy in the Late Nineteenth and Early Twentieth Centuries

John Steele

> What do these facts mean? They simply illuminate one of the ways by which Babylonian mathematics and astronomy, especially the computation of solar, lunar and planetary ephemerides, passed to China for further development.

These words, written by Joseph Needham in the third volume of his *Science and Civilisation in China* published in 1959,[1] represent the apex of a trend in the way that the history of Chinese astronomy has been written in which it is assumed that astronomy began in Babylonia and that Chinese developments in astronomy were founded upon Babylonian knowledge. Statements supporting this view appear throughout Needham's discussion of the astral sciences and stand in stark contrast to his general historiographical approach, which highlights the Chinese origin of scientific and especially technological developments that were then taken up in the West.[2]

Needham was far from the only scholar to claim a Babylonian origin for Chinese astronomy. Several late nineteenth- and early twentieth-century scholars, both in Europe and in Asia, made similar claims that all or part of Chinese astronomy and astrology had its origins in Babylonia. For example, in his *Matériaux pour server a l'Histoire Comparée des Sciences Mathématiques chez les Grecs et les Orientaux*, published in two volumes between 1845 and 1849, only a few years after cuneiform had been deciphered and long before the first astronomical cuneiform texts had been identified, Louis-Pierre-Eugène Amélie

1 Needham, *Science and Civilisation*, 3:205.

2 Finlay, "China."

 | DOI:10.1163/9789004511675_003

Sédillot repeatedly raised the question of Babylonian influence on Chinese mathematical astronomy, comparing passages from Geminos that describe 'Chaldean' astronomy with Chinese texts known to him through the work of the Jesuit Gaubil. Later in the nineteenth century, Terrien de Lacouperie, a French scholar of oriental philology who became professor of Chinese at University College, London, and Joseph Edkins, a British Protestant missionary who spent most of his adult life in China, both claimed a Babylonian origin for Chinese astronomy, although they did not always agree on the details of which aspects of Chinese astronomy were Babylonian. Both scholars saw this Babylonian origin for Chinese astronomy as part of a larger indebtedness of Chinese civilization to Babylonia, for evidence of which they drew on archaeological remains as well as philological arguments concerning both the content of texts and a supposed similarity between the Akkadian and Chinese languages. Despite the evident prejudices that both authors exhibited in their treatment of the Chinese material,[3] both authors were keen to engage with and use recent work in Assyriology, and in particular work on Babylonian astronomy. Edkins corresponded with the Rev. Archibald Henry Sayce about the latter's long paper "The Astronomy and Astrology of the Babylonians," published in the *Transactions of the Society of Biblical Archaeology* in 1874,[4] and Lacouperie refers to the same work by Sayce in the notes to his list of astronomical "elements of culture received by the civilisers of China from Babylonia and Elam" given in his book *Western Origin of the Early Chinese Civilisation from 2,300 B.C. to 200 A.D.*[5] Edkins and Lacouperie's works gave rise to a short-lived movement in European sinology that has become known as Sino-Babylonianism. Although quite quickly abandoned in Europe, Sino-Babylonianism was briefly revived in China and Japan during the first half of the twentieth century.[6] This revival of Sino-Babylonianism in East Asia may have contributed to the approach of some Japanese and Chinese historians who also sought to show that elements of Chinese astronomy originated in Babylonia.

3 Edkins, for example, made statements such as "The early Chinese astronomy looks much more scientific than it probably would do if it had been of purely indigenous origin" ("Babylonian Origin," 90), and Lacouperie remarked that the Chinese are "intelligent, but lacking in originality and creative power, deeply imbued with reverence for the ancients, and specially for those who had introduced civilisation into their land, blindly conservative and respecting precedents and routine, somnolent still in their worship of olden times" (*Western Origin*, x).

4 Edkins' letters to Sayce are preserved among Sayce's papers at the Bodlean library, Oxford. See, in particular, MS Eng. let. d. 62f. 188–189, a letter from Edkins to Sayce dated September 4, 1875, in which he discusses the reaction of Alexander Wylie to Sayce's "Astronomy and Astrology of the Babylonians."

5 Lacouperie, *Western Origin*, 9–14.

6 Hon, "Hierarchy in Time."

Sino-Babylonianism had a counterpart within the field of Assyriology: Pan-Babylonianism. Pan-Babylonianism grew out of an increasing awareness of the similarities between certain stories found in the Hebrew Bible, such as the account of the flood, and stories found in earlier cuneiform sources. The catalyst for the Pan-Babylonianist movement was a series of three lectures by the German Assyriologist Friedrich Delitzsch with the title "Babel und Bibel."[7] These lectures, which quickly became controversial because of their theological implications, were intended to demonstrate the relevance of the study of ancient Babylonia to research on the Bible and on the history of Western civilization more generally. Shortly after the first lecture, Delitzsch's student Hugo Winckler published a short book entitled *Die babylonische Kulture in ihren Beziehungen zur unsrigen* which examined the relationship between Babylonian and Western culture.[8] Winckler's book was the first of several published over the next decade or so by a small group of German Assyriologists: Winckler, Alfred Jeremias, and Heinrich Zimmern, all of whom were former students of Delitzsch and were active in Leipzig, and Jeremias's student Ernst Weidner. The Pan-Babylonians took as their central premise that the mythologies of ancient peoples all over the world were derived from very ancient Babylonian astrological and astronomical knowledge. In order to make this argument, the Pan-Babylonians had to push back the history of Babylonian astronomy to very ancient times, often claiming preposterously early dates for the composition of Babylonian astronomical texts.[9] Sharp criticism of the Pan-Babylonian arguments came from other Assyriologists and from Franz Xaver Kugler, at the time the leading scholar of Babylonian astronomy, who easily disproved the Pan-Babylonians claims for the antiquity of Babylonian astronomy and ridiculed the whole Pan-Babylonian enterprise.[10] Nevertheless, the compre-

---

7 For a detailed study of Delitzsch, his lectures, and their legacy, see Lehmann, *Friedrich Delitzsch und der Babel-Bible-Streit*. For a brief overview of the Pan-Babylonian movement, see Parpola, "Back to Delitzsch and Jeremias."

8 Winckler, *Die babylonische Kulture*.

9 See, for example, Jeremias, *Das Alter der babylonischen Astronomie*, Weidner, "Sum Alter der babylonischen Astronomie", Weidner, "Die Entdeckung der Paäzession", and Weidner, *Alter und Bedeutung der babylonischen Astronomie*. Both Jeremias and Weidner frequently invoked precession in arguments that put the date of various Babylonian star lists in the late fourth millennium BCE. Weidner, for example, dates the compendium MUL.APIN to around 3000 BCE. The earliest known copy of this composition dates to the seventh or perhaps late eighth century BCE. Recently scholarship places its date of composition in the late second or early first millennium BCE. See Hunger and Steele, *Babylonian Astronomical Compendium MUL.APIN*, 16–19 for a review of the arguments for the date of the composition of MUL.APIN.

10 See, in particular, Kugler, *Im Bannkreis Babels*. Kugler showed, for example, that the

hensive nature of the evidence presented by the Pan-Babylonians often meant that for scholars outside of Assyriology, their works were the only access to this material. In particular, Jeremias' *Handbuch der altorientalischen Geisteskultur*, first published in 1913 and then republished in a second edition in 1929, provided the most comprehensive survey of Babylonian religion and sky lore available at the time.[11] Although the Pan-Babylonians themselves did not turn to China for evidence of the spread of Babylonian astral lore, their works almost certainly influenced some early twentieth-century claims for a Babylonian origin of Chinese astronomy.

In making their claims for a Babylonian origin of a Chinese astronomy, Needham, Edkins, Lacouperie, and others sought to establish the origin and subsequent circulation of astronomy and astrology. In doing so, however, what they achieved was not an examination of overlapping knowledge and practices but rather an *overwriting* of one culture's history (the Chinese) with a flawed understanding of another's history (the Babylonian). The reasons for this are many and often contradictory: conscious or unconscious prejudice among Europeans who simply could not accept that the origin of science was in China not the West, a desire to connect China with cultures where there was clear evidence of science and therefore civilization, attempts to extend the antiquity of Chinese civilization back into earlier times, and perhaps even simple intellectual excitement when it was believed that an unexpected connection had been found. Regardless, their effect was the same—a misrepresentation of both Chinese and Babylonian astronomy and astrology.

Most of the arguments that have been made over the years for a Babylonian origin for all or part of Chinese astronomy can easily be disproved on the basis of careful reading of both the Chinese and Babylonian sources.[12] My aim in this chapter, therefore, is not simply to refute such claims but instead to exam-

---

cuneiform tablet CBS 11901, the preserved part of which contains planetary and eclipse data for part of a year whose date is lost in the broken part of the tablet, contained data for the year 425 BCE, despite Weidner's claim that the tablet dated to around 1500 BCE based upon the style of the cuneiform writing. Kugler also demonstrated the absurdity of many of the Pan-Babylonian arguments for parallels between Babylonian astral mythology and the ritual and religious practices of the Egyptians, Greeks, Jews, and Christians, ridiculing the whole enterprise by showing that, if one looks hard enough, one can find parallels between the Babylonian Epic of Gilgamesh and accounts of life of king Ludwig IX, and therefore one could irrationally argue that Ludwig IX must be a reflection of a Babylonian solar hero.

11 Jeremias, *Handbuch der altorientalischen Geisteskultur*, Jeremias, *Handbuch der altorientalischen Geisteskultur. Zweiter*.

12 See Steele, "A Comparison," and Pankenier, "Did Babylonian Astrology."

ine the intellectual context within which those claims were made in order to try to understand why scholars might have hypothesized a Babylonian origin and why such a view gained traction among other scholars. In order to do so, it will be necessary first to consider what was known of Babylonian astronomy itself in the late nineteenth and early twentieth centuries and, equally as importantly, who knew it. In the first section of this chapter, therefore, I recount the rediscovery and reconstruction of Babylonian astronomy in this period. I then present three case studies that examine claims of Babylonian influence on Chinese astronomy and astrology: one made by a German Assyriologist, one made by a Chinese Sinologist, and one made by a number of European and Japanese Sinologists. One motivation for presenting these case studies is to highlight some of the methodological problems we face when trying to study the circulation of knowledge in the ancient world. In doing so, I hope to contribute to a conversation about how we might develop a better methodology for such studies.

1 The Rediscovery of Babylonian Astronomy

Until the decipherment of cuneiform and the recovery of substantial numbers of cuneiform tablets in the middle of the nineteenth century, all that was known about ancient Babylonian astronomy—indeed, all that was known of ancient Babylonia more generally—came from Classical and Biblical sources.[13] These sources include accounts of ten Babylonian observations of lunar eclipses that were reported in Ptolemy's *Almagest* and a brief description of Babylonian lunar theory in Geminos's *Introduction to the Phenomena* as well as a range of general statements of greater and lesser believability found in the writings of classical historians and philosophers.[14] The histories of Babylonian astronomy constructed from these various sources, however, can now be seen in the light of the preserved cuneiform evidence to be not only very incomplete, which is to be expected, but also in most cases very misleading.

---

13 For a description of these sources and their contents and a discussion of how they were used to construct a (largely inaccurate) history of Babylonian astronomy before the discovery of the cuneiform material, see Steele, *Ancient Astronomical Observations*, 52–57, and Steele, "Egypt and Babylon."

14 For Ptolemy's accounts of Babylonian eclipse observations, see Steele, "Re-analysis." For references to Babylonian astronomy in Geminos, see Evans and Berggren, *Geminos's Introduction to the Phenomena* and Steele, "Geminos and Babylonian Astronomy." For general references to Babylonian astronomy and fantastical claims of its great antiquity, see Neugebauer, *History*, 609–612.

The discovery and decipherment of the first cuneiform tablets containing astronomical and astrological texts in the 1870s led to a major revision in how Babylonian astronomy was understood. The first tablets to be identified and studied were recovered from the site of the Assyrian capital Nineveh and date to the seventh century BCE, during the final years of the Neo-Assyrian empire. These tablets include texts containing celestial omens, lists of stars, simple mathematical schemes for the variation in the length of night over the year and the change in the duration of visibility of the moon over the course of a month, and letters and reports sent to the king by his scholarly advisors that describe astronomical observations and their ominous interpretation. The first extensive study of this material was undertaken by the British Assyriologist and Semitic philologist the Rev. Archibald Henry Sayce, fellow and tutor at Queen's College, Oxford and later Professor of Assyriology at Oxford University. In the third volume of the *Transactions of the Society of Biblical Antiquities*, published in 1874, Sayce presented a 195-page study entitled "The Astronomy and Astrology of the Babylonians, with Translations of the Tablets relating to these subjects".[15] Despite its title, this study focuses almost exclusively on the celestial divination series known today as *Enūma Anu Enlil*, which Sayce claims was composed for Sargon of Akkad, whom he dates to the sixteenth century BCE,[16] even though in the introduction Sayce summarizes the Classical evidence for the history of Babylonian astronomy and remarks on its great antiquity. The following year, Sayce published a short paper in the journal *Nature* entitled "The Astronomy of the Babylonians,"[17] which is again based purely upon texts from seventh-century-BCE Nineveh. In it, Sayce claims that it is to the "Accadians" (by which he means the Sumerians) that we owe the origin of Babylonian astronomy, and this knowledge was then passed to the "Semitic Babylonians" (i.e., Akkadian speaking Babylonians) who entered the region sometime in the fourth millennium BCE and conquered it by about 2000 BCE. Sayce claims that it was the Sumerians who developed astronomy and created the texts of celestial omens, as well as the zodiac, the week, the calendar, and the division of the day into twelve double-hours. Sayce and R.H.M. Bosanquet subsequently published three papers on Babylonian astronomy in the *Monthly Notices of the Royal Astronomical Society* during 1879 and 1880.[18] These papers deal with

15 Sayce, "Astronomy and Astrology of the Babylonians."

16 Modern scholarship places Sargon in the twenty-fourth or twenty-third century BCE but does not associate him with the composition of *Enūma Anu Enlil*, which seems to have been standardized as a text only in the early first millennium BCE.

17 Sayce, "Astronomy of the Babylonians."

18 Bosanquet and Sayce, "Preliminary Paper"; Bosanquet and Sayce, "Babylonian Astronomy. No. 2"; Bosanquet and Sayce, "Babylonian Astronomy. No. 3."

the structure of the Babylonian calendar, angular measurement and the so-called planisphere tablet S.162, and the so-called Venus Tablet, a collection of omens based upon the synodic cycle of Venus often assumed to be based upon observations. Sayce and Bosanquet's discussion of S.162 is the first detailed discussion of what is today known as a "Three Stars Each" text (sometimes misleadingly referred to as an "Astrolabe" text), which assigns three stars to each month of the year, one star in each of the three paths of Enlil, Anu, and Ea, which refer to the northern, middle, and southern parts of the sky.[19] Accompanying the star names are numbers that Bosanquet and Sayce took to be angular distances covered by the stars but that we now know to be related to the length of daylight in the relevant month of the year.

Our understanding of Babylonian astronomy radically changed in the 1880s.[20] Whereas Sayce only had access to seventh-century-BCE material from Nineveh, shortly before 1880 the Jesuit Johann Nepomuk Strassmaier identified cuneiform tablets from Babylon that contained astronomical texts. These tablets date to the latter half of the first millennium BCE, and whereas the tablets from Nineveh were mostly either lists of celestial omens or letters containing quite basic descriptions of observed astronomical phenomena together with their astrological interpretation, the newly discovered tablets from Babylon contained something quite different: detailed records of precise and systematic astronomical observations, predictions of future astronomical phenomena (including lunar and solar eclipses and the synodic phenomena of the planets), and texts (mostly tables) of mathematical astronomy. Strassmaier sent his transcriptions of these texts to another Jesuit, Joseph Epping, and over the next decade and a half Epping (often in collaboration with Strassmaier) published a series of papers and a book which elucidated this later phase of Babylonian astronomy.[21] What Epping showed was that Babylonian astronomy developed into a highly precise science that was both observational and theoretical. The primary concern of this astronomy was the occurrence of the regular phenomena exhibited by the moon and the planets: the duration of visibility of the moon on specific days around the new and full moon, the movement of the moon and the planets past a fixed repertoire of reference stars known today as Normal Stars, lunar and solar eclipses, and the synodic phenomena of the planets (first and last appearances, stations, and acronychal rising).

19 For the Three Stars Each texts, see Horowitz, *The Three Stars Each*.

20 For a detailed study of the rediscovery of Babylonian astronomy between 1880 and 1950, see de Jong, "Babylonian Astronomy 1880–1950."

21 See, in particular, Strassmaier and Epping, "Zur Entzifferung," and Epping, *Astronomisches aus Babylon*.

Following Epping's death in 1895, another Jesuit, Franz Xaver Kugler, continued Epping's work on the astronomical cuneiform tablets discovered by Strassmaier. Kugler published several books and numerous articles (all but one in German) between 1900 and the late 1920s in which he further elucidated both late Babylonian observational, predictive, and theoretical astronomy and newly identified compositions from earlier periods in Babylonian history such as the text known by its incipit as MUL.APIN.[22] MUL.APIN is a short compendium of astronomical and astrological texts from the early (pre-eighth century BCE) phase of Babylonian astronomy. It contains several star lists; numerical schemes for the variation in the length of day and night, the duration of the moon's visibility, and the length of the shadow cast by a gnomon; a series of intercalation rules; statements of the intervals between the first and last visibilities of the planets; and a short collection of celestial omens.[23] The text was very popular throughout Babylonia and Assyria from at least the seventh to the second century BCE. From the 1910s onwards, Kugler's work was complemented by that of the Assyriologist Ernst Weidner, who published a series of books and articles on early Babylonian astronomy and astrology, in particular dealing with star lists, the Three Stars Each texts, MUL.APIN, and celestial divination.[24]

By the 1920s, therefore, a fairly comprehensive and self-consistent picture of Babylonian astronomy had emerged. Early Babylonian astronomy was concerned with celestial omens, with listing stars, with the calendar, and with simple numerical schemes for modelling the variation in the length of daylight over the year and the change in the duration of visibility of the moon over a month. In later periods, detailed, precise, and systematic astronomical observations were made and methods for making accurate predictions of certain future astronomical phenomena were developed, along with new mathematical astronomical procedures and new astronomical concepts such as the zodiac. Although this picture of the development of Babylonian astronomy has been subject to refinement and revision, its basic outline has held up to close scrutiny over the last ninety years. However, the history of Babylonian astronomy constructed by Epping, Kugler and Weidner struggled to reach the broader community of scholars. Their work was not as easily accessible as Sayce's: they published in German, not an obstacle in itself but some of Epping's papers and his book were printed using the Gothic script, Epping's work often appeared in

22 See, in particular, Kugler, *Die Babylonische Mondrechnung*, and Kugler, *Sternkunde und Sterndienst in Babel*.

23 For an edition and detailed study of MUL.APIN, see Hunger and Steele, *Babylonian Astronomical Compendium MUL.APIN*.

24 See, in particular, Weidner, *Handbuch* and Weidner, "Ein babylonisches Kompendium."

the fairly obscure Jesuit religious journal *Stimmen aus Maria-Laach* (his book was published as a supplement to this journal), and Kugler and Weidner published their work either in book form or in Assyriology journals. Sayce, in contrast, published in the well-known science journals *Nature* and *Monthly Notices of the Royal Astronomical Society*. More significantly, Epping and Kugler undertook highly technical astronomical analyses of Babylonian astronomical texts and therefore required their readers to have a good command of astronomy in order to follow their arguments. Kugler and Weidner also combined astronomical analysis with discussions of Akkadian terminology, which again made it difficult for outsiders to access their work. Finally, Kugler's books in particular are notoriously hard to navigate: they contain what amount to a series of papers on disparate topics arranged in the order in which he undertook the research rather than presented so that the reader can easily grasp an overarching picture of Babylonian astronomy (something hampered even more by the lack of indexes). Thus, whilst a fairly comprehensive and accurate picture of Babylonian astronomy and astrology had emerged by the 1920s, this history was very hard for anyone outside of the field to access.[25]

2 Case Study 1: Celestial Omens

Probably the most widely cited and accepted claim for Babylonian influence on Chinese astral science has been the German Assyriologist Carl Bezold's conclusion that he found traces of Babylonian influence within Sima Tan's 司馬談 treatise on the celestial offices in Sima Qian's 司馬遷 *Shiji* 史記.[26] This claim was made by Bezold in 1919 in an article entitled "Sze-ma Ts'ien und der babylonische Astronomie," published in the eighth volume of the *Ostasiatische Zeitschrift*. Bezold was not the first scholar to raise the possibility of a Babylonian origin for Chinese celestial omens, however. Already in 1885, Joseph Edkins had written "If the question be asked, is Sï ma Ch'ien's astrology of native origin or not, I should reply that it is a foreign astrology adapted to China", although, he remarks a few paragraphs later that "Sï ma Ch'ien himself had no idea that astrology was foreign in its origins."[27] For Edkins, that foreign origin was Babylonia.

25 For a detailed study of the reception of the history of Babylonian astronomy among historians of science, see Rochberg, "Consideration of Babylonian Astronomy."

26 Bezold, "Sze-ma Ts'ien." For a detailed critical analysis of Bezold's paper, see Pankenier, "Did Babylonian Astrology". See also Brown, *Interactions*, 566–570, who did not know of Pankenier's work but reached similar conclusions.

27 Edkins, "Babylonian Origin."

Edkins' argument for a Babylonian origin of Chinese astrology was based around omens that concern the color of planets and the prominence of Mercury among the omens. Edkins summarized these omens as follows: "A white colour in Mercury indicates drought. Yellow marks good harvests. Red denotes war. Black indicates water. If Mercury and Venus come out together in the east and are both red with singular brightness, foreign powers will be defeated and China will be victorious."[28] Edkins then remarked that he wished "to direct particular attention to the use of the five colours and the position held by Mercury in Chinese astrology as proofs of Babylonian connection."[29] Edkins did not state why this should be proof of a Babylonian origin, nor did he explain why five colors should be significant when above he had only listed four. It seems likely, however, that Edkins was drawing upon a Babylonian text containing lunar eclipse omens that was translated by Sayce in his "Astronomy and Astrology of the Babylonians," which Edkins mentioned earlier in his paper.[30] However, Sayce translated these colors as white, black, dark-blue, greenish-yellow, and pale-yellow, which do not fully agree with the colors given by Edkins for the Chinese omens, which are white, yellow, red, and black. Thus, Edkins' conclusion was probably based more upon his preconceptions, evident in statements such as "Astrology itself is not germane to Chinese thought, and the great minds of China have never worked in this sort of groove",[31] than his comparison of the Chinese and Babylonian texts.

The idea of a possible Babylonian origin of Chinese celestial omens was raised again in 1905, this time by an Assyriologist. Morris Jastrow was born in Poland but spent most of his professional life associated with the University of Pennsylvania. In his *Die Religion Babyloniens und Assyriens*,[32] Jastrow noted some similarities between Babylonian celestial omens and those found in the *Shiji* as he knew it from the French translation by Édouard Chavannes in his *Les Mémoires Historiques de Se-Ma-Ts'ien*.[33]

Jastrow's idea to compare the Babylonian omens with those in the *Shiji* was taken up by Carl Bezold and formed the basis of a 1919 paper. Bezold had trained in Assyriology under Friedrich Delitzsch and spent six years working on cuneiform tablets at the British Museum in order to produce the first catalogue of the museum's Kuyunjik collection of tablets from Nineveh before taking up

28 Edkins, "Babylonian Origin," 93.
29 Edkins, "Babylonian Origin," 93.
30 Sayce, "Astronomy and Astrology of the Babylonians," 223.
31 Edkins, "Babylonian Origin," 93.
32 Jastrow, *Die Religion Babyloniens*, 745–746.
33 Chavannes, *Les Mémoires Historiques* 6, 339–412.

the professorship in oriental philology at the University of Heidelberg in 1894.[34] A mainstream and well-respected Assyriologist, Bezold's interests were wide ranging and included Babylonian celestial divination and the identification of Babylonian stars and constellations, although he did not venture into the history of technical Babylonian astronomy.

Although not a Pan-Babylonian himself, Bezold was certainly familiar with their work through his connection with Deliztsch and his place within the German Assyriological community, and he may well have been subconsciously influenced by the general Pan-Babylonian idea of diffusion from Babylonia to the rest of the world. More significantly, however, one of Bezold's colleagues at Heidelberg was the Classical philologist Franz Boll, who worked extensively on ancient astronomical and astrological texts. Bezold and Boll collaborated on several projects during the 1910s, sometimes with the assistance of the astronomer August Kopff, also a professor at Heidelberg. A common theme of these collaborations was exploring links between Babylonian and Greek astronomy and astrology.

Bezold and Boll's first collaboration led to the publication in 1911 of *Reflexe astrologischer Keilinschriften bei griechischen Schriftstellen* as a volume in the series *Sitzungsberichte der Heidelberger Akademie der Wissenschaften*.[35] In this work, Bezold and Boll searched for traces of Babylonian celestial omens in later Greek texts in order to demonstrate a connection between Babylonian and Greek celestial divination. They compared examples from the Babylonian celestial omen series *Enūma Anu Enlil* as it is preserved in tablets from the Neo-Assyrian capital Nineveh, dating to the seventh century BCE and containing copies of the series and quotations of individual omens in reports sent to the Neo-Assyrian king by his scholarly advisors, with omens contained in a variety of Greek texts dating from the third century BCE to the sixth century CE. Bezold and Boll began by discussing general concepts and individual words before moving on to common descriptions of astronomical phenomena and finally studying complete sentences and groups of sentences that they claimed demonstrate sufficient agreement to show that the Greek texts they examined drew directly from the Babylonian text *Enūma Anu Enlil*. In doing so, Bezold and Boll aimed to build up an argument that progressed from weaker to stronger evidence, leading to their final conclusion that Greek texts of celestial divination were directly dependent upon the cuneiform material.

---

34 Budge, *Rise and Progress*, 226–227.

35 Bezold and Boll, *Reflexe*.

Bezold and Boll's project anticipated later work on the circulation of astronomical and astrological material between Babylonia and the Greek world that has clearly demonstrated the wide-ranging influence of the Babylonian astral sciences on their Greek counterparts.[36] As recently discussed by Misiewicz, however, Bezold and Boll's approach suffered from several methodological problems.[37] In particular, Bezold and Boll were not sufficiently critical in establishing direct influence as opposed to simple similarity of ideas or concepts that could have arisen independently from an interest in celestial divination shared by the Babylonians and the Greeks. Furthermore, Bezold and Boll simply cherry-picked examples from the texts that supported their argument of Babylonian influence on the Greek material and did not discuss the many cases that do not support this conclusion.

Bezold and Boll's study of Babylonian influence on Greek celestial divination provided the model for Bezold's later study of Babylonian influence on early Chinese celestial divination. Indeed, the argument presented in his 1919 paper followed more or less directly the form of argumentation that he and Boll had used in their earlier work. Bezold began by presenting a series of claimed common themes in the type of astronomical phenomena found in Babylonian and early Chinese omens, although he noted that these were not firm evidence of influence in themselves, echoing remarks in his and Boll's publication about the relatively minor significance of basic similarities between general Babylonian and Greek themes within divination texts. Next, Bezold examined a number of examples of specific omens from *Enūma Anu Enlil* that he claimed were sufficiently similar to omens found in Sima Tan's treatise to show evidence of Babylonian influence. Although, according to Budge, Bezold studied Chinese as a schoolboy,[38] for the text of Sima Tan's treatise, Bezold relied purely upon the French translation by Édouard Chavannes in his *Les Mémoires Historiques de Se-Ma-Ts'ien*.[39] As pointed out by Pankenier,[40] there are problems with Chavannes' translations of some of the technical terms, which may have misled Bezold. But, as Pankenier has also pointed out, the bigger problem was with Bezold's overall approach. Just as he had done in his work with Boll, Bezold went looking for parallels and was not sufficiently critical in determining influence as opposed to mere similarity. Furthermore, Bezold again cherry-picked

36 See, for example, Jones, "Babylonian Arithmetical Schemes"; Pingree, *Astral Omens to Astrology*; and Pingree, "Legacies."

37 Misiewicz, "Mesopotamian Lunar Omens."

38 Budge, *Rise and Progress*, 226.

39 Chavannes, *Les Mémoires Historiques* 6, 339–412.

40 Pankenier, "Did Babylonian Astrology."

the evidence, presenting only a few examples that he thought showed evidence for a Babylonian influence and not discussing the many more omens in Sima Tan's text that do not have any similarity with omens known from Babylonia.

3 Case Study II: The Names of the *Tiangan* and the *Dizhi*

The ten *tiangan* 天干 (heavenly stems) and the twelve *dizhi* 地支 (earthly branches) have several important uses in China, including designating the intervals that divide a 24-hour period into twelve double-hours and being combined to form the *ganzhi* sexagenary cycle used to count days and years. The origin of the names for the stems and branches is not certain, however.[41]

An early suggestion of a Babylonian origin for the names of the ten stems was made by Lacouperie in a short note published in *The Academy* in 1883.[42] Lacouperie argued that the names were sufficiently similar to the Akkadian words for the numbers one to ten as to "prove unmistakably that they are part of the stock of scientific notions and elements of culture" borrowed from Babylonia. In support of this assertion, Lacouperie reported that the Assyriologist Theophilus Pinches, at that time a curator at the British Museum, had provided him with the Akkadian words for the numbers one to ten.[43] The following year, Edkins published a reply in *The Academy* in which, while he was supportive of the general idea of the Babylonian of Chinese astronomy, he argued that the Akkadian words were not sufficiently in agreement with the Chinese to support Lacouperie's claim.[44]

A more detailed argument for a Babylonian origin for the names of the branches was put forward by the Chinese scholar Guo Moruo 郭沫若 in 1931.[45] Guo, an expert on ancient Chinese history, archaeology, and paleography who had studied in Japan, undertook a detailed investigation of a possible Babylonian origin for the names of the twelve branches. Guo proposed that the names for the branches were derived from the Babylonian names for the signs of the zodiac. He argued that the process by which the Babylonian names were transformed into the names for the branches varied: sometimes the Chinese name

41 For recent suggestions, see, for example, Pulleyblank, "The *Ganzhi* as Phonograms"; Pankenier, "Getting Straight with Heaven"; and Smith, "*Di Zhi* 地支 as Lunar Phases."

42 Lacouperie, *Western Origin*.

43 Lacouperie, *Western Origin*, 145.

44 Edkins, "Chinese Cycles."

45 Guo, *Jiagu Wenzi Yanjiu*, 243–282. For a further discussion of Guo's work, see Wang, "Guo Moruo."

was simply based upon the sound of the Akkadian word, sometimes the Chinese name was intended to convey the meaning of the Akkadian word, and sometimes their Chinese character was based upon a visual representation of the Akkadian word.

Guo began his discussion by quoting the list of zodiacal constellations given in a text he identified as CT XXXIII, 1–8. This text is a copy of the early astronomical compendium MUL.APIN and was known to Guo through a discussion of its contents by Jeremias in the second edition of his *Handbuch der altorientalischen Geisteskultur* (cited by Guo as "HAOG"). Guo faithfully reproduced the list of seventeen zodiacal constellations presented in MUL.APIN as given by Jeremias and summarized Jeremias' discussion of the list.[46] Guo then presented his analysis of the twelve branches beginning with *yin* 寅 and then cycling through the remaining branches in order. Guo assumed that each branch corresponded to a single sign of the zodiac, with *yin* corresponding to Virgo and then moving backwards through the cycle of the twelve signs. On pages 214–228 of his *Handbuch der altorientalischen Geisteskultur*, Jeremias presented a long discussion of the constellations within each sign of the zodiac and, with one or two odd exceptions (perhaps simply mistakes), Guo used this list as the basis of his discussion. In his discussion, however, Jeremias bundled together the names of zodiacal signs and constellations from a variety of sources, including the stars lists found in MUL.APIN, a text that predated the seventh century BCE, and much later observational texts from the last few centuries BCE. The stars and constellations mentioned by Jeremias, therefore, were simply stars and constellations that he assumed were located within a sign of the zodiac rather than the names of the signs of the zodiac used by the Babylonian themselves.

The concept of the zodiac as a division of the zodiacal band into twelve equal-length parts was developed in Babylonia in the late fifth century BCE. It is worth noting that Guo knew that the *ganzhi* were already in use in China at the time of the Shang oracle bones, nearly a millennium before the development of the Babylonian zodiac. It seems that Guo did not know that the Babylonian zodiac was only developed at this late date.

Three systems for naming the signs of the zodiac are attested in Babylonia: by number, by the name of the month when the sun would be located in the sign, and, most commonly, by the name of a constellation located within that sign. Whilst there was initially some variability in the constellation names

46 The list is found at lines I iv 33–37 in the edition of MUL.APIN by Hunger and Steele, *Babylonian Astronomical Compendium MUL.APIN*.

used to indicate the signs of the zodiac, a more or less standard list emerged by the third century BCE.[47] Table 1.1 compares the Babylonian names for the signs of the zodiac given by Guo with both the standard and alternative names for the signs of the zodiac attested in cuneiform texts and shows that there are some significant discrepancies between what Guo thought were the Babylonian names for the zodiacal signs and their actual names. Most significantly, Jeremias, and hence Guo, placed the constellation the Field within Aries, but in fact this constellation was sometimes used as a name for Pisces. Guo explained the name of the branch *wei* 未 through the meaning grain, which could then be naturally associated with the Field constellation. However, in his system, *wei* corresponded to the zodiacal sign Aries, not the required Pisces. Coupled with other methodological problems in Guo's approach—his arguments based upon similar sounding words often do not distinguish between the Akkadian words themselves and the writing of some of these words using Sumerian logograms (the former are written in italics and the latter in upper case letters in transcriptions)—the incorrect assumptions concerning the Babylonian names for the signs of the zodiac force us to reject Guo's conclusion that the names of the branches were derived from the Babylonian zodiac.

TABLE 1.1 A comparison of the constellations assumed by Guo to refer to the signs of the zodiac with the standard and alternative names found in Babylonian texts

Branch	Zodiacal sign	Constellation names as given by Guo	Standard Babylonian name	Alternative Babylonian names
yin 寅	Virgo	GIŠ.BAN 'The Bow' ŠU.PA 'ŠU.PA'	ABSIN 'The Furrow'	
mao 卯	Leo	UR 'The Lion' *šarru* 'The King'	A 'The Lion'	
chen 辰	Cancer	*allul* 'The Crab' KAK.SI.DI 'The Arrow' *kakku*	ALLA 'The Crab'	
si 巳	Gemini	MAŠ.TAB.BA.GAL.GAL 'The Great Twins' MAŠ.TAB.BA.TUR.TUR 'The Small Twins' ŠIB.ZI.AN.NA 'The True Shepherd of Anu'	MAŠ.MAŠ 'The Twins'	ŠIPA.ZI.AN.NA 'The True Shepherd of Anu'

47 Steele, "Development of the Babylonian Zodiac."

TABLE 1.1 A comparison of the constellations assumed by Guo to refer to the signs of the zodiac (*cont.*)

Branch	Zodiacal sign	Constellation names as given by Guo	Standard Babylonian name	Alternative Babylonian names
		izzazzu		
		tuâmu-rabuti 'The Great Twins'		
wu 午	Taurus	*zappu* 'The Bristle'	MÚL.MÚL 'The Stars/The Bristle'	GU_4.AN 'The Bull'
		GÙ.AN.NA 'The Bull of Heaven'		
		narkabtu		
		šûrê nakabti		
wei 未	Aries	KU.MAL	ḪUN 'The Hired Man'	LU 'The Sheep'
		AŠ.GAN 'The Field'		
		E.KUE		
shen 申	Pisces	ŠIM.MAḪ 'The Swallow'	*zib*.ME 'The Tails'	AŠ.GAN 'The Field'
		ŠA.AM.MAH		
		anunitu 'Anunitu'		
		nûnu 'The Tails'		
		zibbâti		
you 酉	Aquarius	*gula* 'The Great One'	GU 'The Great One'	
xu 戌	Capricorn	SUḪUR.MAŠ 'The Goat-fish'	MÁŠ 'The Goat-fish'	
		zababa		
		ENZU		
hai 亥	Sagittarius	PA.BIL.SAG 'PA.BIL.SAG'	PA 'Pabilsag'	
zi 子	Scorpio	GIR.TAB 'The Scorpion'	GÍR.TAB 'The Scorpion'	
chou 丑	Libra	ZI.BA.AN.NA 'The Balance'	RÍN 'The Balance'	
		zibanitu 'The Balance'		

Guo's analysis raises two interesting points, however: Guo's willingness to consider and then support a Babylonian origin for a fundamental part of Chinese science and the problems faced by scholars using material from another field. I am not competent to discuss Guo's intellectual background and his motivations in detail, but a few of general points are worth making. First, it is probably not insignificant that Guo was based in Japan when he wrote his article. In 1900, two Japanese nonacademic historians, Shirakawa Jirō and Kokubu Tanenori, incorporated detailed summaries of Lacouperie's *Western Origin of the Early Chinese Civilisation* in their *Shina bunmei shi*, a history of Chinese civilization, omitting any mention of the criticisms of Lacouperie's work that had been made by European sinologists.[48] Secondly, in 1921, the Swedish archaeologist Johan Gunnar Andersson discovered prehistoric remains in Henan Province of what became known as Yangshao culture.[49] Andersson suggested that this culture had a Western origin, and Guo knew Andersson's work. Finally, it may also not be without significance that Guo relied on Jeremias's *Handbuch der altorientalischen Geisteskultur* for his knowledge of the Babylonian zodiac. Jeremias was, of course, a leading Pan-Babylonian scholar, and the *Handbuch der altorientalischen Geisteskultur* contains Pan-Babylonian arguments.

Guo's work also provides a clear illustration of the problems faced by scholars using material from another field. The correct names for the Babylonian signs of the zodiac had already long been established and were available in the publications of Epping and Kugler. But Guo apparently did not have access to their publications, which were printed in much smaller numbers and intended for a specialist readership—and as a consequence were less widely known and less accessible to non-specialists than Jeremias' more problematic publication. In short, Guo was as much led astray by the limitations on which Assyriologists he was able to read as he was by the methodological approach he took.

4 Case Study III: The Origin of the *Xiu*

The twenty-eight *xiu* 宿 (lodge) system is fundamental to Chinese astronomy and astrology, serving as a way of dividing the sky into regions that are arranged in sequence along the celestial equator. Each *xiu* spans a range of right ascension and is defined by a particular star at its beginning. The width of the *xiu* varies widely from only a few degrees to more than twenty degrees of right ascension. The Chinese lodge system shares some characteristics with the Indian system of twenty-seven or twenty-eight *nakṣatra* and the Arabic system of twenty-eight *manzil*, and several (but not all) nineteenth- and early twentieth-century scholars argued that the three systems are variations of a single original system. For these early scholars, the difficulty was in establishing which of the three systems, the Chinese *xiu*, the Indian *nakṣatra*, or the Arabic *manzil*, was the original. For example, Biot, in a series of untitled notes published in the *Journal des Savants* of 1840, argued that the Chinese *xiu* system was created in the middle of the third millennium BCE in China and then passed into India, where it became corrupted.[50] Other authors suggested the

48 Hon, "Hierarchy in Time."

49 Andersson, *An Early Chinese Culture*.

50 Biot, untitled notes.

opposite, namely that the Indian *nakṣatra* was the original system and the Chinese learned the system from India. Still others have argued that all three systems are descended from a fourth system. That fourth system was often assumed to have originated in Babylon.

The earliest detailed claim for a Babylonian origin for the *xiu* system was made by A. Weber in his two-part study *Die vedischen Nachrichten von den naxatra (Mondstationen)*, published between 1860 and 1862.[51] Writing more than a decade before the first study of Babylonian astronomy that was based upon cuneiform evidence was published, Weber relied instead upon the description of "Chaldean" astronomy given in the mid-first-century BCE Greek author Diodorus Siculus' *Bibliotheheca Historica*.[52] According to Diodorus, the Babylonians divided the stars into three groups: the twelve stars of the zodiac followed by twenty-four stars, half of which are northern stars and half southern stars. Weber took these twenty-four stars to be an early version of the *xiu*, although there is nothing in Diodorus' text to suggest that the stars are arranged in sequence to define regions of right ascension.

In 1891, Fritz Hommel published a detailed linguistic analysis of Arabic star names, especially those used in the *manzil* system, and compared these with those used in the *nakṣatra* and *xiu* systems.[53] Hommel was an oriental philologist who studied cuneiform under Delitizch and published extensively on Ethiopic and Arabic literature as well as studies of the Sumerian language and ancient near eastern history. Hommel considered the question raised by Weber of whether the *manzil* and other systems were based upon an original Babylonian system. He briefly discussed the twenty-four stars mentioned by Diodorus before turning to the cuneiform tablets for evidence. Citing Epping's recently published book *Astronomisches aus Babylon*, Hommel compared the names and identifications of twenty-four of the so-called Normal Stars with the Arabic *manzil*, finding moderate agreement between the two lists of stars, although significantly less between the Babylonian Normal Stars and the *nakṣatra* and the *xiu*. Nevertheless, Hommel claimed the *manzil*, *nakṣatra*, and *xiu* systems could all be traced back to a twenty-four-star Babylonian version. As was made clear by Epping, however, the Normal Stars are distributed through the zodiacal band, close to the ecliptic, and are unconnected to the celestial equator. Thus, while some of the Normal Stars may appear as determinative stars in the *xiu* and other systems—which is not surprising since both the Normal Stars and

51 Weber, *Die vedischen Nachrichten*.

52 For a recent study of Diodorus' description of Babylonian astronomy, see Jones and Steele, "Diodorus on the Chaldeans."

53 Hommel, "Ueber den Ursprung."

the determinative stars are chosen because they are reasonably bright and the celestial equator and the ecliptic cross one another—the two groups of stars operate quite differently.

Hommel's conclusion was criticized by Thibaut for this reason and because only a very small number of stars in Hommel's supposed twenty-four star Babylonian version agree with those in the Chinese *xiu* system.[54] The German born Thibaut, having worked briefly in England, spent most of his career working at colleges in India. He worked extensively on Sanskrit astronomical and mathematical texts, publishing studies and translations of several key Sanskrit texts. In a long essay published in the *Journal of the Asiatic Society of Bengal* entitled "On the Hypothesis of the Babylonian Origin of the So-called Lunar Zodiac," Thibaut provided a detailed criticism of Hommel's conclusions in which he demonstrated a detailed understanding of Babylonian astronomy—much more detailed than Hommel's—based upon the work of Epping. Thibaut's arguments were clear and to the point. First, he noted there is no evidence that either the Babylonians or any of the supposed descendants of the Babylonian system ever considered there to be twenty-four lunar stations. There were clearly more than twenty-four Babylonian Normal Stars, and there is no evidence for the *manzil*, *nakṣatra*, and *xiu* systems having less than twenty-seven or twenty-eight stations. Furthermore, the idea of twenty-four lunar stations makes no sense because the sidereal period of the moon is between twenty-seven and twenty-eight days, thus we should expect (as indeed we have in the Arabic, Indian, and Chinese systems) either twenty-seven or twenty-eight lunar stations. Secondly, Thibaut said that, as the *manzil*, *nakṣatra*, and *xiu* systems are used as subdivisions of the celestial sphere, we should expect to find a similar subdivision of the path of the sun, moon, and planets into twenty-four parts in cuneiform texts. However, all of the texts published by Epping clearly demonstrate that the path of the sun, moon, and planets was divided only into the twelve signs of the zodiac, never into twenty-four. Finally, Hommel's claim that several of the names of the Babylonian Normal Stars correspond to the names used for the *manzil*, *nakṣatra*, and *xiu* is not convincing—more cases disagree than agree. Thibaut's discussion shows he had a significantly better understanding of both Babylonian astronomy and the *manzil*, *nakṣatra*, and *xiu* systems than any other author up to the second half of the twentieth century.

Despite Thibaut's perceptive criticism of Hommel's arguments, the idea of a Babylonian origin for the *manzil*, *nakṣatra*, and *xiu* systems was revived on

54 Thibaut, "Hypothesis of the Babylonian Origin."

several occasions in the early to mid-twentieth century, both in Europe and in East Asia. In Japan, for example, Iijima Tadao argued that the *xiu* and *nakṣatra* systems originated in Babylon around 400 BCE, at the time when the star α Ari was at the vernal equinox.[55] Iijima did not provide any evidence from Babylonian texts for the existence of a Babylonian lodge system. Indeed, the citation he made in support of his argument was to Weber's *Die vedischen Nachrichten von den naxatra* (*Mondstationen*) of 1860, written before any of the cuneiform material had been studied. Iijima's reliance on Weber's work is surprising given that two years earlier Shinjō Shinzō had severely (and correctly) criticized Weber's work and dismissed the idea of a Babylonian lodge system.[56] Shinjō cited the three articles on Babylonian astronomy by Sayce and Bosanquet (omitting the name of the latter) in the *Monthly Notices of the Royal Astronomical Society* and Jeremias's *Handbuch der altorientalischen Geisteskultur* in his discussion—out of date and not especially reliable sources for the history of Babylonian astronomy, but better than nothing.

As far as I know, the final attempt to revive the theory of a Babylonian origin of the *xiu* appeared in Needham's *Science and Civilisation in China*.[57] Although Needham accepted Thibaut's criticism that only a small number of the stars that appear in Hommel's proposed Babylonian lodge system agree with those in the *xiu* system, he argued that we should not necessarily expect exact agreement between the stars in the two systems and that there are other grounds for concluding a Babylonian origin for the system. Needham then described (and reproduced a drawing of) a part of a circular tablet from Nineveh containing a Three Stars Each text:

> Assyriologists have long been familiar with a number of cuneiform tablets which are preserved in the library of King Assurbanipal (Ashur-bāni-apli, –668 to –626) at Nineveh, but which date as to contents from the late –2nd millennium. These show diagrams of three concentric circles, divided into twelve sections by twelve radii. In each of the thirty-six fields thus obtained there are constellation-names and certain numbers, the exact significance of which has not been yet explained. ... Could they not be regarded as primitive planispheres, showing both the circumpolar stars and the equatorial 'moon-stations' corresponding to them?

---

55 Iijima, *Shina rekihō kigenkō*, 595–599; for an English translation of Iijima's discussion, see Yampolsky, "Origin."

56 Shinjō, *Tōyō temmon gaku shi kenkyū*, 215–224; for an English translation of Shinjō's discussion, see Yampolsky, "Origin."

57 Needham, *Science and Civilisation*, 3:254–257.

> Such a view seems to be supported by the most recent researches. The tablets with 'planispheres' belong to the class now known as that of the 'Enuma (or Ea) Anu Enlil' series. The corpus contains some 7000 astrological omens, and the time of its formation was contemporary with the Shang period, i.e. from about –1400 to –1000.[58]

Needham conflated several different Babylonian texts and concepts in this statement. The 'planispheres' are not directly connected to the omen series *Enūma Anu Enlil*, as is clear from the work of Weidner and van der Waerden, which he cited in a footnote.[59] Instead, these texts, which have the ancient name "Three Stars Each," give the names of three stars, one from each of the paths of Enlil, Anu, and Ea, for each month of the schematic 360-day year. Furthermore, the series *Enūma Anu Enlil* is always known by that name, never Ea Anu Enlil. The Akkadian word *enūma* means "when," and the name of the series translates as "When the gods Anu and Enlil." This name is taken from the first three words of the introduction to the series, which gives a very brief account of the creation. Needham, however, glossed the name of the series in a footnote by saying that Ea, Anu, and Enlil were the gods of Elam, Akkad, and Amurru. Since the three paths mentioned in the Three Stars Each texts are named for these three gods, Needham has clearly conflated the two types of text. Needham then explained that the three paths correspond to stars located near the celestial equator (Anu stars), north of the equator (Enlil stars), and south of the equator (Ea stars) and that a longer list of stars in these three paths is given in the work MUL.APIN, which he dated to about 700 BCE. Needham then came to the core of his argument:

> Now in these texts there is never any mention of any zodiac or of constellations lying along the ecliptic; the earliest documentary evidence of this conception occurs just after –420. On the other hand, the Seleucid Babylonian cuneiform texts of the –3rd and –2nd centuries give great prominence to the zodiac, and use ecliptic coordinates exclusively. … One might fairly surmise, therefore that the equatorial moon-stations of East Asia originated from Old Babylonian astronomy before the middle of the –1st millennium and probably a long time before.[60]

58 Needham, *Science and Civilisation*, 3:254–256.

59 Weidner, *Handbuch*; van der Waerden, "Babylonian Astronomy. II."

60 Needham, *Science and Civilisation*, 3:256.

Whilst Needham was correct that the zodiac as an abstract mathematical system for dividing a band around the ecliptic into twelve equal parts is a late fifth-century BCE invention, early texts do include lists of the zodiacal constellations that lie along the ecliptic. MUL.APIN, for example, contains a list of zodiacal constellations that stand "in the path of the moon," and many omens that mention the position of a planet in a zodiacal constellation are found in *Enūma Anu Enlil*. Furthermore, the existence of three paths centered on the celestial equator is not evidence for a lodge system.

Needham's insistence on a Babylonian origin for the *xiu*, and indeed his wider claims of Chinese astronomy's indebtedness to Babylonia, is puzzling. Not only does it seem to go against the broader goals of his work, which aimed to highlight how advanced Chinese science and technology were compared with their Western counterparts, but in his discussions, Needham often cited authoritative works on Babylonian astronomy by scholars including Kugler, Weidner, Neugebauer, and van der Waerden that do not support the conclusions he drew. Indeed, his presentation of the Babylonian material suggests only a very vague understanding of the work of the scholars he cited, leading one to wonder whether he actually read or understood them. Given the vast scope of Needham's project, it should not come as a surprise that he perhaps did not read works on a tangentially related topic as carefully as he might have. Nevertheless, it remains strange that he then made such strong claims for Babylonian influence on Chinese astronomy and astrology.

5 Final Remarks

As has been pointed out many times, all writing, including all history writing, is influenced by the context in which it is written. The historian's own prejudices, bias, and training constrain the types of questions that are asked, the approach used to answer them, and the answers themselves. I am a British historian of astronomy currently employed at a university in the USA, subject to all the biases that accompany that identity. I am primarily a historian of Babylonian astronomy who also has an interest in Chinese astronomy, but the shortcomings in my knowledge of the latter will be even more apparent to specialists than they are to myself. On a practical level, my background and training mean that I read English and Akkadian cuneiform, can get by with literature in French and German, but do not read Classical Chinese (except at the most basic level) or modern Chinese or Japanese. This chapter itself, therefore, may exhibit many of the problems that I am trying to highlight in how we study the circulation of knowledge between different parts of the world. For example, in some places I have had to rely upon translations and even summaries of the work of others rather than being able to engage with these works themselves. Furthermore, it is much easier to look back to see the biases and methodological problems in the work of earlier scholars than it is to identify them in one's own work. Thus, I intend my remarks to be understood not as a criticism of earlier scholars but rather as an attempt to think about how we can improve our methodology for studying the circulation of knowledge.

It is evident that the problems that can be identified in many late nineteenth- and early twentieth-century arguments for a Babylonian origin of Chinese astronomy and astrology can be ascribed to conscious or unconscious subscription to various ideologies: European Orientalism, nationalism, Pan-Babylonianism, and so on. In some cases, the effect of these ideologies is not direct. As I have discussed, several of the studies examined in this chapter rely for their information about Babylonian astronomy either on Sayce's very early "The Astronomy and Astrology of the Babylonians" or on the Pan-Babylonianist Jeremias' *Handbuch der altorientalischen Geisteskultur*, which had a much wider distribution and was much easier to read than the more reliable works by (non-Pan-Babylonian) specialists of Babylonian astronomy such as Epping, Kugler, and Neugebauer. Reliance on Jeremias' book in particular may therefore have led some authors to subconsciously absorb and be influenced by some of his Pan-Babylonian ideas.

Access to sources is the biggest obstacle to the study of the circulation of knowledge. To my knowledge, there has yet to be a scholar with full command of both the Babylonian and Chinese sources. All of the scholars who have investigated connections between Babylonian and Chinese astronomy and astrology, therefore, have had to rely for at least one side of the study on translations whose accuracy they are not in a position to check. Furthermore, very many sources remain untranslated, and many translations are published in specialist publications and not easily accessible. Thus, comparisons are frequently made on the basis of secondary studies of the material, often incorporated into general surveys rather than being detailed studies themselves. The scholar attempting to compare texts from Babylonia and China, therefore, is almost always several steps removed from either the Babylonian or Chinese texts, or even both. All too often, this can lead to false or partial interpretations of a text being used for comparison.

A related problem is that, in undertaking comparisons of material between a field one knows well and another that one does not know so well, it is easy to project an interpretation from the more familiar field onto a text from the less familiar field, as Needham clearly did in his discussion of the Babylonian Three Stars Each text. Any text needs to be understood within the broader history of the astronomy of which it is part (not to mention its broader cultural context) in order to take into account the conceptual framework that underlies the astronomy that the text is presenting.

A further problem is how to establish a methodology for identifying the transmission of knowledge. How similar must material from one culture be to that of another culture in order for us to claim that one culture has borrowed it from the other? Clearly some similar astronomical ideas and methods are developed independently in different cultures because these cultures are dealing with the same astronomical circumstances (a good example is the Babylonian and Chinese identification of the 19-year cycle for intercalation). We may conclude that the approach of looking for similarities adopted by Bezold is problematic, but what should replace it? In comparing omens, for example, what should be the criteria for concluding that one set of omens is based upon another? A general thematic agenda, for example in linking eclipses and governments, is clearly not sufficient. But is it enough to find a few similar omens? Or do we need a large proportion of similar omens? Is it more significant if we find the same ordering of omens within groups? The answers will inevitably be subjective: at some stage, a scholar will feel that there is enough evidence to assume a connection between the two texts. What is important in these cases, I suggest, is to be up front about what the criteria for identifying transmission are and how certain the result is.

Understanding the circulation of knowledge is an extremely important part of the history of science. My comments here are not meant to discourage its study but rather to call for greater attention to be paid to thinking about the methodology of how we undertake such studies. On a practical level, it may be that this is an area that would benefit from more collaborative studies undertaken by scholars of the individual fields that are being compared. We will never escape our biases nor develop the perfect methodology for the study of knowledge circulation, but we should nevertheless work towards getting as close to these goals as we can.

Acknowledgments

I wish to thank Bill Mak for bringing the work of Guo Morou to my attention and for working through his text with me. I also wish to thank Wang Xianhua and Guan Yuzhen for their comments on an earlier draft of this paper.

Bibliography

Andersson, Johan Gunnar. *An Early Chinese Culture*. Peking: The Geological Survey of China, 1923.

Bezold, Carl. "Sze-ma Ts'ien und der babylonische Astronomie." *Ostasiatische Zeitschrift* 8 (1919): 42–49.

Bezold, Carl, and Franz Boll. *Reflexe astrologischer Keilinschriften bei griechischen Schriftstellen*. Sitzungsberichte der Heidelberger Akademie der Wissenschaften 7. Heidelberg: Carl Winter's Universitätsbuchhandlung, 1911.

Biot, Édouard. Untitled notes. *Journal des Savants* (1840): 27–47, 75–93, 142–152, 227–255, & 264–276.

Bosanquet, Robert H.M., and Archibald H. Sayce. "The Babylonian Astronomy. No. 2." *Monthly Notices of the Royal Astronomical Society* 40 (1880): 105–123.

Bosanquet, Robert H.M., and Archibald H. Sayce. "The Babylonian Astronomy. No. 3. The Venus Tablet." *Monthly Notices of the Royal Astronomical Society* 40 (1880): 565–578.

Bosanquet, Robert H.M., and Archibald H. Sayce. "Preliminary Paper on the Babylonian Astronomy. The Calendar." *Monthly Notices of the Royal Astronomical Society* 39 (1879): 454–461.

Brown, David. *The Interactions of Ancient Astral Science*. Bremen: Hempen Verlag, 2018.

Budge, E.A. Wallis. *The Rise and Progress of Assyriology*. London: Martin Hopkinson, 1925.

Chavannes, Édouard. *Les Mémoires Historiques de Se-Ma-Ts'ien*. Paris: Leroux, 1895–1904.

Edkins, Joseph. "Babylonian Origin of Chinese Astronomy and Astrology." *The China Review* 14 (1885), 90–95.

Edkins, Joseph. "The Chinese Cycles of Ten, Twelve, and Twenty-Eight." *The Academy. A Weekly Review of Literature, Science and Art* 25 (1884): 99–100.

Epping, Jospeh. 1889, *Astronomisches aus Babylon*. Friburg im Briesgau: Herder'sche Verlagshandlung, 1889.

Evans, James, and J. Lennart Breggren. *Geminos's Introduction to the Phenomena: A Translation and Study of a Hellenistic Survey of Astronomy*. Princeton: Princeton University Press, 2006.

Finlay, Robert. "China, the West, and World History in Joseph Needham's *Science and Civilisation in China*." *Journal of World History* 11 (2000): 265–303.

Guo Moruo 郭沫若. *Jiagu Wenzi Yanjiu* 甲骨文字研究. Shanghai: Dadong Shuju, 1931.

Hommel, Fritz. "Ueber den Ursprung und des Alter der arabischen Sternnames und insbesondere der Mondstationer." *Zeitschrift der Deutschen Morganländischen Gesselleshaft* 45 (1891): 592–619.

Hon, Tze-ki. "From a Hierarchy in Time to a Hierarchy in Space: The Meanings of Sino-Babylonianism in Early Twentieth-Century China." *Modern China* 36 (2010): 139–169.

Horowitz, Wayne. *The Three Stars Each: The Astrolabes and Related Texts*. Horn: Berger & Söhne, 2014.

Hunger, Hermann, and John M. Steele. *The Babylonian Astronomical Compendium MUL.APIN*. Abingdon: Routledge, 2019.

Iijima Tadao. *Shina rekihô kigen-kô*, Toyko: Oka shoin, 1930.

Jastrow, Morris. *Die Religion Babyloniens und Assyriens*. Giessen: Ricker'sche Verlagsbuchhandlung, 1905.

Jeremias, Alfred. *Das Alter der babylonischen Astronomie*. Leipzig: Hinrichs'sche Buchhandlung, 1908.

Jeremias, Alfred. *Handbuch der altorientalischen Geisteskultur*. Leipzig: Hinrichs'sche Buchhandlung, 1913.

Jeremias, Alfred. *Handbuch der altorientalischen Geisteskultur. Zweiter, völlig erneuerte auflage*. Berlin: de Gruyter, 1929.

Jones, Alexander. "Evidence for Babylonian Arithmetical Schemes in Greek Astronomy." In *Die Rolle der Astronomie in den Kulturen Mesopotamiens*, edited by H.D. Galter, 77–94. Graz: Kult, 1993.

Jones, Alexander and John M. Steele. "Diodorus on the Chaldeans." In *The Scaffolding of Our Thoughts: Essays on Assyriology and the History of Science in Honor of Francesca Rochberg*, edited by C. Jay Crisostomo, Eduardo A. Escobar, Terri Tanaka, and Niek Veldhuis 333–352. Leiden: Brill, 2018.

de Jong, Teije. "Babylonian Astronomy 1880–1950: The Players and the Field." In *A Mathematician's Journeys: Otto Neuegbauer and Modern Transformations of Ancient Science*, edited by Alexander Jones, Christine Proust, and John M. Steele, 365–402. Dordrecht: Springer, 2016.

Kugler, Franz Xaver. *Die Babylonische Mondrechnung*. Friburg im Briesgau: Herder'sche Verlagshandlung, 1900.

Kugler, Franz Xaver. *Im Bannkreis Babels. Panbabylonisctische Konstruktionen und Religionsgeschichtliche Tatsachen*. Münster im Westfalen: Aschensorffsche Verlagsbuchhandlung, 1910.

Kugler, Franz Xaver. *Sternkunde und Sterndienst in Babel*. Münster im Westfalen: Aschensorffsche Verlagsbuchhandlung, 1907–1924.

Lacouperie, Terrien de. "The Affinity of the Ten Stems of the Chinese Cycle with the Akkadian Numerals." *The Academy. A Weekly Review of Literature, Science and Art* 24 (1883): 144–145.

Lacouperie, Terrien de. *Western Origin of the Early Chinese Civilisation from 2,300 B.C. to 200 A.D.* London: Asher & Co., 1894.

Lehmann, Reinhard. G. *Friedrich Delitzsch und der Babel-Bible-Streit*. Frieburg: Universitäts-Verlag, 1994.

Misiewicz, Zöe, "Mesopotamian Lunar Omens in Justinian's Constantinople." In *The Circulation of Astronomical Knowledge in the Ancient World*, edited by John M. Steele, 350–395. Leiden: Brill, 2016.

Needham, Joseph. *Science and Civilisation in China. Volume 3: Mathematics and the Sciences of the Heavens and the Earth*. Cambridge: Cambridge University Press, 1959.

Neugebauer, Otto. *A History of Ancient Mathematical Astronomy*. New York: Springer, 1975.

Pankenier, David W. "Did Babylonian Astrology Influence Early Chinese Astral Prognostication *Xing Zhan Shu* 星占術?" *Early China* 37 (2014): 1–13.

Pankenier, David W. "Getting Straight with Heaven and the Origins of Writing in China." In *Writing and Literacy in Early China: Studies from the Columbia Early China Seminar*, edited by Li Feng and D.P. Branner, 13–48. Seattle: University of Washington Press, 2011.

Parpola, Simo. "Back to Delitzsch and Jeremias: The Relevance of the Pan-Babylonian School to the MELAMMU Project." In *Schools of Oriental Studies and the Development of Modern Historiography*, edited by Antonio Panaino and Andrea Piras, 239–247. Milan: IsIAO, 2004.

Pingree, David. *From Astral Omens to Astrology, from Babylon to Bīkāna*. Rome: Instituto Italiano per l'Africa e l'Oriente, 1997.

Pingree, David. "Legacies in Astronomy and Celestial Omens." In *The Legacy of Mesopotamia*, edited by Stephanie Dalley, 125–137. Oxford: Oxford University Press, 1998.

Pulleyblank, Edwin G. "The *Ganzhi* as Phonograms and their Application to the Calendar." *Early China* 16 (1991): 39–80.

Rochberg, Francesca. "A Consideration of Babylonian Astronomy Within the Historiography of Science." *Studies in the History and Philosophy of Science* 33 (2002): 661–684.

Sayce, Archibald H. "The Astronomy and Astrology of the Babylonians, with Translations of the Tablets Relating to These Subjects." *Transactions of the Society of Biblical Antiquities* 3 (1874): 145–339.

Sayce, Archibald H. "The Astronomy of the Babylonians." *Nature* 12 (1875): 489–491.

Sédillot, Louis-Pierre-Eugène Amélie. *Matériaux pur server a l'Histoire Comparée des Sciences Mathématiques chez les Grecs et les Orientaux. Tome Second.* Paris: Firmin Didot Freres, 1845–1849.

Shinjō Shinzō. *Tōyō temmon gakushi kenkyū*. Tokyo: Kōbundō, 1928.

Smith, Jonathan M. "The *Di Zhi* 地支 as Lunar Phases and their Coordination with the *Tian Gan* 天干 as Ecliptic Asterisms in China Before Anyang." *Early China* 33–34 (2010–2011): 199–228.

Steele, John M. *Ancient Astronomical Observations and the Study of the Moon's Motion (1691–1757)*. New York: Springer, 2012.

Steele, John M. "A Comparison of Astronomical Terminology, Methods and Concepts in China and Mesopotamia, With Some Comments on Claims for the Transmission of Mesopotamian Astronomy to China." *Journal of Astronomical History and Heritage* 16 (2013): 250–260.

Steele, John M. "The Development of the Babylonian Zodiac: Some Preliminary Observations." *Mediterranean Archaeology and Archaeometry* 18/4 (2018): 97–105.

Steele, John M. "Egypt and Babylon in Eighteenth Century European Histories of Astronomy." In *The Allure of the Ancient: Receptions of the Ancient Middle East, ca. 1600–1800*, edited by Margaret Geoga and John M. Steele, 312–344. Leiden: Brill, 2022.

Steele, John M. "Geminos and Babylonian Astronomy." In *Instruments—Observations—Theories: Studies in the History of Early Astronomy in Honor of James Evans*, edited by Christián Carman and Alexander Jones, 205–212. New York: ISAW, 2000.

Steele, John M. "A Re-analysis of the Eclipse Observations in Ptolemy's *Almagest*." *Centaurus* 42 (2002): 89–108.

Strassmaier, Johan N. and Joseph Epping. "Zur Entzifferung der astronomischen Tafeln der Chaldäer." *Stimmen aus Maria-Laach* 21 (1881): 277–292.

Thibaut, George. "On the Hypothesis of the Babylonian Origin of the So-called Lunar Zodiac." *Journal of the Asiatic Society of Bengal* 63 (1894): 144–163.

van der Waerden, Bartel L. "Babylonian Astronomy. II. The Thirty-Six Stars." *Journal of Near Eastern Studies* 8 (1949): 6–26.

Wang Xianhua 王献华. "Guo Moruo 'shi zhigan' yu fanbabilunzhuyi 郭沫若《释支干》与泛巴比伦主义." *Guo Morou Xuekan* 郭沫若学刊 115 (2016): 41–44.

Weber, Albrecht. *Die vedischen Nachrichten von den naxatra (Mondstationen)*. Berlin: Königl. Akademie der Wissenschaften, 1860–1862.

Weidner, Ernst F. *Alter und Bedeutung der babylonischen Astronomie und Astallehre, nebst Studien über Fixsternhimmel und Kalendar*. Leipzig: Hinrichs'sche Buchhandlung, 1914.

Weidner, Ernst F. "Ein babylonisches Kompendium der Himmelskunde." *The American Journal of Semitic Languages and Literatures* 40 (1923): 186–208.

Weidner, Ernst F. "Die Entdeckung der Paäzession, eine Geistestat babylonischer Astronomen." *Babyloniaca* 7 (1913): 1–19.

Weidner, Ernst F. *Handbuch der babylonischen Astronomie. Erster Band*. Leipzig: Hinrichs'sche Buchhandlung, 1915.

Weidner, Ernst F. "Sum Alter der babylonischen Astronomie." *Babyloniaca* 6 (1912): 129–134.

Winckler, Hugo. *Die babylonische Kulture in ihren Beziehungen zur unsrigen*. Leipzig: Hinrichs'sche Buchhandlung, 1902.

Yampolsky, Philip. "The Origin of the Twenty-Eight Lunar Mansions." *Osiris* 9 (1950): 62–83.

CHAPTER 2

Greek Astral Sciences in China

Bill M. Mak

1 Introduction: Early Contacts up to the Early Tang

During the first millennium CE, three ancient civilizations, China, Greece, and India, all had a unique and highly developed body of astral science that exerted great influences on the rest of the ancient world. The interaction among these bodies of knowledge is well documented in some cases and less so in others. In the cases of Greece and India, the Greeks or *yavanas* in India brought about the development of Greco-Indian astral science that became the foundation of classical Indian *siddhānta* astronomy and genethliacal astrology.[1] For China and India, Indian astral science mediated by Indian and Central Asian Buddhists led to the formation of the East Asian Buddhist astral science, reaching as far as Japan in the form of *sukuyōdō* 宿曜道.[2] As for Greece and China, the interaction between Greek and Chinese astral sciences is less well understood and awaits a thorough investigation.[3] Given the scope of the topic and the technicalities involved, I would give only examples of interaction from the periods that I consider to be representative.

When talking about the interaction between Greek and Chinese astral sciences, a number of methodological issues should first be considered. Firstly, Greek cosmologies especially during the early period were pluralistic, to the extent that in the words of Lloyd, "for almost every idea that was put forward, the antithetical view was also proposed."[4] Although some major trends did emerge in the later Hellenistic astral tradition, one should avoid the assumption that there was ever a singular, unadulterated, and coherent body of Greek astral science. Secondly, since the eastward transmission of Hellenistic astral science was inevitably mediated by other cultures, it was constantly shaped

1 Pingree, *The Yavanajātaka of Sphujidhvaja*. For more recent discussions, see Mak, "Last Chapter"; Brown, *Ancient Astral Science*, 712–733.

2 Yano, *Buddhist Esoteric Astrology*.

3 For a brief history of comparative studies of the two cultures, see Beecroft, "Comparisons of Greece and China." An enlightening discussion on the scientific and technological developments of the two civilizations with comparative insights, though without consideration of their historical interaction, may be found in Lloyd and Sivin, *The Way and the Word*.

4 Lloyd, "Greek Cosmologies," 205.

DOI:10.1163/9789004511675_004

by those who learned, translated, and taught it. It is not always possible to discern a particular variety of Greek practices from a local adaptation or accretion. Thirdly, the sources that attest to the historical interaction between the Greeks and the Chinese, especially during the earlier period, are remarkably scarce.

The paucity of sources is doubtless due to the vast distance that separated Greece and China.[5] Until the arrival of the Jesuits in China in the sixteenth century, the Chinese had no apparent knowledge of the Greeks as a distinct people nor any awareness of their unique culture. Nevertheless, from the beginning of the first millennium, elements of Hellenism percolated eastwards and reached China through the intermediary cultures on both land and maritime routes. As one of the foremost sciences in both cultures, astral science had considerable influences on neighbouring cultures. It is only to be expected that there would be some form of interaction between the two bodies of knowledge. To begin, it may be instructive to understand how the Chinese perceived Greek astral science, where it came from, and who transmitted it.

The Chinese word for Greece as both the civilization and the nation, *xila* 希臘 (Canontese: hei^{1}laap^{6}, from Ἑλλάς), is a modern usage unknown to the Chinese before the nineteenth century.[6] For much of the premodern period, this distant land belonging to the westernmost region of the world was known to the Chinese as Daqin 大秦, literally the "Great Qin," a reference to the Roman Empire or at least the Syrian part of it. Culturally speaking, Daqin is the Hellenistic world, however vague it was in the Chinese mind. No later than the first century CE, the Chinese envoy Gan Ying 甘英 was said to have reached as far as the Parthian Empire.[7] Thereafter, other descriptions of Daqin continued to find their ways into Chinese historiographical works, providing faint impressions of the ethnography and material culture of the Hellenistic world throughout the first millennium.[8] One of the difficulties for the early Chinese must have been that the geographical concept of Daqin was tainted with legends from the very

5 For a broad survey of the extant sources, see Leslie and Gardiner, *Roman Empire in Chinese Sources*; see also the critical review by Pulleyblank, "Roman Empire."

6 I thank Liu Jinyu for pointing out to me that the Jesuits rarely referred to Greece and Rome in their discussion of the classical authors, whom they described in indigenous terms such as "lofty sages" (*mingzhe* 明哲) or "ancient sages of the Western Land" (*xituguxian* 西土古賢), evincing geographical vagueness not uncommon in pre-modern Chinese sources (Personal communication, 11 Mar 2021). The picture of Greco-Roman antiquity the Jesuits portrayed to the Chinese was not a cohesive one as noted in Müller-Lee, "Jesuit Mission to China," 44.

7 *Hou Hanshu* 88, 2918.

8 Tian, "Handai Zhongguo yu Tuolemi Aiji de diandi jiaowang" (glassware from Hellenistic Egypt); Zhang, "Jingjiao dongjian ji chuanru Zhongguo de Xila"; "Zhengti lishi shiye zhong de Zhongguo yu Xila" (Greek medicine, metallurgy, and architecture during the later Byzantine period).

beginning, when it was conceived as a utopian counter-China situated in the extreme west of the known world. Later, it was conflated with the imaginary homeland of the Persian Christians.[9]

2 Indian Intermediaries

Among the earliest mediators of Hellenistic culture and its astral knowledge in China, unbeknownst to themselves and perhaps somewhat surprisingly, were the Indian and Central Asian Buddhists.[10] In the *Naxian biqiu jing* 那先比丘經 (T1670), translated sometime before the early fifth century CE, a conversation between the Buddhist philosopher Nāgasena and the Indo-Greek King Milinda of Bactria (Μένανδρος, 155–30 BCE) is presented. The king claims to have come from Alisan (Alexandria), which was 2,000 *yojana* or 80,000 *li* away.[11] Greek astral knowledge, as part of the eastward spread of Hellenistic culture, reached as far as Bactria and likely also Gandhāra by the third century BCE, as revealed by objects such as two astronomical sundials found in Ai Khanoum.[12] Early records of Hellenistic astral science mediated by Indian and Central Asian Buddhists show signs of local adaptation, as in the case of the description of the planets, which were presented in the Greco-Indian order (Sun, Moon, Mars, Mercury, Jupiter, Venus, Saturn) followed by the two Indian pseudoplanets, Rāhu and Ketu.[13] Examples of such transformation are found in the extant Sanskrit fragments of the Buddhist narrative *Śārdūlakarṇāvadāna* and one of the two Chinese translations.[14] By the sixth century, Greco-Indian horoscopy had

9 See Barrett, "Buddhism, Taoism," 557–560.

10 On the Greek elements in the Indian astral science transmitted to East Asia, see Yabuuti, "Indian and Arabian Astronomy," 586–591; Mak, "*Yusi Jing*," 121n78. For the evolution of Greco-Buddhism in the Hellenistic East, see Halkias, "When the Greeks Converted the Buddha."

11 T(1670A)32.702a. The Pali recension gives the place name as *Alasanda*. See Pulleyblank, "Roman Empire," 77.

12 Mairs, *The Hellenistic Far East*, 184; Savoie, "Le Cadran Solaire."

13 Neugebauer, *Exact Sciences in Antiquity*, 168–170 (rationale for the Hellenistic planetary order and its Egyptian and Greek elements); Yano, "Planet Worship," 336–337 (evolution of the descriptions of planets in Indian sources).

14 The preliminary discussions found in both Yano, *Esoteric Buddhist Astrology*, 25–26 (following Zenba), and Mak, "Transmission," 62–63, ought to be updated. The Chinese translation of the *Śārdūlakarṇāvadāna/Mātaṅgasūtra* (T1300), traditionally attributed to the two third-century translators and where the planets in Hellenistic order are found, should be dated no earlier than the second half of fifth century CE (Giebel, *The Mātaṅga Sutra*, 31). The second translation, T1301, attributed to Dharmarakṣa of the third century, contains

become a part of mainstream Indian astral science. Traces of this unique body of astral knowledge gradually found their way into the Buddhist texts compiled during this period. Descriptions of the Greco-Indian zodiac signs both transcribed and translated from Sanskrit are found in the Chinese translations of a recension of the *Mahāsaṃnipāta* (T397) by Narendrayaśas 那連提耶舍 of the late sixth century, the earliest among the extant Chinese sources.[15] Since no explanation of these concepts was included, we do not know how proficient these early Buddhist mediators were in terms of their technical knowledge and to what extent they were aware of the Hellenistic origin of the astral materials they translated.[16]

Techniques of both Hellenistic and Indian astral sciences in the extant Buddhist texts may be described as fragmentary at best. As a result, Chinese Buddhists never fully acquired the rich body of Hellenistic astral knowledge from their Indic sources. However, one should be cautious not to construe this incomplete transmission as an indication of deficiency on the side of the Chinese as a result of the latter's inferiority or incapability of scientific understanding. As Needham had noted, with further insights by Graham, Cullen, and others, the Chinese were capable of tackling all kinds of problems in mathematical astronomy since the Han period in their own terms. Chinese mathematical astronomy resembles Babylonian astronomy during the Hellenistic period in their similar use of algebraic methods.[17] The Chinese throughout the first millennium never saw their science as inferior to the foreigners'. This is in contrast to the medieval Europeans, who recognized their own inferiority and readily borrowed the Arabic sciences wholesale, and later, as Graham described, became the "possessor of the all-important combination

---

no reference to the planets, suggesting that the reference to the planets in T1300 is likely a later interpolation.

15 On the presence of astral materials in this Buddhist text, see Mak, "Transmission," 64; "Matching Stellar Ideas," 151–152 (Chinese translations of the zodiac signs). The Chinese translations for the first two signs in T397 should be *teyang zhi shen* 特羊之神 and *teniu zhi shen* 特羊之神, from the Sanskrit *meṣa* (ram) and *vṛṣa* (bull), with the character *te* 特 to specify the male. Capricorn, *makara* in Sanskrit, is left untranslated in Chinese as *mojie* 磨竭 (MC *ma-gjet/-gjot*).

16 Judging from the rise of Brahmanic orthodoxy and the high degree of Sanskritization during this period in India, such awareness was likely to be fairly low. Thus Varāhamihira, commenting on the etymology of the astral term *horā* (from Greek ὥρα, see below) in his *Bṛhatsaṃhitā*, wrongly suggested that it was an abbreviation of *ahorātra*, the Sanskrit word for day (lit. day-night).

17 Needham, *The Grand Titration*, 42; Graham, "China, Europe, and the Origins of Modern Science," 60. For a comparative study of the example of intercalation cycle, see Cullen, *Heavenly Numbers*, 39–49.

of Greek and Indian mathematics."[18] Only in the exceptional cases of some highly technical Greco-Indian astronomical and mathematical vocabulary do we identify traces of Greek elements in the extant Chinese sources. One such example is the sexagesimal Greek unit λεπτόν (arc minute), which is *liptā* in Sanskrit and *liduo* 立多 (MC *lipta*) in Chinese, mentioned in Gautama Siddhārtha's (Qutan Xida 瞿曇悉達, fl. late seventh to early eighth century CE) *Nine Seizers Canon* (*Jiuzhi li* 九執曆), an eighth-century Chinese translation of an unknown Indian astronomical treatise, commissioned by the Chinese emperor Xuanzong 玄宗 in 718 CE.[19] It should be noted that the term *liduo* was translated from Sanskrit rather than from Greek, just as the term *bojia* 薄伽 (MC *bak-gja*, arc degree) found in the same passage was derived from Sanskrit *bhāga* and not from Greek μοίρα. Gautama Siddhārtha described these terms in his linguistic musings and did not actually use any of these terms in the mathematical sections of his treatise. Instead, the terms he used for minute and degree were the two Chinese near equivalents: *fen* 分 and *du* 度.[20]

Just as obscure was another Chinese astral term, *huoluo* 火羅, which was probably derived from the Greco-Indian term *horā* (horoscopy or ascendent), from the Greek word ὥρα.[21] The term is found in the title of an esoteric Buddhist astral text called *Fantian huoluo jiuyao* 梵天火羅九曜 (T1311) and in the expression *huoluotu* 火羅圖 (diagram of *huoluo*) that appears in the text. The term *huoluo* may originally refer to a horoscope diagram, such as the Hellenistic natal chart first transmitted to India during the early centuries of the Common Era, known generally in the later periods as *rāśikuṇḍalī* (circle of signs), *janmapattra* (birth-document), and so on. The extant transmission of this text in Japan (Jp. *karazu*), however, suggests that the expression refers to an astrological illustration of zodiac signs, together with other celestial entities such as the nine Indian planets and the twenty-eight Chinese lunar lodges. Although this term was widely adopted outside India in the Indianized polities in Southeast

18 Graham, "China, Europe, and the Origins of Modern Science," 67–68.

19 Wylie, "On the Knowledge," 42; Yabuuti, "Researches." The algorithms used in this work bear some resemblance to those found in Brahmagupta's *Khaṇḍakhādyaka* (665 CE) and also other earlier works cited in Varāhamihira's *Pañcasiddhāntikā*.

20 Yabuuti, "Researches," 16. These are near equivalents since the Chinese divided the celestial sphere not into 360 parts, as the Babylonians and the Greeks did, but into 365¼ parts corresponding to the approximate number of days in a year.

21 Pelliot and Chavannes, "Un Traité Manichéen," 160, 167. Some useful insights on connection between the Indic *horā* and the Greek ὥρα may still be gleaned from Jacobi's pioneering Latin dissertation on the subject, "De Astrologiae Indicae 'Hora' Appellatae Originibus."

Asia to refer to horoscopy or even astrologers in general, it has no presence in the Chinese astral corpus or the Chinese language, unlike other Sanskrit terms that were enthusiastically sinicized and adopted by East Asians with the spread of Buddhism.

Besides some preliminary Greco-Indian astral concepts and terminology, more technical mathematical techniques involving trigonometry and a sine table of ultimately Greek origin were also transmitted to China, as evinced in the *Nine Seizers Canon*. Once again, one ought not to overemphasize the Greekness of these techniques, since the mode of transmission and the technical expressions were thoroughly Indian, and neither the Chinese text nor any Indian texts that may be associated with them has any known Greek exemplar. The astronomical formulae and the tabulated values were expressed by pithy Sanskrit verses in manners that can only be deciphered by learned Brahmins, who for generations shaped this hybrid body of Greco-Indian astral science into what they are. By the time of Āryabhaṭa in the late fifth century CE, there was already a very active community of indigenous Indian astronomers and mathematicians who showed no awareness of Hellenistic astral science.

One may inquire into whether the spirit and some key elements of Greek science did ever reach China through the convolutions of transmission, adaptation, and translation. Among the greatest strengths of Greek science is the application of mathematical and especially geometrical methods to the explanation of natural phenomena, as is evident in Eudoxus' doctrine of concentric spheres from the fourth century BCE, Apollonius' doctrine of epicycles and eccentrics, and a variety of techniques using trigonometry and Euclidian geometry.[22] Some of these techniques were indeed adapted, mostly algorithmically, in Indian astronomical texts. As a result, trigonometry and epicycles became part of the repertoire of classical Indian astronomical techniques, which are also found in the *Nine Seizers Canon*. In the latter, it may be noted that no explanation of the underlying geometrical principle was ever given. In fact, Greek mathematics and philosophy, aspects of Hellenistic culture that are considered foundational to its other achievements, were either unknown or deliberately ignored by both Indians and Chinese until over a millennium after the initial contacts.[23] In the case of the *Nine Seizers Canon*, its author

22 Lloyd, *Greek Cosmologies*, 210.

23 It is of interest to note that, for example, Euclid's *Elements* was translated in full into both Chinese and Sanskrit only in the second half of the second millennium: *Jihe yuanben* 幾何原本 by Matteo Ricci in 1607 and *Rekhāgaṇita* by Jagannātha Samrāt in 1719.

focused exclusively on computation, following the Chinese model of *li* 曆 and reducing natural phenomena to long sequences of algebraic derivations.[24]

The *Nine Seizers Canon* was included in Gautama Siddhārtha's *Kaiyuan zhan jing* 開元占經, a major astral compendium that was lost by the tenth century and rediscovered only in the seventeenth century. Despite its lack of explanation of the mathematical and astronomical principles that underlay the work, the resultant techniques were in fact new, and in many ways superior to those available to the Chinese astronomers of the eighth century CE. However, the text had little impact on the Chinese, especially after a dramatic demonstration that it was inferior to other Chinese methods in making accurate predictions of eclipses based on historical records. This apparent failure of the application of the *Nine Seizers Canon* was not due to any intrinsic deficiency of its techniques, such as the use of trigonometry, precession, and epicycle. Rather, it was partly politically motivated and partly due to the less accurate astronomical constants used in the Indian text.[25] It should be borne in mind that the reasons why foreign astronomical theories had little impact on Chinese astronomy throughout the first millennium were not limited to issues of transmission or the different theoretical concerns of the Chinese. Chinese astronomers had a long tradition of astronomical observation, having produced a star atlas, an accurate system of time measurement, calendars, and methods of astronomical prediction that, as Sivin put it, "had met the needs of agriculture and government" by 85 CE.[26]

The reception of Greco-Indian astrology was much more favorable. This exotic astral science was presented in the *Xiuyao jing* 宿曜經 (T1299), Amoghavajra's "Buddhist" astral compendium in Chinese. The Greco-Indian horoscopy described there is only fragmentary at best. It appears that in the case of both Greco-Indian astronomy and astrology, there was insufficient effort to demonstrate the superiority of these new foreign methods against their much more well-established Chinese counterparts. They were seen by the contemporary Chinese as objects of curiosity, or in some cases, of contempt, rather than genuine learning that could seriously rival their own. While such an attitude may be easily dismissed as parochialism or cultural chauvinism, one should bear in mind that there have always been enthusiastic supporters and promoters of foreign knowledge in China. The Gautama astronomers occupied various

24 On the *li* tradition in China, with ample illustrations of the methods of computation, see Cullen, *Foundations of Celestial Reckoning*.

25 Chen, "Qutanxida he ta de tianwen gongzuo," 326–327.

26 Needham, "Astronomy in Classical China," 90–92, 95–96; Sivin, *Granting the Seasons*, 41–56.

high-profile roles in the Tang Astronomical Bureau for four generations. Perhaps even more importantly they heralded a tradition of foreign astronomers playing a highly nuanced role in the Chinese political and intellectual world, that is, as both scientific advisors and counterweights to their Chinese colleagues in the Astronomical Bureau.[27]

3 First Wave: Christians of the Church of the East during the Tang and Beyond

Knowledge of Greek astral science in China remained fragmentary until the advent of Christianity in the Tang Dynasty. Formerly known as Nestorians, and better known by the Chinese as the followers of the Luminous Teaching (*jingjiao* 景教), the Christians of the "Church of the East" were said to have arrived in China from Daqin, in other words, Persia and other parts of the Near East. They brought to the Chinese not only their religious texts in Syriac but also scientific knowledge of a largely Hellenistic lineage that was already circulating in the Near East.[28]

In 1980, a double tombstone was discovered in Xi'an that provides a glimpse of the hitherto unknown astronomical activities of Persian Christians in Tang China. The inscription on the tombstone gives the biographies of a Persian couple, Li Su 李素 (743–817 CE) and his wife Bei Shi 卑失,[29] and informs us that Li Su, known also by his style name Wenzhen 文貞, came from a royal lineage of the "western country of Persia" (*xiguo bosi* 西國波斯). His ancestors had been dispatched to China during the mid-eighth century as hostages (*zhizi* 質子), in exchange for Tang's protection of their government in exile against the threats posed by Western Turks, Arabs, and Tibetans.[30] The Li family had previously been settled for generations in Guangzhou 廣州, one of the maritime gateways of southern China. Sometime during the years of Dali 大歷 (766–779 CE), Li Su was recruited as an astronomical officer by the Sitiantai 司天台 (Astronomical Bureau) in Chang'an, where he eventually worked for nearly half a century, earning the title of Sitianjian 司天監 (Director of the Astronomical Bureau) before he passed away at the age of seventy-five in 817 CE.

27 Sen, "Gautama Zhuan."

28 Takahashi, "Syriac as a Vehicle."

29 Chen, "Xi'an dongjiao sanzuo tangwu qingli ji"; Rong, "Yige rushi Tangchao de Bosi Jingjiao jiazu."

30 Ge, "Tangdai Chang'an yige Sute jiating de Jingjiao xinyang," 182–183; English translation in Ge and Nicolini-Zani, "Christian Faith," 187.

Li Su's remarkable achievement in the Astronomical Bureau is comparable to that of Gautama Siddhārtha and other members of the Gautama family who preceded him.[31] Rather remarkably, his style name, Wenzhen, is noted among the names of Christian clergy in the Xi'an stele dated 781 CE, with an additional name in Syriac script identified as *Lūqā*.[32] It appears that Li Su, Wenzhen, or *Lūqā*, moved from Guangzhou to Chang'an at the age of thirty-eight to take up a prestigious position at the Astronomical Bureau and became a member of the elite Christian community in Chang'an. The exact circumstances that led to the meteoric rise of Li Su in the Astronomical Bureau are unknown. A number of favorable factors may nonetheless be considered. Firstly, throughout the early and mid-Tang periods there were channels for special talents to be recruited into the Tang court that bypassed the traditional examination system; this was how Li Su and other foreigners, along with talented locals of often humble origin, entered the court.[33] Secondly, the Persian Christians appear to have been given favorable treatment by the Chinese ruler after the disastrous An-Shi Rebellion (755–763 CE) due to their efforts in suppressing the rebels, as evinced by Li Su's subsequent secondment outside the Astronomical Bureau and the military and political careers of his children.[34] Last but not least, Li Su's own abilities as an outstanding astronomer deserves our attention. Although none of his works survived, his astronomical knowledge must have rivaled that of his Chinese colleagues. Along with possibly other political factors yet to be explored, Li Su had probably filled a vacuum left by Indian astronomers, whose activities were already in decline and were disrupted during the chaotic period in Central China.[35]

No further records of Li Su the Perso-Chinese astronomer are extant. Two pieces of evidence, however, point to the presence and indigenization of Hellenistic astral science in the extant Perso-Chinese sources: the calendrical

---

31 Among the last Gautama astronomers who attained the position of Sitianjian was Gautama Zhuan 瞿曇譔 (Qutan Zhuan, 712–776 CE), the fourth son of Gautama Siddhārtha. Gautama Zhuan was disgraced in his youth when he lost in a court case to sue posthumously the astronomer monk Yixing 一行 for plagiarizing his father's *Nine Seizers Canon*. He was later reinstated and promoted to become the Director of the Astronomical Bureau in 763/764 CE immediately after the An-Shi Rebellion. On Gautama Zhuan, see Sen, "Gautama Zhuan."

32 Pelliot, *L'inscription Nestorienne de Si-ngan-fou*, 502. For the identification of Li Su with *Lūqā* (*Luka*), see Rong, "Yige rushi Tangchao de Bosi Jingjiao jiazu," 256.

33 Lai, "Tangdai de hanlin daizhao he sitiantai."

34 Wang, "Tang Bosi Jingjiao Li Su chushi hezhong jin zhou de shijian ji yuanyou."

35 See also the insightful discussions by Godwin (*Persian Christians*, 187–194) on the reconfiguration and rise of the Persian Christians in Tang China especially after the Abbasid revolution on one hand and the An-Shi rebellion on the other.

nomenclature of the bilingual Chinese-Syriac Xi'an stele and the Chinese astral treatise *Duliyusi jing*. The Xi'an stele is an important testimony of the Hellenistic culture introduced to China by Persian Christians. Evidence of such influences includes reference to the Greco-Persian era in Syriac ("in the year one thousand and ninety-two of the Greeks [of the Seleucid era = 781 CE]") and the use of seven planetary weekdays in Pahlavī as discussed below.[36] According to the Chinese inscription, the stele was erected on the great day of *yaosenwen* 耀森文, which is the Chinese transcription of the Pahlavī expression *ēw-šambih*, literally "first day of the week," namely Sunday, the most important weekday in Christian worship.[37] This sinicized expression suggests the presence of a multilingual Christian community that consisted of Persian expatriates and Chinese converts who adhered to a Persian tradition with a Hellenistic outlook. To what extent was this community visible to Chinese society at large? As far as their astral knowledge is concerned, some forms of exchange between the Christians and the much more dominant Buddhists certainly took place. In the second fascicle of the *Xiuyao jing*, Amoghavajra's Buddhist astral compendium compiled by his disciple Shi Yao 史瑤 in 759 CE,[38] readers were told of the method to determine the day of the week. The notion of a seven-day week cycle was apparently a novelty to the Buddhists of the Esoteric Sect, who required some explanations. The solution given by Shi Yao was surprisingly simple—one should ask the foreigners! Five years later in the second redaction of the same text by Yang Jingfeng 楊景風, another disciple of Amoghavajra and a court astrologer of some status of his own, the same passage was replaced by an apparently more rational method of weekday computation excerpted from Gautama Siddhārtha's *Nine Seizers Canon*. Doubtless this change was motivated by Yang's desire to upgrade the text with the prestige that came with a text commissioned by the Emperor himself, composed in the Astronomical Bureau, and contained mathematical formulae in sophisticated Chinese mathematical language. Shi Yao's earlier account attests nonetheless to the popularity of the seven-day Hellenistic planetary week among the foreigners known to the Chinese at the time. To help readers to decipher the answers given by the foreigners, Shi Yao gave the names for the seven planetary weekdays in three

36 Zhang, "Jingjiao dongjian ji chuanru Zhongguo de Xila," 76–88; "Zhengti lishi shiye zhong de Zhongguo yu Xila—Luoma shijie," 229–234. For translation, see Eccles and Lieu, "Stele."

37 Pelliot, *L'inscription Nestorienne de Si-ngan-fou*, 308n281. Based on the Chinese date given, the date has been suggested to be 4 February, 781 CE (Saeki, *Nestorian Documents and Relics*, 45–46).

38 His name in the text appears as *Sima Shi Yao*, the first part of which, *Sima* 司馬, I consider as the court position rather than his last name.

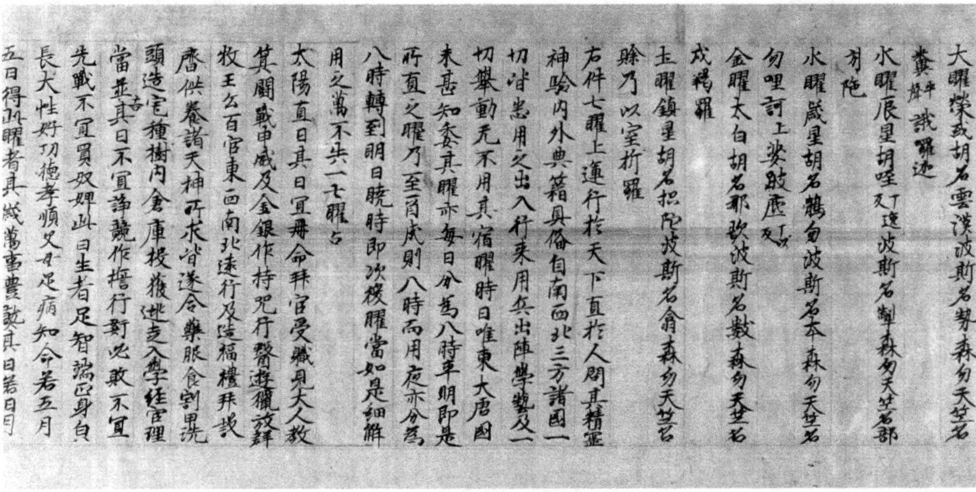

FIGURE 2.1 Multilingual glosses of the seven planetary weekdays in the *Xiuyao jing*

languages besides Chinese, namely Sogdian, Middle Persian, and Sanskrit, all transcribed in Chinese characters (fig. 2.1).[39] The presence of Middle Persian words in this Buddhist astral treatise suggests that some form of Greco-Persian astral science was in circulation in China by the mid-eighth century, to the extent that Shi Yao, an official based in Duanzhou 端州 in Southern China,[40] knew about it. In 714 CE, the authority in Guangzhou was alarmed by the "Persian monks" (*Bosi seng* 波斯僧) making "unusual instruments" and gave order to ban them.[41] From these accounts, one gets the impression that the Persian expatriate community in major cities across China possessed a body of technical knowledge, rising in prestige and gaining influences in Chinese society at large.

The second piece of evidence that points to the presence and indigenization of Hellenistic astral science among the extant Perso-Chinese sources is the *Duliyusi jing* 都利聿斯經, a Chinese astral treatise likely associated with Persian Christians in China. Until recently, this text was thought to exist only in fragmentary citations. According to the *Xin Tang shu* 新唐書, the *Duliyusi jing* from its Syriac or Pahlavī original in five fascicles (mistaken as Sanskrit by the bibliographer) was translated into the *Duliyusi jing* in Chinese in two fascicles during the years of Zhenyuan 貞元 (785–805 CE). Rather remarkably, this coin-

39 Saeki, *Nestorian Documents and Relics*, 45; Yano, *Esoteric Buddhist Astrology*, 78–81.

40 Now near Shaoxing 肇興, about 100 km west of Guangzhou.

41 This refers most likely to the Persian Christian clergy. *Tang huiyao* 唐會要, 62:6; Kuwabara, "Suitō jidai-ni raiōshita Saiikijin-ni tsuite."

cided with the tenure of Li Su in the state capital.[42] This text was purportedly brought to China by Li Miqian 李彌乾 and was translated by a certain Officer Qu 璩公 whom we know nothing about.[43] On the basis of fragmentary citations of this work, Japanese scholars suggested that the title *Duliyusi* was a corruption of Πτολεμαῖος (Ptolemy) and identified the text as a work of Hellenistic genethliacal astrology.[44] My recent study of the *Xitian yusi jing* 西天聿斯經, an abridgement of the *Yusi jing*, shows that it contains highly idiosyncratic materials that resemble those found in the five-chapter astrological treatise of Δωρόθεος Σιδώνιος (Dorotheus of Sidon, first century CE).[45] *Duliyusi* appears to be a derivation of a Persian or Syriac transcription of Dorotheus instead of Ptolemy.

The *Duliyusi jing* is distinguished from other Greco-Indian *horā* texts by virtue of a number of features not found in the latter. These include an emphasis on planetary conjunction, the configuration of trine, and the use of "lots," an astrological concept first attested in Greek sources.[46] As mentioned earlier, the Indians never succeeded in transmitting to China a complete body of astral knowledge. The *Duliyusi jing*, being one of the first complete treatises on Hellenistic horoscopy, thus appears to have filled the gap, introducing to Chinese readers concepts such as astrological places (*topoi*) and the precise methods of horoscope casting (fig. 2.2).[47] This also explains why Esoteric Buddhists kept this non-Buddhist, extra-canonical work in their library, which was eventually transmitted to Japan, but practically forgotten in China.[48]

---

42 Niu ("On the Dunhuang Manuscript P.4071," 541) points out that the *Yusi jing* was cited in the Buddhist astral manual *Fantian huoluo jiuyao* 梵天火羅九曜 dated to 751 CE, and hence Li Su, being only of the age of eight, is unlikely to be the translator of the *Yusi jing*. The argument is however problematic since the *Fantian huoluo jiuyao* contains remarks dated as late as 873/874 CE as Niu remarked. I thus tend to agree with Niu's earlier suggestion ("Fantian huoluo jiuyao," 327) that this text was composed not by Yixing, and was compiled instead by his disciples with interpolations made before the text arrived in Japan, where it has since been preserved since the ninth century CE.

43 See Rong, "Yige rushi Tangchao de Bosi Jingjiao jiazu," 251. Nothing further is known about these two figures, other than the fact that the *Xin Tang shu* identified Li Miqian as a "Duli" necromancer 都利術士, possibly an astrologer known for his skill in the technique of *Duli*[*yusi jing*]. It is possible that both *Miqian* and *Qu* are the corrupt Chinese transcriptions of some Syriac names. Tentatively, I had suggested *Micā* and [*Lū*]*qā* to be their original forms. At any rate, Li Su must have been aware of this work and might have even played a role in its translation and subsequent dissemination.

44 Ishida, "Durietsusikyō-to sono itsubun"; Yabuuti, *Chūgoku chūsei kagaku gijutsushi-no kenkyū*, 169–172; Momo, "Sukuyōdō-to sukuyō kanbun."

45 Mak, "*Yusi Jing*," 124.

46 On these technical terms and concepts, see Pingree, "Astrology," 120 (aspects, opposition, quartile, trine, sextile, lots), 123 (conjunctions and triplicities).

47 Mak, "*Yusi Jing*," 125–127.

48 Yano, *Esoteric Buddhist Astrology*, 119–121.

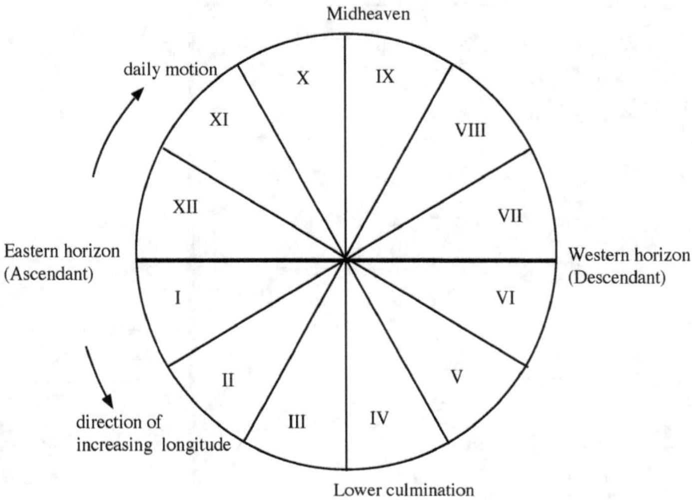

Place	Branch
I	卯 mǎo
II	寅 yín
III	丑 chǒu
IV	子 zǐ
V	亥 hài
VI	戌 xū
VII	酉 yǒu
VIII	申 shēn
IX	未 wèi
X	午 wǔ
XI	巳 sì
XII	辰 chén

FIGURE 2.2 Twelve places and the Chinese earthly branches according to the *Xitian yusi jing*

Although the contributions of scientific knowledge by these early Christians have never been recognized by the Chinese, and their overall impact on the trajectory of the history of science in China appears negligible, the legacy of Hellenistic astral science of this period survived in a number of ways. The *Jingjiao* communities in China, though greatly declined and dispersed, survived well into the Ming Period, as attested by inscriptions in Uighur script up to the fifteenth century.[49] The tradition of Persian astronomical learning in China also continued until the thirteenth century, when the Christian astronomer Isaac was employed by Mongol officials in the Yuan observatory. Some features of the *Yusi jing* also survived in a number of Chinese divinatory systems. A sinicized form of Hellenistic horoscopy exhibiting features of the *Yusi jing*, containing Indian and Persian elements, was practiced by some Chinese up to the Ming Dynasty, as shown in Wan Minying's 萬民英 astral compendium, the *Xingxue dacheng* 星學大成 (Grand achievement of astral learning, 1563 CE, fig. 2.3).[50] More closely connected to their Hellenistic predecessors are the radial variety of East Asian horoscopes, which are standard in Esoteric Buddhism in Japan and preserved in the Chinese astral treatise *Zhangguo xingzong* 張果星宗 (fig. 2.4).[51]

49 Malek, *Jingjiao*, 209–242. These inscriptions bear distant memory of the Persian Christian's Greek connection, preserving the Seleucid era in their calendar and referring to the "era of Alexander, son of King Philippe, native of the town of Macedonia."

50 Mak, "*Yusi jing*," 110–111, 127.

51 Yabuuti, *Chūgoku-no tenmon rekihō*, 190; Needham, *Science and Civilisation*, 2:352, Plate XVII.

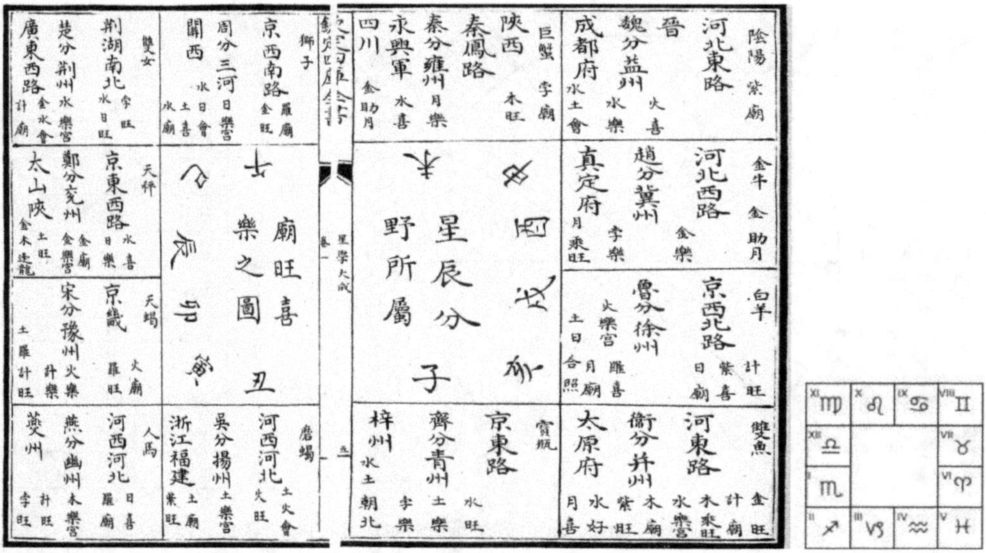

FIGURE 2.3 Quadrate Chinese horoscope in the *Xingxue Dacheng* with corresponding zodiac signs (right)

It is unknown whether this hybridized form of Hellenistic horoscopy was practiced by the Chinese after the sixteenth century. Far more popular and widely circulated was a form of fate calculation such as the later widely popular *Ziweidoushu* 紫微斗數 that absorbed certain elements of Hellenistic horoscopy.[52] These texts share the same basic representation of the cosmos, namely, the celestial sphere divided into twelve equal portions corresponding to the basic form of the Hellenistic horoscope, i.e., twelve signs of the zodiac and the twelve astrological places. Curiously, the Chinese traditions inspired by horoscopy ultimately became completely numerological and non-astronomical. Thus, unlike in Europe or India where horoscopy drove the development of mathematical astronomy in order to determine the accurate positions of the planets,[53] in China only its form and certain technicalities were appropriated. As a result, Hellenistic horoscopy did not play any significant role in the history of Chinese astronomy in the manner that it did in medieval Europe.

---

52 Mak, "Astral Science," 90. The Hellenistic and Indic elements of the *Ziweidoushu* were first discussed in Ho, "Ziweidoushu yu xingzhanxue de yuanyuan"; *Chinese mathematical astrology*, 71–82.

53 As Colebrooke noted, citing Bhāskara, in fact the Indians considered astronomy to be in the service of astrology ("On the Indian," 373). The same attitude was held by Ptolemy and other Greek astronomers/astrologers, who considered the excellence of an astrologer may only be measured by his knowledge of astronomy.

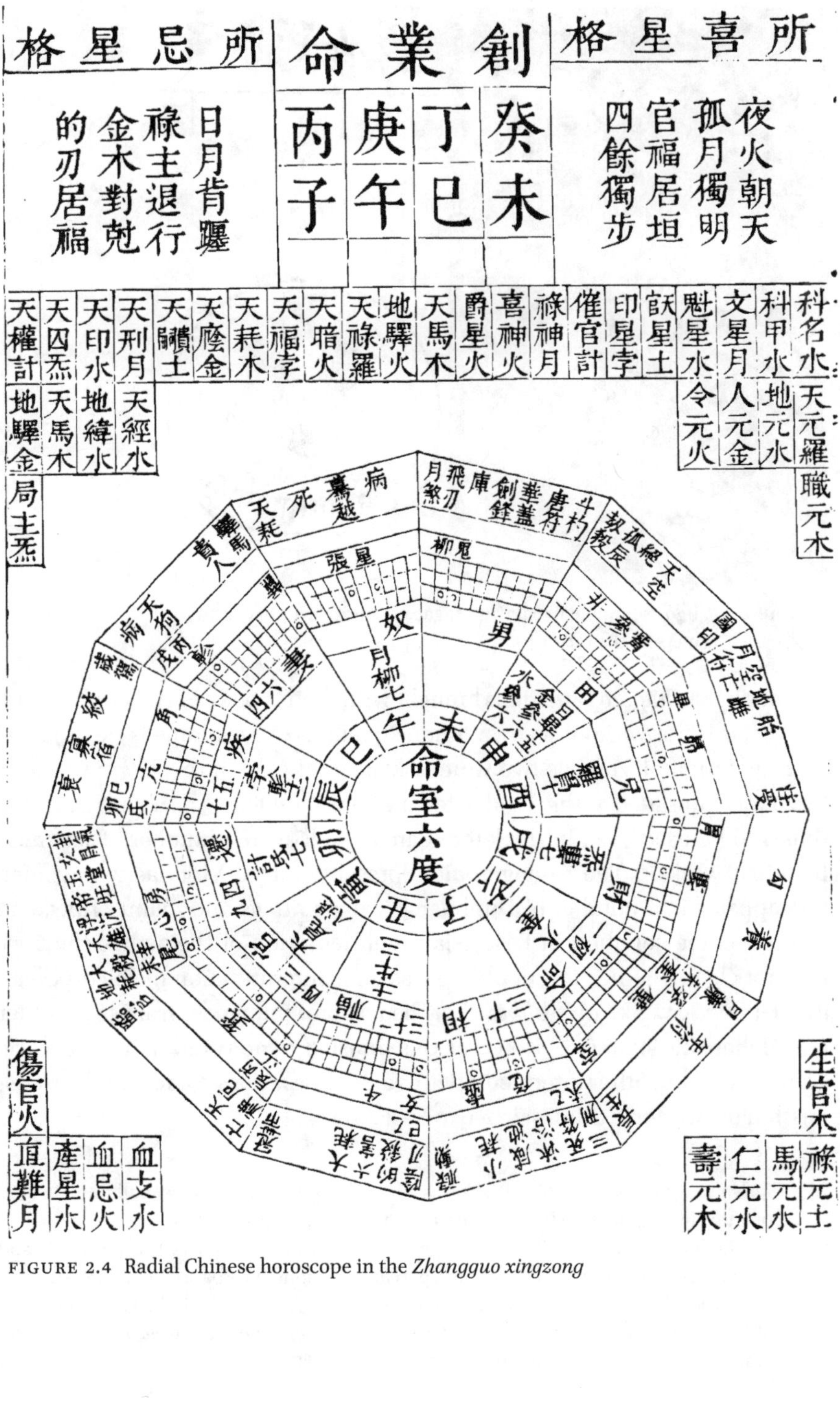

FIGURE 2.4 Radial Chinese horoscope in the *Zhangguo xingzong*

FIGURE 2.5 Virgo as "double female" in Mogao cave 66, eleventh century (left) and the *Bukkyōzu'i*, 1690 (right)

Although the *Jingjiao* Christians brought with them a body of Hellenistic astral science, it is not without its own unique qualities. An example is the unusual representation of Virgo as two women instead of one. In the *Xitian yusi jing*, Virgo was translated as the "double female," or *shuangnü* 雙女, a term that continues to be used even in the later Yuan, as well as in a number of Japanese Buddhist works.[54] The iconographic representation of Virgo as two women also appears in a number of Dunhuang cave paintings and manuscripts, and it has become common in East Asian representation of the zodiac (fig. 2.5). It is not clear whether this doubling was a misinterpretation of "bicorporeality" (Gk. δίσωμον, Skt. *dvisvabhāva*) or the result of influences from some other non-Hellenistic sources.[55] However this unusual reinterpretation took place, Central Asia and other intermediaries had doubtless played a role in the transmission of "western" astral science to East Asia.

54 *Xitian yusi jing* 7.41 (v). This variant was attested also in P. 4091 and was adopted in later practically all Chinese astral treatises, including the Buddhist translations.

55 On the bicorporeal signs (Gemini, Virgo, Sagittarius, and Pisces), see *Tetrabiblos* 1.11 and Vettius Valens 1.2. This concept of bicorporeality is adopted verbatim in Indian astral science, known as *dvisvabhāva* in Sanskrit. Other variants include the Indian representation of Gemini as a male-female couple, as seen in the *Yavanajātaka*, the *Bṛhajjātaka*, and naturally all Buddhist works later transmitted to East Asia.

4 Second Wave: Islamic Astronomers from Yuan to Ming Period

Unlike Gautama Siddhārtha or "Lūqā" Li Su, who both had meteoric but soon forgotten careers in the Astronomical Bureau of the Tang Court, Islamic astronomers, especially those from the Ming period, played a much more enduring role in Chinese history, as demonstrated in the works of Yabuuti, Chen, Shi, and others.[56] Not only did Islamic astronomers establish themselves much more successfully within Chinese society in comparison with other, earlier groups of foreigners, the Hellenistic astral science they transmitted had both greater clarity and precision, a reflection of the erudition and innovation of the Islamic astronomers. Particularly noteworthy are also their abilities to absorb astronomical knowledge from other non-Hellenistic cultures, such as the Indian and even the Chinese.[57]

As early as the Song Dynasty, Islamic astronomers began to occupy positions at the Astronomical Bureau, starting with the Islamic-Chinese astronomer Ma Yize 馬依澤 (ca. 910–1005), who for the first time introduced the weekday system in an official Chinese calendar, the *Yingtian li* 應天曆, in 963.[58] The epoch of this calendar was set to Friday (*jinyaori* 金曜日, *ādīnah*), July 16, 622 CE in the Julian calendar, in other words, the beginning of the Hijri calendar. The circulation of this weekday system reached as far as Dunhuang, as shown by the annotated calendar *Song Taipingxingguo sannian maoyin sui yingtian juzhu liri* 宋太平興國三年戊寅歲應天具注曆日 of 978 CE.[59] Such Islamic influence may be seen as a continuation of the eastward transmission of Hellenistic science which had started almost a millennium earlier with the Indians and the Persians. The Chinese translation of the names of the seven planetary weekdays or planets as 七直 or 七曜 found in the *Yingtian li* followed the earlier usage of the Chinese Buddhists (for example, *jinyaori* and so on). In fact, the astronomical formulae associated with the epoch used in the *Yingtian li* took into consideration of the seven-day week by including multiples of seven.[60] The substrata of Indian and Persian astral science may be detected in the astral nomenclature of the later *Huihui li*, where the Chinese names of the twelve zodiac signs include *mojie* 磨羯 (Capricorn, from Skt. *makara*), *yinyang* 陰陽

56 Yabuuti, "Indian and Arabian Astronomy," 580–603; Chen, *Huihui tianwenxueshi yanjiu*; Shi, "Islamic Astronomy"; also, Weil's chapter in this volume.

57 See Isahaya's chapter in this volume.

58 Chen, "Madeludi fuzi he huihui tianwenxue," 29. See CMD 431–432 for possible precursors with similar use of proximate epoch in works such Cao Shiwei's 曹士蒍 *Futian li* 符天曆 (c. 780), an astronomical work of Central Asian origin.

59 British Library S612.

60 Chen, *Huihui tianwen xueshi yanjiu*, 56–60.

(Gemini, from Skt. *mithuna*), and *shuangnü* 雙女 (Virgo, lit. "double female," see above).[61] Other astral works of Hellenistic lineage introduced by Ma Yize to China include his translations of a number of astronomical works by al-Battānī (858–929): a star table from the *Kitāb al-Zīj*, astronomical computations from the *al-Zīj al-Sābī*, and treatises on horoscopy. All these works of the latter are much more technically demanding in terms of mathematical skills and knowledge of astronomy than similar works transmitted by the earlier Indian and Persian astronomers.

The concept of solar months is particularly prominent in the *Huihui li*, as adapted from the Solar Hijri calendar, in which the twelve months correspond to the twelve signs starting from Aries. In the Chinese military text *Wujing zongyao* 武經總要 (1044), one encounters for the first time in a Chinese text a description of the entrance of the sun into the twelve zodiac signs in relation to Chinese solar terms:[62]

> *yushui* 雨水—[sun] enters Pisces one day after
> *chunfen* 春分 (spring equinox)—[sun] enters Aries three days after …

The application of solar longitude in determining the true solar terms in an official Chinese calendar must wait nearly another five centuries until a proposal by Jesuit astronomers was finally accepted (see below). It is of interest to note that some of these Arabic works by al-Battānī, which exerted great influence on medieval European astronomy, were in fact first transmitted to China before they were translated into Latin in the twelfth century.[63]

During the Yuan period, a large number of Islamic astronomical texts and instruments were brought by Jamāl ad-Dīn (*Zhamaluding* 扎馬魯丁, ca. 1255–1291) and others to the Islamic Astronomical Bureau, founded by Khubilai Khan in 1271 and operating in parallel to its Chinese counterpart.[64] The establishment of the Islamic Astronomical Bureau was a response to the large influx of Muslims arriving in China during the latter half of the thirteenth century, resulting also in the creation of the Sino-Islamic calendar *Huihui li* 回回曆, established in 1267. Among the titles found in the private library of Jamāl ad-Dīn were also Ptolemy's *Almagest* and Kūshyār ibn Labbān's astral treatise *al-Madkhal fī Ṣināʿat Aḥkām al-Nujūm* (henceforth *Madkhal*), which was based mostly on Ptolemy's *Tetrabiblos* or one of its versions.[65] Jamāl ad-Dīn's son and grandson

---

61 *Mingshi*, *Huihui lifa*, 746–747.
62 Chen, *Huihui tianwen xueshi yanjiu*, 65.
63 Dalen, "Battānī."
64 Dalen, "Zhamaluding."
65 See Weil's chapter in this volume.

continued to occupy important positions in the Astronomical Bureau. Large-scale Chinese translation of Islamic astral works took place only when Chinese rule was restored in the Ming Period, when skilled Islamic astronomers were employed alongside Chinese ones. The *Madkhal* was translated into Chinese as *Mingyi tianwen shu* 明譯天文書 (c. 1383), together with another Islamic astronomical work called *Huihui lifa* 回回曆法.[66] These two works are Ptolemaic in character, as they were based largely on the *Tetrabiblos* and the *Almagest*. The astrological treatise *Mingyi tianwen shu* demonstrates some of the continuity of the transmission of Hellenistic astral science in China (fig. 2.6). Its vocabulary reveals a mixture of influences from different periods, including those of the Sino-Indian Buddhists (*mojie* 磨羯, from Sanskrit *makara*), Sino-Persian Christians (*shuangnü* 雙女, from the *Yusi jing*),[67] and some other additions by Islamic astronomers.[68] It was through these multiple waves of Arab astronomers that Greek astronomy was introduced to the Chinese in a systematic manner by the early Ming period, albeit in a Sino-Islamic guise.

5 Third Wave: Jesuit Astronomers from Late Ming and Qing Periods

Despite the intense activities and effort of Islamic astronomers during the Yuan and early Ming periods, the Greek astral science introduced to the Chinese remained marginal in the Chinese intellectual milieu. The Huihui 回回 astral system was exclusively associated with the Sino-Islamic community or the Hui 回 ethnicity, although serious Chinese astronomers in general were acquainted with it as an alternative system to their own, unlike the earlier systems of the Indians and the Persians, which were forgotten. A much closer encounter between Chinese and Greek astral systems took place when the

66 The translation was made by four Islamic astronomers, Haida'er 海達兒, Adawuding 阿荅兀丁, Mashayihei 馬沙亦黑, and Mahama 馬哈麻, officers of the Ming Royal Observatory under the supervision of Wu Bozong 吳伯宗. See Chen, "Madeludi fuzi he huihui tianwenxue" on the background of the *Mingyi tianwenshu*; for the *Madkhal*, see Yano, *Kūšyār Ibn Labbān's Introduction*, v–xxv.

67 Given the similarity of Chinese expressions noted in the technical vocabulary of the *Xitian yusi jing* and the *Mingyi tianwenshu*, it is possible that the translators of the latter were familiar with and influenced by the *Yusi*-cluster of texts.

68 Yabuuti *Chūgoku-no tenmon rekihō*, 235–242; Yano, *Kūšyār Ibn Labbān's Introduction*, vi–vii, xvii. Some of Kūshyār's materials are not found in any extant version of the *Tetrabiblos* and are likely taken from the astral works by his Islamic predecessors Abū Maʿshar and Māshāʾallāh. There are furthermore differences between the *Mingyi tianwenshu* and the extant version of the Arabic *Madkhal*. It is not certain whether the *Mingyi tianwenshu* was translated from the Arabic *Madkhal* or its Middle Persian version (Yano, xxi).

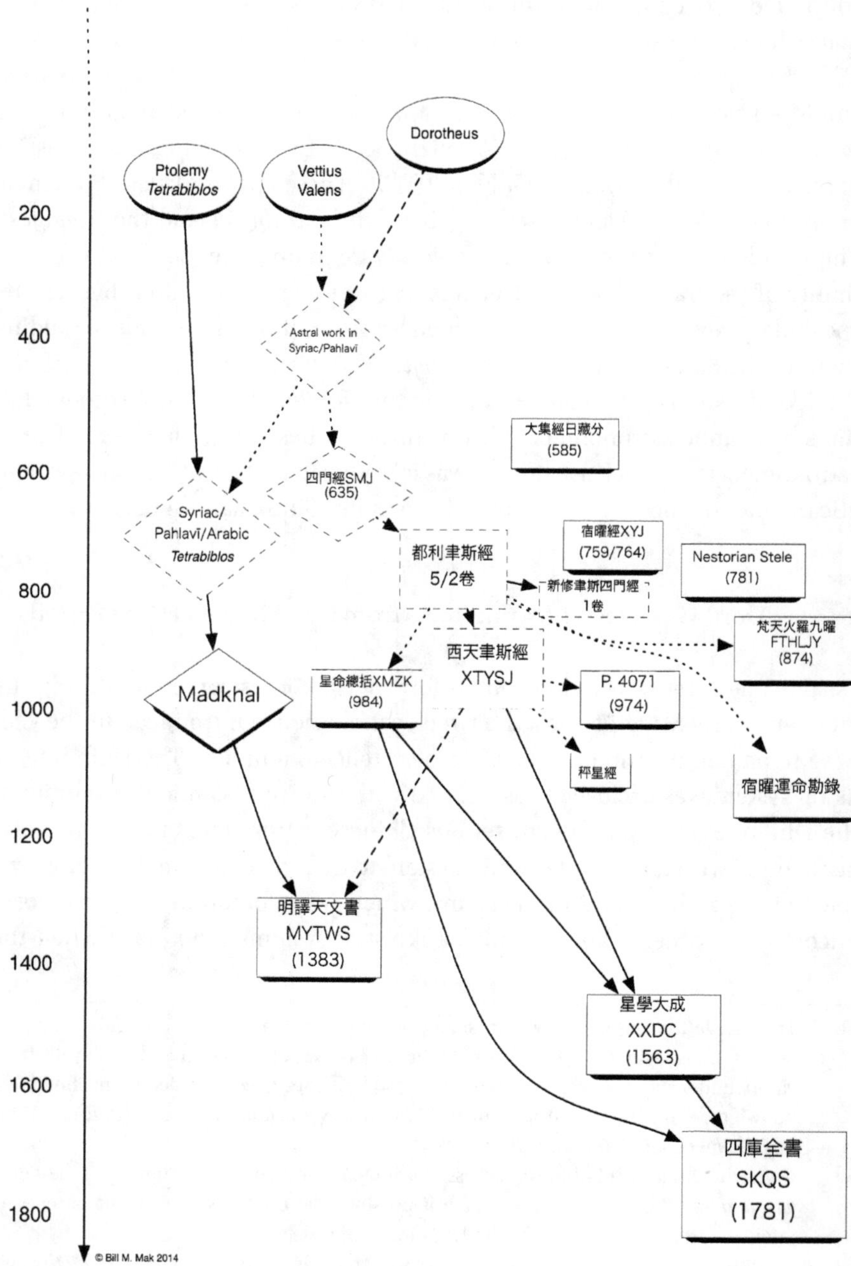

FIGURE 2.6 Genealogy of the *Yusi jing* in China

Jesuits arrived in the late sixteenth century.[69] The Jesuits, beginning with Matteo Ricci (1552–1610), who arrived in Ming China in 1583, were keen to display their scientific knowledge as a strategy to gain recognition from Chinese rulers and scholars. With the Collegium Romanum in Rome as their resource, they introduced to the Chinese both the latest European scientific theories and the ancient Greek or Latin astral texts which those theories were believed to be founded. Matteo Ricci translated, with the help of Xu Guangqi 徐光啓 (1562–1633) and others, a number of key scientific works of his teacher Christopher Clavius (1537–1612).[70] Similarly, the German Jesuit Johannes Schreck (1576–1630) and his successor Adam Schall von Bell (1592–1666) used the latest European astronomical theories in their attempts to revise the Chinese calendar and astronomical system.[71] Although only after the Ming Dynasty collapsed were their works officially adopted by the new Manchurian rulers, astronomical concepts of ultimately Greek origin (such as the inequality of seasons expressed in terms of true motion of the Sun, a geometric method for the computation of planetary motion, and spherical trigonometry) were formulated in Chinese for the first time in the *Xiyang xinfa lishu* 西洋新法曆書 (c. 1634).[72] They became the basis of the official calendar *Shixian li* 時憲曆 (1645) at the beginning of the Qing Period. It should be noted that Greek astral science proper was by then considered outdated by European scholars, who gradually embraced the new astronomical theories of Copernicus, Tycho Brahe, Galileo, and Kepler.[73] A number of Hellenistic astrological treatises were also transmitted by the Jesuits during this period. Two such texts associated with Ptolemy's *Tetrabiblos* may be highlighted. In 1637, having completed the compilation of *Chongzhen lishu*, Schall von Bell, together with Giacomo Rho (1592–1638) and their Chinese collaborators, began to compile a text titled *Tianwen shiyong* 天文實用 (Practical application of astral knowledge).[74] In one extant fascicle, there is a description of the astrological effects of the seven planets, an apparent adaptation of

69 Han, *Tongtian zhi xue*.

70 An example would be *In Sphaeram Ioannis de Sacro Bosco Commentarius*, which later became the basis of the *Qiankun tiyi* 乾坤體義.

71 Schreck, Schall von Bell and other Jesuit and Chinese collaborators compiled the *Chongzhen lishu* 崇禎曆書 (1629–1634) toward the end of the Ming Period as an attempt to replace the outdated *Shoushi li* 授時曆, which had been used for over three hundred and forty-eight years, since the Yuan Period.

72 CMD 302, 643–648.

73 Jiang, "Lun Yesuhuishi meiyou zhurao Gebaini xueshuo zaihua chuanbo—Xifang tianwenxue zaoqi zaihua chuanbo zhi zaipingjia."

74 Han, "Adam Schall von Bell," 486.

the *Tetrabiblos*.[75] Whatever the motivation behind the translation of this text was, it did not appear to be considered a work of great importance for either the Jesuits or Chinese officials, as the text was not circulated in the Qing Astronomical Bureau. The text was, however, known to the leading scholars of the time, including Mei Wending 梅文鼎 (1633–1721), Liu Xianting 劉獻廷 (1648–1695), and Lü Liuliang 呂留良 (1629–1683), and it was cited in works up to the eighteenth century.[76] Works such as the *Tianwen shiyong* and other similar less-known titles reached as far as Korea and Japan in the seventeenth century, suggesting a widespread interest among East Asian intellectuals, many of whom for the first time engaged directly with astrology, an aspect of Hellenistic astral science that interested them most—whether out of sympathy or antipathy towards it.[77]

Another attempt to introduce Greek astral science was made by the Polish Jesuit astronomer Nikolaus Smogulecki (1610–1656). Together with his Chinese colleague Xue Fengzuo 薛鳳祚, Smogulecki compiled a work titled *Tianbu zhenyuan* 天步眞原 (True origin of the movement of the heavens).[78] The astrological section of the work has been identified as a translation of Girolamo Cardano's (1501–1576) commentary on Ptolemy's *Tetrabiblios*.[79] In blatantly presenting Western astrology to the Chinese, Smogulecki might have been considered a rival or even a dissident among the Jesuits in China.[80] Smogulecki was summoned by Schall von Bell to Beijing in 1653 as the "great mathematician" but was not offered a position in the Astronomical Bureau, and was instead dispatched to Hainan as a missionary.[81] In Xue's other astral works, the Chinese author who was sympathetic to Western learning attempted to reconcile Western astral ideas with Chinese ones.[82]

---

75 Han, "Adam Schall von Bell," 487.

76 Han, "Adam Schall von Bell," 487–488. Among texts which quoted *Tianwen shiyong* are You Yi 游藝 's *Tianjing huowen* 天經或問 and Zhang Yongzuo's 張永祚 *Tianxiang yuanwei* 天象源委. On *Tianjing huowen*, see Hiraoka, "Tenkyōwakumon-no kanpon-to shahon."

77 Han, "Adam Schall von Bell," 489. For the case in Korea, see Jun, "Western Horoscopic Astrology."

78 Scholars such as Wylie and Pfister have suggested the work to have been falsely attributed to Smogulecki, but recent scholarship shows that the translation could only be made with someone who had a thorough mastery of Latin (Standaert, *Handbook of Christianity*, 340). See also Shi, "Nikolaus Smogulecki," 82–102.

79 Standaert, "European Astrology."

80 About Tycho, Smogulecki said that he was a "second-rate astronomer and his method has not been imported into China in a complete version," and about Schall von Bell that he "is not expert [in some aspects of astronomy]" and that he supported Lansberge's heliocentricism, which was not adopted by other Jesuits (Shi, "Nikolaus Smogulecki," 101).

81 Kosibowiez, "Un Missionaire Polonaise Oublié," citied in Shi, "Nikolaus Smogulecki," 102.

82 Song, "Xue Fengzuo zhongxi zhanyan huitong yu lifa gaige."

The close encounter in the field of astral science between the Jesuits and the Chinese during the late Ming and the Qing periods, and to some extent also the Sino-Islamic astronomers who saw the Jesuits as their rivals, led to some genuine attempts from all parties to rationalize each other's systems in their own terms on the Chinese soil. For the Chinese traditionalists, the attempt was to interpret or even subsume foreign astral concepts within the Chinese intellectual framework.[83] By the seventeenth century, Chinese astronomers came to the realization that the Islamic astronomical theories transmitted to China centuries before shared many similarities with the Jesuits' and belonged to the same system. A consensus then emerged among Chinese scholars that all forms of foreign astral knowledge had their origin in China. This curious, ethnocentric view, though by no means unique among the learned in the premodern world, is evident in the detailed comparative studies of Mei Wending, such as his *Lixue quanshu* 曆學全書 (Complete work on the study of astronomy).[84] Later, in Japan, the Buddhist scholar Entsū 圓通 carries a similar argument in his *Bukkokurekishōhen* 佛國曆象編 (An astronomical and astrological compendium of the Buddhist country, 1810).[85] Hence, Mei wrote:[86]

> The Huihui li (Islamic astronomy) and the Ouluoba li (European astronomy) are different astronomical systems of the same origin, differing only in their precision … Therefore I said that the way Western (i.e., European) astronomy and Islamic astronomy are related is like [that of the Chinese astronomical systems] Jiyuan 紀元 and the Tongtian 統天.[87] Their [difference in] precision is obvious … Those good methods are identical to the Western ones. Now, if one uses the Western method just like the Islamic one, how could there be any difference? Note: Huihui was known in the past as the Western Region (i.e., Central Asia and India). This is because Zheng He of the Ming [Period], who sailed the oceans under the Imperial Decree, realized that there was more than one country and called

83 See Hiraoka's chapter on the *Nanban Unkiron* 南蛮運気論 (*Yunqi* theory of the southern babarians) in this volume.

84 CMD 686–695.

85 On the *Bukkokurekishōhen* and how Entsū was indebted to Mei in his positioning of foreign astral sciences in the East Asian context, see Mak and Ueda, "Bukkokurekihenshō-ni okeru indo tenmongaku-ni tsuite," 445, 456. See also Moerman's chapter in this volume.

86 *Lixue yiwen* 曆學疑問, 1:15a–b. Cited in Shi, "Cong kexue lijie dao wenhua yixiang," 261. Translation mine.

87 *Era Epoch* and *Concord with Heaven* are two Chinese astronomical canons composed in the early and late twelfth centuries respectively. See Sivin, *Granting the Seasons*, 50–51.

> them collectively as the "Western Ocean" (*xiyang* 西洋). This astronomical treatise is titled "New Methods of Western Ocean." This is because Islamic Astronomy belongs to the old method of "Western Ocean."

The work of Islamic, Jesuit, and Chinese authors who were sympathetic to the former were often discounted because of the traditionalist desire to demonstrate the antiquity and superiority of Chinese (and in some cases, Indian) astronomy. Furthermore, the Chinese traditionalists developed the idea that foreign, Hellenistic systems could be more accurate but were ultimately unoriginal derivatives of the Chinese's. The demise of the Jesuit mission in China by the late eighteenth century put an abrupt end to their burgeoning endeavors, before the Chinese realized the deficiencies of their parochial interpretations. By the nineteenth century, the introduction of European sciences to the Chinese fell into the hands of Protestant missionaries and other secular scholars, whose interest in transmitting ancient Hellenistic science was far less significant, and the Chinese engagement with Greek astral science came to a definite end.

6 Conclusion

Greek astral sciences were introduced to China by different groups of foreigners with varying success. By examining the patterns of dissemination and reception of this foreign knowledge in the *longue durée*, a few remarks may be made with regard to also the character of Chinese astral science and its development. Throughout the long history of Chinese astral science, there were several moments of close encounters with its foreign counterparts. Outstanding Indian, Persian, Islamic, and Jesuit astronomers were promoted to senior positions in the Chinese astronomical bureau, where they were given opportunities to demonstrate their abilities and commissioned to improve existing astronomical systems. While such employment was driven by pragmatic motives, namely to bolster the Chinese rulers' Mandate of Heaven and to keep the power of the Chinese astronomers in check, it attested also to the Chinese rulers and the astronomers' openness, their efforts to create room for dialogue and mutual learning with the foreigners. The results, as shown, were mixed. As astral science was largely a state-controlled enterprise, such high-level engagements took place almost exclusively within the confines of the court, and under limiting and highly sensitive circumstances. As a result, even the most learned Chinese scholars were often unable to fully grasp and appreciate the new astral concepts and theories, such as the geometry and trigonometry introduced to them.

In secular society, Hellenistic astral knowledge spread in a much more liberal manner, though without the court astronomers' sophistication. This often resulted in curiously hybridized Sino-Hellenistic astral systems. In astrology, Buddhist and Central Asian planetary weekdays and Taoist fate calculation inspired by Persian horoscopy have survived up to the present day in Japan and China. In astronomy, the Huihui astronomical system incorporated the seven-day cycle and in the Qing calender true solar longitudes are applied to the Chinese solar terms analogous to zodiac divisions.

The intense attempts by Ming and Qing scholars to evaluate and compare Hellenistic astral concepts and theories with their Chinese counterparts further suggest that this encounter between Chinese and Hellenistic science was an ongoing project that came to a premature end as a result of the demise of the Jesuit mission and the collapse of the Qing dynasty.

Acknowledgments

Drafts of several versions of this paper were read at the following conferences: "The Reception of Greek and Roman Culture in East Asia: Texts & Artefacts, Institutions & Practices" (Freie Universität Berlin, Germany, July 2013), the 19th Asian Studies Conference Japan (Meiji Gakuin University, Tokyo, Japan, June 2015), and "Eurasian Connections—Annual Conference of the Center for Global Asia, Shanghai New York University" (Shanghai, China, August 2018). I thank Eric Huntington, Mark Wainwright, Ronnie Dreyer, and the anonymous reviewers for their generous comments and suggestions.

Bibliography

Barrett, T.H. "Buddhism, Taoism and the Eighth-Century Chinese Term for Christianity: A Response to Recent Work by A. Forte and Others." *Bulletin of the School of Oriental and African Studies* 65, no. 3 (October 2002): 555–560.

Beecroft, Alexander. "Comparisons of Greece and China." *Oxford Handbooks Online*, 2016, http://doi.org/10.1093/oxfordhb/9780199935390.013.14.

Brown, David. *The Interactions of Ancient Astral Science*. Bremen: Hempen Verlag, 2018.

Chen Guoying 陈国英. "Xi'an dongjiao sanzuo tangwu qingli ji" 西安东郊三座唐墓清理记, In *Kaogu yu wenwu* 考古与文物 2 (1981): 25–31.

Chen Jiujin 陈久金. *Huihui tianwenxueshi yanjiu* 回回天文学史研究. Nanning: Guangxi kexue jishu chubanshe, 1996.

Chen Jiujin 陈久金. "Madeludi fuzi he huihui tianwenxue" 马德鲁丁父子和回回天文学, *Ziran kexueshi yanjiu* 自然科学史研究 8, no. 1 (1989): 28–36.

Chen Jiujin 陈久金. "Qutanxida he ta de tianwen gongzuo" 瞿昙悉达和他的天文工作. *Ziran kexueshi yanjiu* 自然科学史研究 4, no. 4 (1985): 321–327.

Chen Jiujin 陈久金. "Tang kaiyuan zhan jing / Tianzhu jiuzhi li jing" 唐開元占經/天竺九執歷經. In *Zhongguo zaiji zhong nanya shiliao huibian* 中國載籍中南亞史料匯編, 280–293. Shanghai: Shanghai Guji Chubanshe, 1994.

Colebrooke, H.T. "On the Indian and Arabian Divisions of the Zodiac." *Asiatic Researches* 9 (1807): 323–376.

Cullen, Christopher. *Heavenly Numbers: Astronomy and Authority in Early Imperial China*. Oxford: Oxford University Press, 2017.

Cullen, Christopher. *The Foundations of Celestial Reckoning*. London: Routledge, 2017.

Dalen, Benno van. "Battānī: Abū 'Abd Allāh Muḥammad ibn Jābir ibn Sinān al-Battānī al-Ḥarrānī al-Ṣābi'." In *The Biographical Encyclopedia of Astronomers*, eds. T. Hockey et al., 1270. New York: Springer, 2007.

Dalen, Benno van. "Zhamaluding: Jamāl Al-Dīn Muḥammad Ibn Ṭāhir Ibn Muḥammad Al-Zaydī Al-Bukhārī." In *The Biographical Encyclopedia of Astronomers*, eds. T. Hockey et al., 1262–1263. New York: Springer, 2007.

Eccles, L. and S. Lieu. " Stele on the Diffusion of the Luminous Religion of Da Qin (Rome) in the Middle Kingdom," 2016. https://www.mq.edu.au/__data/assets/pdf_file/0007/55987/Xian-Nestorian-Monument-27-07-2016.pdf (Accessed 29/6/2020).

Jiang Boqin 姜伯勤. "Dunhuang yu Bosi" 敦煌与波斯. *Dunhuang Yanjiu* 敦煌研究 3 (1990): 1–15.

Jiang Xiaoyuan 江晓原. "Lun Yesuhuishi meiyou zhurao Gebaini xueshuo zaihua chuanbo—Xifang tianwenxue zaoqi zaihua chuanbo zhi zaipingjia." 论耶稣会士没有阻挠哥白尼学说在华传播—西方天文学早期在华传播之再评价. *Xueshu yuekan* 学术月刊 12 (2004): 101–110.

Ge Chengyong 葛承雍. "Tangdai Chang'an yige Sute jiating de Jingjiao xinyang." 唐代长安一个粟特家庭的景教信仰. *Lishi yanjiu* 历史研究 3 (2001): 181–186.

Giebel, Rolf. "The Mātaṅga Sutra." In *Esoteric Texts*, 31–104. Moraga, CA: Bukkyō Dendō Kyōkai America, 2015.

Godwin, R. Todd. *Persian Christians at the Chinese Court: The Xi'an Stele and the Early Medieval Church of the East*. London: I.B. Tauris, 2018.

Graham, A.C. "China, Europe and the Origins of Modern Science: Needham's *The Grand Titration*." In *Chinese Science. Exploration of an Ancient Tradition*, eds. N. Sivin et al., 45–69. Cambridge, MA: MIT Press, 1973.

Han Qi. "From Adam Schall von Bell to Jan Mikołaj Smogulecki: The Introduction of European Astrology in late Ming and Early Qing China." *Monumenta Serica* 59, no. 1 (2011): 485–490.

Han Qi 韩琦. *Tongtian zhi xue: Yesuhuishi he tianwenxue zai zhongguo de chuanbo* 通天之学: 耶稣会士和天文学在中国的传播. Beijing: Sanlian shudian, 2018.

Halkias, Georgios T. "When the Greeks Converted the Buddha: Asymmetrical Transfers of Knowledge in Indo-Greek Cultures." In *Religions and Trade: Religious Formation, Transformation and Cross-Cultural Exchange between East and West*, edited by P. Wick and V. Rabens, 65–115. Leiden: Brill, 2014.

Hiraoka Ryūji 平岡隆二. "Tenkyōwakumon-no kanpon-to shahon" 『天経或問』の刊本と写本. *Kagakushi kenkyū* 科学史研究 (Series III) 58, no. 289 (2019): 2–21.

Ho Peng Yoke [He Bingyu] 何丙郁. "Ziweidoushu yu xingzhan de yuanyuan" '紫微斗数' 与星占学的渊源. *Lishi yuekan* 歷史月刊 68 (1993), 38–50. Reprinted by Ho Peng Yoke. In *He Bingyu Zhongguo kejishi lunji* 何丙郁中国科技史论集 [*Science and Culture in Ancient China: Selected Works of Ho Peng Yoke*], 239–255. Shenyang: Liaoning Jiaoyu Chubanshe, 2001.

Hou Hanshu 後漢書 (Latter Han History). Compiled by Fan Ye 范曄. Beijing: Zhonghua shuju, 1973.

Ishida Mikinosuke 石田幹之助. "Durietsusikyō-to sono itsubun" 都利聿斯經とその佚文. In *Toyōshi ronsō: Haneda hakashi juju kinen* 東洋史論叢：羽田博士頌壽紀念, 49–62. Tokyo: Tōyōshi kenkyukai, 1950.

Jacobi, Hermann. "De Astrologiae Indicae 'Hora' Appellatae Originibus." PhD dissertation, University of Bonn, 1872.

Jun Yong Hoon. "Western Horoscopic Astrology in Korea." In *Coping with the Future: Theories and Practices of Divination in East Asia*, edited by Michael Lackner, 486–498. Leiden: Brill, 2018.

Kosibowiez, K. "Un Missionaire Polonaise Oublié: Le Père Jean Nicolas Smogulecki S.J., Missionaire en Chine au XVIIe Siècle." *Revue d'Histoire des Missions* 22 (1929): 335–360.

Kuwabara Jitsuzo 桑原隲藏. "Suitō jidai-ni raiōshita Saiikijin-ni tsuite" 隋唐時代に支那に來往した西域人に就いて. In *Shinagaku Ronsō: Naitō Hakashi Kanreki Jukuga* 支那學論叢：内藤博士還暦祝賀, 565–660. Kyoto: Kōbundō, 1926.

Lai Ruihe 賴瑞和. "Tangdai de hanlin daizhao he sitiantai: Guanyu 'Li Su Muzhi' he 'Beishi Shi Muzhi' de zaikaocha" 唐代的翰林待詔和司天臺－關於《李素墓誌》和《卑失氏墓誌》的再考察. *Tang yanjiu* 唐研究 9 (2003): 315–342.

Leslie, Donald, and K.H.J. Gardiner. *The Roman Empire in Chinese Sources*. Roma: Bardi, 1996.

Lloyd, G.E.R. "Greek Cosmologies." In *Ancient Cosmologies*, edited by Carmen Blacker and Michael Loewe, 198–224. London: George Allen & Unwin, 1975.

Lloyd, G.E.R., and Nathan Sivin. *The Way and the Word: Science and Medicine in Early China and Greece*. New Haven: Yale University Press, 2002.

Mairs, Rachel. *The Hellenistic Far East: Archaeology, Language, and Identity in Greek Central Asia*. Berkeley: University of California Press, 2014.

Mak, Bill M. "Astral Science of the East Syriac Christians in China during the Late First Millennium AD." *Mediterranean Archaeology and Archaeometry* 16, no. 4 (2016): 87–92.

Mak, Bill M. "Matching Stellar Ideas to the Stars: Remarks on the Translation of Indian Jyotiṣa in the Chinese Buddhist Canon." In *Cross-Cultural Transmission of Buddhist Texts: Theories and Practices of Translation*, edited by D. Wangchuk, 139–158. Hamburg: Department of Indian and Tibetan Studies, Universität Hamburg, 2016.

Mak, Bill M. " The Last Chapter of Sphujidhvaja's Yavanajātaka Critically Edited with Notes." *SCIAMVS* 14 (2013): 59–148.

Mak, Bill M. "The Transmission of Buddhist Astral Science from India to East Asia: The Central Asian Connection." *Historia Scientiarum* 24, no. 2 (2015): 59–75.

Mak, Bill M. "*Yusi Jing*: A Treatise of 'Western' Astral Science in Chinese and its Versified Version *Xitian yusi jing*." *SCIAMVS* 15 (2014): 105–169.

Mak, Bill M. 麥文彪, and Masahiro Ueda 上田真啓. "Bukkokurekihenshō-ni okeru indo tenmongaku-ni tsuite" 『仏国暦象編』におけるインド天文学について. In *East-West Encounter in the Science of Heaven and Earth* (*Ten-to chi-no kagaku* 天と地の科学), edited by Bill M. Mak and Takeda Tokimasa, 441–457. Kyoto: Institute for Research in Humanities, Kyoto University, 2019.

Malek, Roman, ed. *Jingjiao: The Church of the East in China and Central Asia.* Sankt Augustin: Institut Monumenta Serica, 2006.

Mingshi 明史 (Ming History). Compiled by Zhang Tingyu 張廷玉 et al. Beijing: Zhonghua shuju, 1974.

Momo Hiroyuki 桃裕行. "Sukuyōdō-to Sukuyō Kanbun" 宿曜道と宿曜勘文. *Risshō Shigaku* 立正史学 39 (1975): 1–20.

Müller-Lee, Andreas. "The Jesuit Mission to China and the Reception of Ancient Greek and Roman Culture in China and Korea." In *Receptions of Greek and Roman Antiquity in East Asia.* Leiden: Brill, 2018.

Needham, Joseph. "Astronomy in Classical China." *Quarterly Journal of the Royal Astronomical Society* 3 (1962): 87–98.

Needham, Joseph. *Science and Civilization in China*, vol. 2. Cambridge: Cambridge University Press, 1956.

Needham, Joseph. *The Grand Titration: Science and Society in East and West.* London: Allen & Unwin, 1969.

Neugebauer, O. *The Exact Sciences in Antiquity*. Second unabridged and corrected edition. New York: Dover, 1969.

Nicolini-Zani, Matteo. "The Christian Faith of a Sogdian Family in Chang'an During the Tang Dynasty." *Annali* 64, no. 1/4 (2004): 181–196.

Niu Weixing 钮卫星. "Fantian huoluo jiuyao kaoshi ji qi zhuanxie niandai he zuozhe wenti tantao" 《梵天火罗九曜》考释及其撰写年代和作者问题探讨. *Ziran kexueshi yanjiu* 自然科学史研究 24, no. 4 (2005): 319–329.

Niu, Weixing. "On the Dunhuang Manuscript P.4071: A Case Study on the Sinicization of Western Horoscope in Late 10th Century China." In *The Circulation of Astronomical Knowledge in the Ancient World*, edited by John M. Steele, 527–558. Leiden: Brill, 2016.

Pelliot, Paul. *L'inscription Nestorienne de Si-ngan-fou*, edited by Antonio Forte. Kyoto: Scuola di studi sull'Asia Orientale, 1996.

Pelliot, Paul, and Éd. Chavannes. "Un Traité Manichéen Retrouvé En Chine II." *Journal Asiatique* 11 (1913): 99–199, 261–392.

Pingree, David. "Astrology." In *Dictionary for the History of Ideas*, vol. I, edited by P.P. Wiener, 118–126. New York: Scribner, 1973.

Pingree, David. *The Yavanajātaka of Sphujidhvaja*, 2 vols. Cambridge: Harvard University Press, 1978.

Pulleyblank, Edwin G. "The Roman Empire as Known to Han China," *Journal of the American Oriental Society* 119, no. 1 (1999): 71–79.

Rong Xinjiang 荣新江. "Yige rushi Tangchao de Bosi Jingjiao jiazu" 一个入仕唐朝的波斯景教家族. In *Yilangxue zai Zhongguo lunwen ji di er ji* 伊朗学在中国论文集第二集, edited by Ye Yiliang, 82–90. Beijing: Beijing daxue chubanshe, 1998. Reprinted by Rong Xinjiang. In *Zhonggu zhonguo yu wailai wenming* 中古中国与外来文明, 238–257. Beijing: Sanlian shudian, 2001. Page references are to the 2001 edition.

Saeki, P.Y. *The Nestorian Documents and Relics in China*. Tokyo: Toho Bunka Gakuin, 1937.

Savoie, Denis. "Le Cadran Solaire Grec d'Aï Khanoum: La Question de l'exactitude des Cadrans Antiques." *Comptes Rendus Des Séances de l'Académie des Inscriptions et Belles-Lettres* 151, no. 2 (2007): 1161–1190.

Sen, Tansen. "Gautama Zhuan: An Indian Astronomer at the Tang Court." *China Report* 31, no. 2 (1995): 197–208.

Shi Yunli 石云里. "Cong kexue lijie dao wenhua yixiang: Ming Qing xuezhe dui Alabo tianwenxue de renshi" 从科学理解到文化臆想——明清学者对阿拉伯天文学的认识. *Zhongguo Kejishi Zazhi* 中国科技史杂志 40, no. 3 (2019): 253–270.

Shi Yunli. "Islamic Astronomy in the Service of Yuan and Ming Monarchs." *Suhayl* 13 (2014): 41–61.

Shi Yunli. "Nikolaus Smogulecki and Xue Fengzuo's True Principles of the Pacing of the Heavens: Its Production, Publication and Reception." *EASTM* 27 (2007): 63–126.

Sivin, Nathan. *Granting the Seasons: The Chinese Astronomical Reform of 1280*. New York: Springer, 2009.

Song Zhiye 宋芝业. "Xue Fengzuo zhongxi zhanyan huitong yu lifa gaige" 薛凤祚中西占验会通与历法改革. *Shandong shehui kexue* 山东社会科学 6 (2011): 38–43.

Standaert, Nicolas. "European Astrology in Early Qing China: Xue Fengzuo's and Smogulecki's Translation of Cardano's Commentaries on Ptolemy's *Tetrabiblos*." *Sino-Western Cultural Relations Journal* 23 (2001): 50–79.

Standaert, Nicolas. *Handbook of Christianity in China*. Leiden: Brill, 2001.

Takahashi, Hidemi. "Syriac as a Vehicle for Transmission of Knowledge across Borders of Empires." *Horizons* 5 (2014): 29–52.

Tang huiyao 唐會要. Compiled by Wang Pu 王溥. Beijing: Zhonghua shuju, 1955.

Tian Ming 田明. “Handai Zhongguo yu Tuolemi Aiji de diandi jiaowang” 汉代中国与托勒密埃及的点滴交往. *Journal of Inner Mongolia University for Nationalities (Social Sciences)* 内蒙古民族大学学报(社会科学版) 6 (2004): 6–11.

Xin Tang shu 新唐書. Compiled by Ouyang Xiu 歐陽修 et al. Beijing: Zhonghua shuju, 1975,

Wang Changming 王长命. “Tang Bosi Jingjiao Li Su chushi Hezhong Jinzhou de shijian ji yuanyou” 唐波斯景教徒李素出仕河中晋州的时间及缘由. *Wenwu Shijie* 文物世界 2 (2018): 39–41.

Wylie, Arthur. “On the Knowledge of a Weekly Sabbath in China.” *Chinese Recorder and Missionary Journal* 4 (1871): 40–45.

Yabuuti Kiyoshi 薮内清. *Chūgoku chūsei kagaku gijutsushi-no kenkyū* 中国中世科学技術史の研究. Tokyo: Kadokawa Shoten, 1963.

Yabuuti Kiyoshi 薮内清. *Chūgoku-no tenmon rekihō* 中国の天文暦法. Tokyo: Heibonsha, 1969.

Yabuuti Kiyoshi. “Indian and Arabian Astronomy in China,” trans. Leon Hurvitz. In *Silver jubilee volume of the Zinbun-Kagaku-Kenkyusyo Kyoto University*, 585–603. Kyoto: Nissha Print, 1954.

Yabuuti Kiyoshi. “Researches on the *Chiu-chih li*: Indian Astronomy under the T’ang Dynasty.” *Acta Asiatica* 36 (1979): 7–48.

Yano, Michio. *Esoteric Buddhist Astrology*, trans. Bill M. Mak. New Delhi: Aditya Prakashan, 2019.

Yano Michio. *Kūšyār Ibn Labbān’s Introduction to Astrology*. Tokyo: Institute for the Study of Languages and Cultures of Asia and Africa, 1997.

Yano, Michio. “Planet Worship in Ancient India.” In *Studies in the History of the Exact Sciences in Honour of David Pingree*, edited by Charles Burnett, 331–348. Leiden: Brill, 2004.

Zhang Xushan 张绪山. “Jingjiao dongjian ji chuanru Zhongguo de Xila—Baizhanting wenhua” 景教东渐及传入中国的希腊—拜占庭文化. *Shijie lishi* 世界历史 6 (2005): 76–89.

Zhang Xushan 张绪山. “Zhengti lishi shiye zhong de Zhongguo yu Xila–Luoma shijie—Han-Tang shiqi wenhua jiaoliu de jige dianli” 整体历史视野中的中国与希腊—罗马世界—汉唐时期文化交流的几个典例. *Quanqiushi pinglun* 全球史评论 1 (2008): 215–235.

CHAPTER 3

Eclipse Dragons, Seasonal Change, and the Salvation of Light: A Case of Overlapping Cosmologies in Manichaeism

Adrian C. Pirtea

1 Introduction

Few concepts became as popular in the traditional astral sciences of Asia as that of the eclipse dragon. This astrological and cosmological concept resulted from the combination of at least two distinct sets of ideas: (a) the myth of a celestial snake, dragon, or demon who devours, seizes, or covers the sun and the moon, thereby causing eclipses; (b) an astronomical understanding of the two lunar nodes[1] combined with a belief in their astral influence on human affairs. The merging of these two elements, resulting in what will be referred to here as the *eclipse-dragon theory*, occurred sometime in the first centuries CE and often led to the inclusion of the two nodes, called the head and tail of the dragon, in the conventional list of planets as additional dark planets. In some cases, one finds a pair of dragons fulfilling the same function.

Varieties of this theory are attested in countless texts on cosmology and astral science from across premodern Eurasia and North Africa. Thus, the eclipse demon Rāhu (the Vedic Svarbhānu) is found in Sanskrit treatises on astronomy,[2] in early Khmer inscriptions containing astrological data,[3] as well as in Buddhist literature and art from China, Tibet, Mongolia, and Japan.[4] Rāhu

---

1 The two lunar nodes (the ascending and descending node) are the imaginary points of intersection between the moon's orbit and the ecliptic. Their apparent retrograde movement can be astronomically computed, allowing the prediction of eclipses.

2 The earliest extensive treatment (and refutation) of the theory is found in Varāhamihira's sixth-century *Bṛhatsaṃhitā*, V.1–13 (Kern, 23–25). On Rāhu and Ketu in Hindu astronomy and astrology in general, see Billard, *L'astronomie indienne*, 125–132; Markel, *Planetary Deities*, 55–67.

3 See, for example, the Sanskrit-Khmer bilingual inscriptions K. 115 (665 CE, Coedès, *Inscriptions*, 6:10), K. 926 (624 CE, Coedès, *Inscriptions*, 5:20–21); K. 726 (eighth century CE, Coedès, *Inscriptions*, 5:75–80).

4 On the role of astrology in Chinese Esoteric Buddhism, see Sørensen, "Astrology." On Rāhu and Ketu in Mongol astrology and divination, see Baumann, *Divine Knowledge*, 68 and 125–128.

 | DOI:10.1163/9789004511675_005

is often depicted together with the five planets, the two luminaries, and Ketu (initially a personified comet, later also the descending node)[5] in a composition known as the *navagraha*, i.e., the nine planets.[6] A very similar notion, the celestial snake or dragon Gōzihr,[7] appears in Zoroastrian writings on cosmology and astrology, where it is closely associated with two dark planets: the Dark Sun and the Dark Moon.[8] Beginning with the sixth or seventh century CE, Syriac texts on astrology refer to the eclipse dragon as Aṯālyā.[9] Finally, the Arabic al-Jawzahr, a derivate of Middle Persian Gōzihr, features in virtually all Islamic treatises on astrology from Māshā'allāh (d. 815 CE) and Abū Maʿshar (d. 886 CE) onwards, from where it eventually reached medieval Jewish and Latin astrology.[10]

While the earliest sources to include an eclipse demon in the list of planets hail from India and Central Asia,[11] the overall development of this eclipse theory and its transmission to different cultural areas of the Eurasian landmass are difficult to establish. In light of the constant interactions and overlaps between

5 The earliest clear identifications of Ketu with the descending node in Sanskrit texts date to the twelfth century (Billard, *L'astronomie indienne*, 127). However, Billard also notes that al-Bīrūnī already describes Ketu as the descending node in the early eleventh century (cf. Markel, *Planetary Deities*, 64–66). At least in one Chinese treatise from the Tang period, the *Qiyao rangzai jue* 七曜攘災決, Ketu represents the lunar apogee (see Yano, "The *Ch'i-yao jang-tsai-chüeh*"). I thank Bill M. Mak for this reference.

6 See Hartner, "Pseudoplanetary Nodes"; Pingree, "Astronomy and Astrology"; Pingree, "Planetary Images"; Gail, "Planets and Pseudoplanets"; Markel, *Planetary Deities*; Mak, "Transmission of the *Grahamātṛkādhāraṇī*."

7 Or possibly *Gawčihr*. See Panaino, "Pahlavi Gwcyhl"; cf. Soudavar, "Significance"; Hintze, "Avestan Expression."

8 *Bundahišn* v.4–5, Pakzad, *Bundahišn*, 73–74; Agostini and Thrope, *The Bundahišn*, 33–42. On the astrological passages in this work, see MacKenzie, "Zoroastrian Astrology"; MacKenzie, "Gōzihr"; MacKenzie, "Bundahišn"; Henning, "Astronomical Chapter"; Pingree, *Astral Omens*, 39–40; Raffaelli, *L'oroscopo del mondo*.

9 Villey, "Textes astronomiques"; Pirtea, "Eclipse Dragon in Manichaeism."

10 Kennedy and Pingree, *Astrological History*; Hartner, "Djawzahar"; Kuehn, *Dragon*; Kuehn, "Eclipse Demons." On the Jewish, Byzantine and medieval Latin reception, see Sharf, *Universe*; Mancuso, *Sefer Ḥakhmoni*; Cùscito, "Il tly nella cosmologia"; Pirtea, "From Lunar Nodes"; Draelants, "Le *Liber Nemroth*."

11 The earliest Indian sources that apparently include Rāhu in the list of planets are the *Gārgīyajyotiṣa* (ca. first century CE) and the *Śārdūlakarṇāvadāna*, at least as preserved in a fourth-century Sanskrit manuscript found in Central Asia (St. Petersburg, SI 1942, fol. 16r7, Miyazaki et al., "The *Śārdūlakarṇāvadāna*," 6) and a Chinese translation (*Modengjia jing* 摩登伽經, T1300, 405b15). On these texts, see Pingree, "Empires," 268–269; Pingree, "Astronomy and Astrology," 240–241; Pingree, *Jyotiḥśāstra*, 69–72; Mitchiner, *The Yuga Purāṇa*; Raffaelli, *L'oroscopo del mondo*, 16; Geslani et al., "Garga"; Mak, "Indian *Jyotiṣa*," 6–7; Mak, "Buddhist Astral Science," 62–63; Mak, "Vedic Astral Lore."

the multiple cosmological systems of premodern Eurasia, one has to take into account all the variations and reinterpretations of the theory that these interactions facilitated. Moreover, since explanations of eclipses are always given within a larger cosmological framework and/or religious worldview, any transfer of a specific eclipse theory from one system to another will transform, at least to some degree, both the theory and the systems that adopt it. In the case of the eclipse dragon, this transformation is often restricted to iconographic or terminological variations, or to a few adjustments to the overarching cosmological system. However, in some instances it may involve changes on a more fundamental level that are motivated by philosophical or religious concerns.

The elaboration of a cosmological explanation of eclipses in Manichaeism represents an exceptional case study of precisely such a religiously motivated reinterpretation of the eclipse-dragon theory. Manichaeism was a world religion founded by Mani in third-century Mesopotamia, which in various periods spread from North Africa and the Eastern Mediterranean to Central Asia and China. The structure of the visible cosmos, as understood by Mani and his followers, was inextricably linked to the Manichaean view of salvation, seen as the liberation of light entrapped in matter. Therefore, Mani's adoption of specific theories regarding eclipses and other natural phenomena from the surrounding cultures was primarily determined by soteriological considerations rather than by an interest in cosmology or astrology per se. As will become clear later on, the Manichaean view of eclipses was from the outset the result of an overlap between two different cosmological concepts, which Mani subsumed under an overarching religious concern: defending the divinity of the sun and the moon and explaining the mechanics of the redemption process.

With the spread of Mani's new religion from Sasanian Mesopotamia to other parts of Eurasia and the Mediterranean, the Manichaean understanding of eclipses inevitably collided with competing cosmological systems. Thus, in the Eastern Roman Empire Manichaeans met with the Ptolemaic model of the universe prevalent there. A study of the Coptic Manichaean sources from Late Roman Egypt can show how the Manichaean view of eclipses changed as a result of this encounter and indicate the extent to which early Manichaeans interacted with the body of Hellenistic astronomical and astrological knowledge.

Approaching these problems from the perspective of overlapping cosmologies[12] is not only appropriate and necessary given the diffusion of Manichaeism across multiple linguistic and cultural boundaries, but it could also explain

12 See the introduction to this volume.

some puzzling or seemingly contradictory statements found in the Manichaean writings. Moreover, this approach may prove useful beyond the case study examined here, as it could provide methodological guidelines for addressing other particularities of Manichaean cosmology.

2 Eclipses in Manichaean Cosmology

Mani's original writings in Aramaic are lost, but an extensive corpus of Manichaean writings have survived in Greek, Latin, Coptic, Middle Iranian, Chinese, and Old Uyghur. Mani is also credited with a *Book of Pictures* (*Eikōn* or *Ardhāng*), in which he supposedly offered a detailed visual representation of his teachings. Although this work is also lost, a ca. fourteenth-century Chinese Manichaean painting discovered in Japan appears to be a later and partial reflection of Mani's *Eikōn*.[13] Through a careful comparison of these textual and iconographic sources, the central tenets of Manichaeism, including many elements of Mani's cosmology, can be reconstructed with a fair amount of certainty.[14]

Thanks to their accommodative missionary strategy, Manichaeans were very efficient in adopting different theological, cosmological, and anthropological ideas and beliefs, while also reinterpreting them in a distinctive Manichaean key. To give one well-known example, Manichaeans could claim that Jesus Christ, Zarathustra, and the Buddha were all preachers of the same true religion and the forerunners of Mani, who was both the Paraclete announced in the *Gospel of John* and the Maitreya of Buddhist eschatology.[15]

Manichaeans applied the same syncretistic principle to cosmology and formulated their views on the zodiac, the planets, and their effect on humans, plants, and animals by interacting with other cosmological traditions.[16] Nevertheless, Mani's fundamental insight about the nature of the material cosmos remained pivotal for all the later developments of the Manichaean system. According to Mani, the entire visible universe is a giant mechanism that filters

13 On the Manichaean cosmology painting, see Yoshida, "Shinshutsu Manikyō"; Kósa, "Translating the *Eikōn*"; Yoshida and Furukawa, *Manikyō Kaiga Kenkyū*; Gulácsi, *Mani's Pictures*.

14 See, for example, Kósa, "Manichaean Attitude"; Kósa, "Ships and Ferries"; Kósa, "Atlas and Splenditenens."

15 See Klimkeit, *Gnosis on the Silk Road*, 134; Oort, "Paraclete Mani"; Sundermann, "Der Paraklet."

16 For an overview of Manichaean astrology, see von Stuckrad, *Ringen um die Astrologie*, 696–766.

and redeems divine light-elements that had been trapped and mixed with darkness after the primordial battle between the First Man (Ohrmazd) and the King of Darkness (Ahriman). The Chinese *Manichaean Treatise*, an eighth- or ninth-century précis of Manichaean doctrine, describes the fashioning and purpose of the universe as follows:

> The five kinds of Demons adhered to the Five Light-Bodies like a fly stuck in honey, like a bird caught in lime, like a fish that had swallowed a hook. For this reason, the Pure Wind (i.e., the Living Spirit) Envoy of Light, *by combining the two powers of the five kinds of Demons and the Five Light-Bodies, created the Ten Heavens and Eight Earths of the cosmos. Thus, the cosmos is the hall of healing where the Light-Bodies are cured, and it is also the prison where the dark demons are incarcerated.*[17] The two of them, Pure Wind and Good Mother, by skillful means *established the Ten Heavens; then they positioned the Wheel of Karma, together with the palaces of the Sun and the Moon*, the Eight Earths beneath these [...].[18]

The wheel of karma (Ch. *ye lun* 業輪)[19] mentioned here is almost certainly identical to what other Manichaean sources call "sphere,"[20] the "circle (of the zodiac) with planets and stars,"[21] the "rolling wheel and zodiac,"[22] or the "the wheel of the stars."[23] Since this sphere is where the most evil demons are imprisoned,[24] the zodiac, the stars, and planets are consequently regarded as evil agents. Their maleficent influence on human affairs is further explained by the fact that the demoness Āz (Greed) created human beings as an exact replica of the universe, but with the opposite purpose: while the universe imprisons demons and is designed to save the light-elements, the bodies of living beings

17 A close parallel is found in one of the Manichaean Coptic Bema-Psalms: "He made it a prison too for all the powers of Darkness, it is also a place of purification for the Soul [...]" (Allberry, *Psalm-Book*, 10.27–28).

18 *Manichaean Treatise*, II.10–15 (Lieu and Mikkelsen, *Tractatus Manichaicus*, 4–5; modified, italics mine).

19 Mikkelsen, *Dictionary*, 84; Ma, "Wheel of Karma."

20 Pth. ʿyw ʿspyr (Turfan frag. M 384 + M 2067 + M 4517 + M 5190 + M 5682; Sundermann, *Der Sermon vom Licht-Nous*, 62–63).

21 MP gyrd'sm'n ʿyw pd 'xtr''n 'wd ʿst'rg'n (Turfan frag. M 7984 II R/i/27–28; Hutter, *Šābuhragān-Texte*, 28).

22 Sgd. cxryy 'ty 'nxrwzn (Turfan frag. M 178; Henning, "Sogdian Fragment," 312–313).

23 Cpt. *p-trochos n-nsiou* (*Kephalaia*, Ch. 48; Böhlig and Polotsky, *Kephalaia I* (*Lieferung 1–10*), 121.33).

24 For a recent study on the imprisonment of evil in the cosmos according to Manichaeism, see Kósa, "Evil forces."

are built to imprison the light and impede its redemption.[25] Āz is also said to have bound the human body with invisible chords to the planets and zodiac, in order to subjugate the body to their evil influence.[26] Human life is thus completely determined by the movements of the celestial sphere. Humans are "marked out by the stars and the signs of the zodiac in the sphere. They are appointed for them; in them are their (i.e., the humans') births. And their root is bound up with their zodiacal signs [...]."[27]

On the other hand, Mani considered the sun and the moon to be divine and therefore of a different nature than the other planets. Described as light palaces or boats, the role of the two luminaries was to continuously gather the redeemed particles of light from the material cosmos and to transfer them to the Realm of Light.[28] Against this background, the phenomenon of eclipses posed a vexing metaphysical problem: could the forces of evil seize or engulf the divine Sun and Moon and stop, even if only temporarily, the unceasing process of redemption? In other words, could an eclipse dragon even be conceivable within the Manichaean worldview?

As several scholars have observed, a series of terms denoting the eclipse dragon(s) or the lunar nodes does appear in Manichaean sources: Aṯālyā, Gōzihr, the two nodes (Cpt. *anabibazōn*), and the two dragons (MP *dō azdahāg*).[29] At first sight, the presence of these terms, well known from the Syriac, Iranian, and Greek astrological vocabulary, seems to suggest that Manichaeans did accept the validity of the eclipse-dragon theory.[30] However, a closer look at the extant evidence rather suggests that an important distinction needs to be made between Aṯālyā and Gōzihr on the one hand, and the two nodes and dragons on the other. As I have argued elsewhere, Manichaeans understood Aṯālyā to be a "veil" or "curtain" that the luminaries could draw in order to protect themselves against the onslaught of darkness.[31] This type of clothing or covering metaphors to describe the phenomenon of eclipses, which probably reflects an older Babylonian conception, is consistently used in the

25 *Manichaean Treatise*, III.21–27 (Lieu and Mikkelsen, *Tractatus Manichaicus*, 8–9).
26 On this topic, see Panaino, *Tessere il cielo*, 87–130, with numerous references.
27 *Kephalaia*, Ch. 46, Böhlig and Polotsky, *Kephalaia I (Lieferung 1–10)*, 118.1–3, Gardner, *Kephalaia of the Teacher*, 124. See also Sundermann, *Parabeltexte der Manichäer*, 29n58.
28 On the role of the sun and moon in Manichaeism, see Richter, "Sonnenhymnus"; Kósa, "Ships and Ferries."
29 Pirtea, "Eclipse Dragon in Manichaeism," 548–549; Reck and Sundermann, "Omen-Text," 10; Hutter, *Šābuhragān-Texte*, 10; Böhlig and Polotsky, *Kephalaia I (Lieferung 1–10)*, 168–169.
30 See, for example, Beck, "Anabibazontes"; Reck and Sundermann, "Omen-Text," 13–14.
31 Pirtea, "Eclipse Dragon in Manichaeism." The same case can be made for the use of Gōzihr in the Turfan fragment M 556 ("Eclipse Dragon in Manichaeism," 535–536, 548).

Manichaean sources that explicitly mention eclipses.[32] Conversely, one never finds in these instances an explicit link between eclipses and dragons.

But if Mani regarded eclipses as solar and lunar protective veils, what role could the two dragons and the two nodes have played in the Manichaean system? Is this peculiar overlap of eclipse theories an inconsistency on Mani's part, or does each theory play a distinct role in Mani's cosmology? A careful reading of the two Manichaean texts where the two dragons and the nodes are mentioned may provide an answer to these questions.

3 The Two Dragons in Mani's *Šābuhragān* (M 98)

The Manichaean text preserved in the Turfan manuscript M98 is part of a larger group of fragments which transmit Mani's *Šābuhragān*, a treatise dedicated to the Sasanian king of kings Šāpur I (r. 240–270 CE), whom Mani was hoping to convert. The work thus represents the earliest layer of Manichaean teachings on cosmology, salvation, and eschatology, as formulated in third-century Sasanian Mesopotamia (some possible later interpolations notwithstanding).[33] The reference to the two celestial dragons is found at the beginning of the fragment and is embedded in the same cosmogonical context as the passage from the Chinese *Manichaean Treatise* quoted above:

> [And the Living Spirit and the Mother of Life ...] fixed the seven planets (to the firmament) *and hung up the two dragons* (*dō azdahāg*) and bound them fast (there). And they hung them up on that lowest firmament, and *in order to make them rotate unceasingly upon the Call*, they appointed two angels, one male and one female.[34]

Some scholars have identified the two dragons with the eclipse dragon Gōzihr and the comet-planet Muš Parīg, which are mentioned together with the planets in the astrological section of the Zoroastrian treatise *Bundahišn*.[35] However, since Zoroastrian sources never describe Muš Parīg as a dragon or snake, this

32 Pirtea, "Eclipse Dragon in Manichaeism," 537–538, 545–549.
33 See Hutter, *Šābuhragān-Texte*, 124–134.
34 Frg. M98/I/R/1–6: [...] hpt 'b'xtr przyd · 'wd dw / 'zdh'g ''gwst 'wd gyšt · / 'wd pd h'n ʿy 'yrdwm 'sm'n / 'wl ''gwst · 'wš'n pd w'ng / 'n'spyn grdnydn r'y · nr u m'yg / prystg dw 'br gwm'rd (Hutter, *Šābuhragān-Texte*, 10); Klimkeit, *Gnosis on the Silk Road*, 225–226 (italics mine).
35 See, for example, Skjærvø, Khaleghi-Motlagh, and Russell, "Aždahā."

identification is not fully convincing.[36] Closer parallels to the dragons in M 98 are found in later astrological sources in Middle Persian and Hebrew which report the existence of two eclipse dragons. The Middle Persian compendium *Dādēstān-ī dēnīg* composed by the Zoroastrian scholar Manuščihr (ninth century CE) speaks of "two Gōzihrs" (MP *dō gōčihrān*) as causing eclipses,[37] as does the tenth-century Jewish astrologer Shabbatai Donnolo, who mentions that some books posit the existence of two eclipse dragons (Heb. *telayyim*).[38] Both texts claim that the dragons each cover the extent of six zodiacal signs, and that they follow each other's tails. A somewhat similar description of "two Rāhus" (Skt. *rāhudvayaṃ*) is found in Varāhamihira's *Bṛhatsaṃhitā*, albeit only as a hypothetical argument that the author refutes.[39]

How these later sources relate to Mani's *Šābuhragān* is not entirely certain, but Manuščihr's account in particular may reflect an older Iranian astrological tradition that was also available to Mani. If this is the case, then Mani's attempt to integrate Iranian lore on eclipses in his *Šābuhragān* could be seen as an effort to attune his exposition of the Manichaean worldview to the Zoroastrian audience at Šāpur's court.[40] Yet even so, it remains unclear what specific role Mani attributed to these two dragons and if he understood them to be identical to the lunar nodes. Unfortunately, the two dragons only appear in this passage and Mani never mentions dragons or eclipses elsewhere in the preserved parts of the *Šābuhragān*. Furthermore, the few explicit references to eclipses in other Manichaean sources always describe them as veils or curtains drawn by the luminaries as a protection against evil.

The only clear activity ascribed to the dragons in M98 is that they rotate unceasingly together with the stars and planets at the command of the two appointed angels. The obvious consequence of this unceasing rotation of the dragons, the stars, and the zodiac is the creation of measurable time (years, months, days, etc.) and the alternation of the seasons. Indeed, the subsequent sections of the *Šābuhragān* confirm that Mani was primarily concerned with these latter aspects, i.e., the year and the seasons.[41] The fragments M 7980I+II,

36 On Muš Parīg, see Raffaelli, *L'oroscopo del mondo*, 51–52, 108; Grenet, "Mythological Figures."

37 *Dādēstān-ī dēnīg*, question 69 (Anklesaria, "Dādestān-ī Dīnīk," 137).

38 Mancuso, *Sefer Ḥakhmoni*, 346–347.

39 Varāhamihira, *Bṛhatsaṃhitā* V.7 (Kern, 24). I thank Bill M. Mak for pointing out that Varāhamihira is not reporting here a real belief in two Rāhus.

40 On Mani's interactions with Zoroastrianism at the Sasanian court, as described in the Dublin *Kephalaia*, see the collection of studies in Gardner, BeDuhn, and Dilley, *Mani at the Court*.

41 On the Manichaean solar year, see Tubach, "Sonnenjahr."

M 506 and M 7981 (also part of the *Šābuhragān*) describe in great detail the effect of the sun's annual journey through the "thresholds" (MP *'st'ng*) of the months and through each sign of the zodiac. On its path, the sun encounters evil demonic forces and meteorologic calamities that affect the natural world:

> Then the Sun also ascends from the second threshold to the first threshold, which is higher, larger and wider than the others. And then thirty days of the month of Ādur are brought about by those thirty rotations and 360 double-hours of just this first threshold, where the month of Ādur has its time (lit., place). *And in that (threshold) there are, in the month of Ādur, accordingly the burning dark she-demons, which are like hot winds, and the constellation Gemini. And he frees this heat to descend. Then plants and fruits begin to ripen.* [...] And thereafter the cosmos experiences summer and noon, (that is) *rapih* (i.e., midday heat, AP), just as when the sun existed in the beginning. And then he divides the year into twelve months, in accordance with the twelve zodiacal signs, (and) into Spring and Summer, Fall and Winter, and he calls them into being (successively). Then the trees become green, and herbs and grass (grow) and fruits and plants ripen and animals bring forth their young. [...] And in the same manner the constellation Virgo gains a day each month, (that is) that (month) *when cold starts being released from it (the constellation) and starts to descend. Then the trees wither, it is considered Fall and the evening of the cosmos sets in.* [...] And in the same manner the constellation Pisces gains a day every month, (that is) that month when heat starts being released from it (the constellation), and starts to descend. Then the trees become green, it is considered Spring and the morning of the cosmos dawns.[42]

Mani describes here the sun's passage through the zodiacal signs over the course of a year as a conflict between the beneficent influence of the sun and the extreme heat and cold that the fettered demons, the stars, and the zodiac emit towards earth. Mani likens some months of the year to the essential moments of the cosmic day. Thus, the summer solstice in the month Ādur (May–June)[43] represents the noon of the universe, while the vernal and autumnal equinoxes symbolize dawn and evening respectively. Because of the dualistic nature of the cosmic structure, as devised by the demiurgic Living Spirit, the

42 Hutter, *Šābuhragān-Texte*, 62–68; Klimkeit, *Gnosis on the Silk Road*, 230–231 (italics mine).

43 On the names of the months in M 98, see Henning, "Henochbuch"; Boyce, *Reader*, 67–68.

initially harmful qualities of the zodiac and stars (severe cold, extreme heat, etc.) are providentially diverted towards a good end through the rotation of the sphere. The *Šābuhragān* also insists that the cycle of seasons enables the growth of plants and the ripening of fruit.

The underlying soteriological reasoning that drives Mani's account is that the ripe fruit and vegetables must be ritually consumed by the Manichaean elect in order for the light-elements trapped within them to be liberated and transferred to the Realm of Light. The annual agricultural cycle and the success of harvests are not only essential for a community's earthly subsistence, but also for its eternal salvation.[44]

It is within the framework of this larger cosmic process that the function of the two dragons can be elucidated. Although it is unlikely that Mani already made an exact correlation between the "rotation of the dragons" in M98 and the retrograde movement of the lunar nodes,[45] his description implies at least the basic notion that the movements of these dragons/nodes are somehow causally linked to the occurrence of eclipses. Yet instead of interpreting the dragons' movement as an attack on the luminaries, whom he regards as invulnerable and protected by veils, Mani focuses here on the negative effects of these rotations on terrestrial life. This concern fits well with the overall aim of the *Šābuhragān*. In other words, Mani's juxtaposition of two distinct eclipse theories (eclipses as veils vs. eclipses caused by dragons) is designed to explain the same ominous event, but from two different perspectives: from the point of view of the luminaries, the veil explains how the two light ships remain unharmed during an eclipse. From a human perspective, the unceasing rotation of the dragons and the sphere must be judged in terms of its harmful consequences for life on earth, i.e., earthquakes, droughts, and other phenomena commonly associated with eclipses.

Although this overlap of two different concepts of eclipses may seem inconsistent, having the two perspectives side by side helps Mani solve the problem of solar and lunar divinity and invulnerability: while the eclipse dragons/nodes cannot harm the sun and the moon, they do have a harmful effect on the sublunar world. Keeping the eclipse dragon terminology in this latter context was

---

44 See also the Parthian *Sermon on the Soul*, § 36–39 (Sundermann, *Sermon von der Seele*, 76–77). For a detailed study on the Manichaean view of plant life and the salvation of light through the ritual meals of Manichaean elect, see BeDuhn, *The Manichaean Body*, 163–233.

45 Such a correlation is found e.g., in the *Bundahišn* v A.5 (Pakzad, *Bundahišn*, 352; Agostini and Thrope, *The Bundahišn*, 39), where half of Gōzihr's orbital period is approximated to ten years. The modern value is ca. 9.3 years.

important, since the term *azdahāg* (dragon) suggested to a Zoroastrian audience an association with the various harmful *aži* (dragons) of ancient Iranian lore.[46] It is also worth noting that the *Bundahišn* elsewhere mentions winged and wingless serpents together with the winter season as the evil creations of Ahriman.[47] Although these serpents do not play any role in cosmology, their common evil origin and their association with the winter season might also have inspired Mani in the elaboration of his cosmological account. In both Zoroastrianism and Manichaeism, summer is regarded as a beneficent period (lasting five months)[48] during which fruits ripen and the days are longer than the nights, whereas the autumnal equinox marks the beginning of the cold season and the prevalence of the Ahrimanian forces in the universe. Thus, Mani links this dualistic interpretation of the yearly seasons inherited from Zoroastrianism to the cosmic drama of the liberation of light.

4 The Lunar Nodes in the Coptic Manichaean *Kephalaia*

As far as can be judged from the extant fragments, Mani never elaborates on how exactly the movement of the two dragons and the occurrence of eclipses affect this annual cycle of plant life. Later Manichaeans were thus called to refine Mani's teachings on the subject. In the Eastern Roman Empire, this was done by appealing to the principles of Hellenistic astrology.

The first Manichaean missionaries reached Roman Syria and Egypt already during Mani's lifetime. By the end of the third century, Manichaean presence in Egypt was already strong enough to elicit the first polemical reactions against the new religion by Christians and non-Christians alike. The newly established Manichaean communities, who soon turned to Coptic as a literary language, readily began to integrate the theory and practice of Greco-Egyptian astrology into Mani's cosmological system. The most extensive treatment of astrological topics is found in the Coptic Manichaean *Kephalaia*, a collection of Manichaean teachings preserved in a fourth- or fifth-century manuscript from

46 For an overview, see Skjærvø, Khaleghi-Motlagh, and Russell, "Aždahā."

47 *Bundahišn* XXXI.3, Pakzad, *Bundahišn*, 352; Agostini and Thrope, *The Bundahišn*, 163; cf. Christensen, *Premier chapitre du Vendidad*, 26–27. The passage in question is in fact a paraphrase of the *Widēwdād*, where Ahriman creates a "red dragon" (Av. *aži raoiδita*) and the "dēw-created winter" (Av. *ziiąmca daēuuo.dātəm*) to counter Ahura Mazda's creation of Ērān-wēz (the ancestral Iranian homeland). See *Widēwdād* I.2–3, Christensen, *Premier chapitre du Vendidad*, 23.

48 Hutter, *Šābuhragān-Texte*, 67. A later addition to *Widēwdād* I.3 also insists that Iranian summers last five months (Christensen, *Premier chapitre du Vendidad*, 24).

Medinet Madi.[49] The following passages from *Kephalaia* Ch. 69, which discuss the role of the planets and the zodiac, provide the most detailed description of the lunar nodes in Manichaean literature:

> [The Teacher, i.e., Mani, speaks:] Know, also about these Five Stars, the wanderers, how they are put in place. The star of Zeus is generated by the World of Smoke, which is the Mind. Aphrodite, in turn, came to be from the World of Fire. Ares, however, belongs to the World of Wind. Hermes belongs to the World of Water, while Kronos belongs to the World of Darkness. *The Two Ascendants* (Copt. *n-anabibazōn sneu*), *however, belong to fire and lust, which are dryness and moisture, they are the father and mother of all these things.* [...] *These seven, which we have named—the Five Stars and the two Ascendants—they are the evil-doers who perpetrate every wickedness and evil in every single land, above and below, in every creation, in dryness and moisture, in tree and flesh.*[50]

At the end of the chapter, Mani recapitulates his doctrine concerning the nature of the planets:

> I have also taught you about the *Five Stars*, that they too have come to be from the Five Worlds of Darkness. *I have proclaimed to you, also, about the Two Ascendants, that they stand (according) to the mystery of fire and lust, which is dryness and moisture, the father and mother.* I have also revealed to you about the Sun and the Moon, that they are foreign to them. But because of the (irresistible) force of the planned arrangement according to which one thing takes on the other, and they are being plundered by them, that is why they were counted among them according to the numerical value, *although the Sun and Moon come from the Greatness, (and) do not belong to the stars and the zodiacal signs.*[51]

The recurrent expression "two Ascendants" (Cpt. *n-anabibazōn sneu*) for the lunar nodes requires some explanation. The word *anabibazōn* is borrowed from

---

49 Böhlig and Polotsky, *Kephalaia I (Lieferung 1–10)*. Further fascicles of the text were edited in Böhlig, *Kephalaia I (Lieferung 11/12)*; Funk, *Kephalaia I (Lieferung 13/14)*; Funk, *Kephalaia I (Lieferung 15/16)*.

50 *Kephalaia*, Ch. 69 (Böhlig and Polotsky, *Kephalaia I (Lieferung 1–10)*, 168.1–16; Pettipiece, *Pentadic Redaction*, 180; italics mine). For an alternative translation, see Gardner, *Kephalaia of the Teacher*, 177–178.

51 *Kephalaia*, Ch. 69 (Böhlig and Polotsky, *Kephalaia I (Lieferung 1–10)*, 169.11–22; Pettipiece, *Pentadic Redaction*, 181; italics mine). See also Gardner, *Kephalaia of the Teacher*, 179.

the Greek technical term for the ascending node, i.e., ἀναβιβάζων σύνδεσμος. It is however odd that the Coptic text does not mention the corresponding descending node (καταβιβάζων σύνδεσμος), but simply reduplicates the first term, even though having two ascending nodes is astronomically impossible. This suggests that the Manichaean author of the *Kephalaia* was either ignorant of, or not particularly interested in a precise scientific terminology when he chose this expression.

Scholarly discussions of these passages have generally focused on the motives behind the substitution of the sun and the moon with the two nodes.[52] Indeed, the entire chapter can be read as a Manichaean attempt to respond to competing (Gnostic, Mandean, etc.) views that *all* the seven planets are evil, and to argue instead that the sun and the moon are divine beings that purify and lead the liberated light-elements back to the Realm of Light. The introduction of the two nodes would thus be purely motivated by the need to restore the number of planets to the canonical number of seven.[53] While this explanation seems straightforward, it remains unlikely that in the *Kephalaia* the two nodes were understood to be real planets (as, for example, the Indian *navagraha*). The title and main text of the chapter consistently give the number of planets as five. The planetary names are Greek (Ζεύς, Ἀφροδίτη, Ἄρης, Ἑρμῆς, Κρόνος, in this order), with no traces of the Babylonian, Syriac, Iranian, or for that matter Egyptian planet names. Moreover, the association of the planets with the five evil aeons (smoke, fire, wind, water, darkness) is in agreement with other fivefold divisions in Manichaeism and with what Timothy Pettipiece calls the pentadic redaction of specific elements of the Manichaean pantheon.[54] The same interest in pentads is obvious in the attempt to link the twelve zodiacal signs to the same five aeons.[55] Thus, although the two ascendants are listed together with the five planets, they are never called stars or planets in their own right.

So far, little attention has been given to the description of the qualities and functions ascribed to the two nodes in the Coptic text. The nodes are twice said to belong to the domains of “fire and lust,” which represent “dryness and moisture” and the “Father and Mother of them all.”[56] The implications of attributing

52 Sundermann, *Parabeltexte der Manichäer*, 45n13; Beck, “Anabibazontes”; Demaria, *Kephalaia copti manichei*, 57.

53 Cf. Panaino, *Tessere il cielo*, 101–102, who argues that, in astrological contexts, Manichaeans did include the sun and the moon among the seven planets, despite the theological conflict this standpoint entailed.

54 Pettipiece, *Pentadic Redaction*.

55 *Kephalaia*, 69, Böhlig and Polotsky, *Kephalaia I* (*Lieferung 1–10*), 167.22–30.

56 *Kephalaia*, 69, Böhlig and Polotsky, *Kephalaia I* (*Lieferung 1–10*), 167.7–9, 169. 14–16.

the pairs fire/lust, dry/humid, father/mother to the lunar nodes are not immediately clear, but they seem to involve an astrological reasoning based on the Greek astral sciences practiced in Late Antique Egypt. While it is true that the distinction between a dry and a humid earth is well attested in Manichaean sources,[57] the pair dry/humid attached to the planets and the lunar nodes is more likely to derive from Greek astrology. In Ptolemy's *Tetrabiblos*, the nature of the different planets is described in terms of the four fundamental qualities: dry, humid, hot, and cold. According to this scheme, the nature of Saturn is cold and dry, the nature of Mars is hot and dry, etc.[58] Because the four qualities are further said to be either active/creative (the hot and the humid) or passive/destructive (the dry and the cold), the specific combination of qualities that characterizes each planet also determines its beneficent or maleficent nature.[59] The same reasoning applies for the masculine or feminine nature of the planets, another idea rooted in Ptolemaic astrology.[60] Thus, as Serena Demaria observes, the authors of the *Kephalaia* employ the same Ptolemaic principles to describe the qualities of the two nodes: the ascending node is characterized by heat, dryness, and masculinity, whereas the descending node embodies the opposite qualities (coldness, humidity, femininity).[61]

It is however striking that beyond this detailed description of the nodes' qualities, the *Kephalaia* nowhere discusses their role in producing eclipses. Furthermore, the only passage in Coptic Manichaean literature that mentions a solar eclipse uses the same clothing metaphors as the Eastern Manichaean sources and does not allude to the nodes or to dragons.[62] What function, therefore, do the two nodes play in the astrological context of the Coptic *Kephalaia*?

An answer to this question can be found if we compare the description of the two nodes in the Coptic text and the expositions on cosmology and the seasonal cycle in Mani's *Šābuhragān*. In Mani's Middle Persian treatise, the evil influence of the two dragons and the planets is somehow linked to the alternation of seasons and to their effects on terrestrial (especially plant) life, although Mani does not give any specific details. Notably, the Coptic *Kephalaia* express a very similar concern. Here, "the Teacher" (i.e., Mani) says that the five planets and the two nodes are the evil-doers "who perpetrate wickedness ... in *dry-*

57 See Sundermann, "Die vierzehn Wunden." I thank Gábor Kósa for this reference.

58 Ptolemy, *Tetrabiblos* 1.4 (Robbins, 35–38).

59 Ptolemy, *Tetrabiblos*, 1.5 (Robbins, 38–40).

60 Ptolemy, *Tetrabiblos*, 1.6 (Robbins, 40).

61 Demaria, *Kephalaia copti manichei*, 58. Note also the male and female angels who rotate the sphere according to Mani's *Šābuhragān* (Hutter, *Šābuhragān-Texte*, 10); see above, n34.

62 Allberry, *Psalm-Book*, 196.6–8, see Pirtea, "Eclipse Dragon in Manichaeism," 544–545.

ness and moisture, in *tree and flesh*," that is, on plants, animals, and humans.[63] This echoes the description in the *Šābuhragān* where the sun's passage through the thresholds of each month and zodiacal sign is marked by the descent of extreme heat or cold upon the earth, which affects the life of plants and animals.

Ptolemy's *Tetrabiblos* treats at length the meteorological effects of each zodiacal sign (destructive winds, extreme head and cold, etc.) and describes them in terms reminiscent of the cited passages in the *Šābuhragān*.[64] On the basis of these broad similarities, the Manichaean author(s) of the *Kephalaia* could easily adapt the peculiarities of Mani's cosmology to the Greek astrological tradition of Late Antique Egypt and even elaborate on Mani's terse account using an updated (Ptolemaic) terminology. In this way, the various astrological aspects of the five planets, the two nodes, and the twelve zodiacal signs described in the same chapter (opposition, conjunction, etc.)[65] could be used to explain in more precise terms how the rotation of the sphere affected "tree and flesh" and in what way it influenced the salvation of the light-elements trapped in the world.

Another important similarity is the connection between eclipses and equinoxes, on one hand, and Mani's understanding of the sun's annual journey through the zodiac, on the other. In the chapter dealing with the effects of eclipses, Ptolemy warns that the most severe damage and destruction to vegetation comes about when eclipses occur in the equinoctial and solstitial signs (Aries/Libra and Cancer/Capricorn):

> Likewise stars in the solstitial or equinoctial signs have significance in general for the conditions of the air and the seasons related to each of these signs, and in particular they concern the spring and things which grow from the earth. For when they are at the spring equinox they affect the new shoots of the arboreal crops, such as grapes and figs, and whatever matures with them; at the summer solstice, the gathering and storing of the crops, and in Egypt, peculiarly, the rising of the Nile; at the autumn solstice they concern the sowing, the hay crops, and such; and at the winter equinox the vegetables and the kinds of birds and fish most common at this season.[66]

63 *Kephalaia*, 69 (Böhlig and Polotsky, *Kephalaia I* (*Lieferung 1–10*), 168.13–16; italics mine).
64 See, for example, Ptolemy, *Tetrabiblos*, II.11 (Robbins, 200–205).
65 On this topic, see Stegemann, "Zu Kapitel 69"; Demaria, *Kephalaia copti manichei*, 56–57.
66 Ptolemy, *Tetrabiblos*, II.7 (Robbins, 174).

This potentially destructive effect on vegetation is explained by Ptolemy as a consequence of the close relationship between the four equinoctial and solstitial signs and the four seasons: the vernal equinox (Aries) marks the beginning of the moist season (spring), the autumnal equinox (Libra) that of the dry season, etc.[67] Here again, the author(s) of the *Kephalaia* could easily connect Ptolemy's account with Mani's discussion of the seasons and the equinoxes in the *Šābuhragān*. For example, Mani and Ptolemy use similar analogies when they compare the seasons with the various moments (dawn, noon, evening) of the cosmic day (Mani), or with the four stages (infancy, youth, adulthood, old age) of human life (Ptolemy).[68] But more importantly, Ptolemy's observation that eclipses particularly affect vegetation when they occur at the equinoxes could be used to expand upon Mani's simple mention of the two dragons in M98. Put differently, Ptolemy's theory can help explain, using Greek astrological terminology, how Mani understood the astral effects of the planets, the zodiac, and the two dragons to unfold. One should stress again that the underlying soteriological framework in this astrologized version of Mani's cosmological myth had to remain intact. The interest in the astral effects on vegetation was driven by the daily Manichaean practice of ritual consumption, through which the light-elements trapped in the plants could be released.

In conclusion, the overlap between Ptolemaic and Manichaean cosmological views in the Coptic *Kephalaia* resembles in many ways the approach taken by Mani in the *Šābuhragān*. Mani had appealed both to an eclipse-veil and an eclipse-dragon theory in order to elucidate two different aspects of his cosmology: the invulnerability of the luminaries to eclipses (the veils), and the negative effects of eclipses on vegetative, animal, and human life (the two dragons). As stated above, the former view was likely a Babylonian remnant, while the latter concept was possibly adopted from contemporary Zoroastrian texts on astral science. Going a step further, the Manichaeans in Egypt spelled out the details of these astral effects by adding yet another layer to Mani's theory, i.e., the Greek astrological tradition.

Discerning between these different layers in the Coptic *Manichaica* can also help explain some apparent incongruities: why are, for instance, the two Ascendants in *Kephalaia* Ch. 69 never described as dragons or serpents? A likely answer is that since in Greco-Roman astrology the nodes were never associated with any such creatures, the equivalence between the two dragons in M98 and

67 Ptolemy, *Tetrabiblos*, 1.10 (Robbins, 58–59).

68 Ptolemy, *Tetrabiblos*, 1.10 (Robbins, 60–61).

the two Ascendants in the *Kephalaia* remained strictly functional. It mattered less if these entities were described in mythical (Zoroastrian) or in astronomical (Hellenistic) terms; it was more important to preserve intact their specific role in the Manichaean cosmological system.

5 Conclusion

In their encounter with the various cosmological traditions in Mesopotamia, Sasanian Iran, and Egypt, Mani and his followers intentionally and inventively adapted specific teachings of these surrounding cultures for their own ends. In the case of the eclipse dragon(s), Mani interpreted the concept with respect to his own religious concerns, specifically the liberation of the light-elements trapped in the natural world (plants, animals, humans). In this way, Mani subordinated a primarily cosmological and astrological concept to the strict soteriological framework of his religious system. Consequently, in their own later modifications and elaborations of Mani's teachings in the context of Late Antique Egypt, the Manichaean author(s) of the *Kephalaia* used Ptolemaic astrology to describe in more detail the evil effects of the nodes, the planets, and the zodiac on the life of plants, animals, and human beings. Since the Sun and the Moon were seen as invulnerable to evil, the eclipse dragons were used to explain the destructive effects of eclipses on "tree and flesh" and on the light-elements trapped inside them, which can only be redeemed through the rotation of the heavenly sphere and the sun's yearly path through the signs and seasons.

The overlap of different eclipse theories in Mani's cosmological writings and in the Coptic Manichaean corpus offers a methodologically relevant case study that can hopefully serve as an incentive for further research. For example, the extent to which the Manichaean communities on the Silk Road and in China continued this accommodative strategy in their encounter with Buddhist cosmology still needs to be explored in detail. Although Eastern Manichaean sources on eclipses or eclipse dragons are very scarce, the overall wealth of textual and iconographic material discovered in Turfan, Dunhuang, and southeast China (especially the Chinese cosmology painting) can certainly contribute to a better understanding of Mani's worldview and the different ways in which Manichaeism interacted with the various cosmological traditions of Central and East Asia.

Acknowledgments

A version of this paper was presented at the *International Conference on Traditional Sciences in Asia: East-West Encounter in the Science of Heaven and Earth* (October 25–27, 2017, Kyoto University, Japan), with the financial support of the MPIWG project "Visualization of the Heavens and Their Material Cultures in Eurasia and North Africa (4000 BCE–1700 CE)." I am grateful to Dr. Sonja Brentjes (MPIWG, Berlin) for the kind invitation and for her useful comments on an earlier draft of this paper. I also thank Gábor Kósa, Bill M. Mak, Eric Huntington, and the two reviewers for their helpful remarks. All remaining errors and inaccuracies are my own. I carried out a significant part of the preliminary research on this subject at the Warburg Institute (University of London) in 2016, with the generous support of a Henri Frankfort/Grete Sondheimer Fellowship. I wish to express my gratitude to all the members and colleagues at the Institute, first and foremost to Prof. Charles Burnett, who supervised and guided my research during my stay in London.

Bibliography

Agostini, Domenico and Samuel Thrope, *The Bundahišn. The Zoroastrian Book of Creation*, Oxford: Oxford University Press, 2020.

Allberry, Charles R.C. *A Manichaean Psalm-Book, Part II*. With a contribution by Hugo Ibscher. Stuttgart: Kohlhammer, 1938.

Anklesaria, Peshotan Kavashaw. "A Critical Edition of the Unedited Portion of the Dādestān-i dīnīk." PhD diss., University of London, 1958.

Baumann, Brian. *Divine Knowledge: Buddhist Mathematics According to the Anonymous Manual of Mongolian Astrology and Divination*. Leiden: Brill, 2008.

Beck, Roger. "The Anabibazontes in the Manichaean Kephalaia." *Zeitschrift für Papyrologie und Epigraphik* 69 (1987): 193–196.

BeDuhn, Jason D. *The Manichaean Body in Discipline and Ritual*. Baltimore: John Hopkins University Press, 2000.

Billard, Roger. *L'astronomie indienne: Investigation des textes sanskrits et des données numériques*. Paris: Adrien-Maisonneuve, 1971.

Böhlig, Alexander. *Kephalaia I, 2. Hälfte, Lieferung 11/12*. Stuttgart: Kohlhammer, 1966.

Böhlig, Alexander and H.J. Polotsky. *Kephalaia I, 2. Hälfte, Lieferung 1–10*. Stuttgart: Kohlhammer, 1940.

Boyce, Mary. *A Reader in Manichaean Middle Persian and Parthia: Texts with Notes*. Leiden: Brill, 1975.

Christensen, Arthur. *Le premier chapitre du Vendidad et l'histoire primitive des tribus iraniennes*. Copenhagen: Munksgaard, 1943.

Coedès, George. *Inscriptions du Cambodge*. Vol. 5. Paris: Boccard, 1953.

Coedès, George. *Inscriptions du Cambodge*. Vol. 6. Paris: École française d'Extrême-Orient, 1954.

Cùscito, Giuseppe. "Il *tly* nella cosmologia del *Sefer ḥaḵmônî*." *Sefer Yuḥasin* 8 (2020): 105–135. doi: 10.6092/2281–6062/7229.

Demaria, Serena. *I capitoli* LXIX *e* LXX *dei Kephalaia copti manichei: Traduzione e commento*. Imola: La mandragora, 1998.

Draelants, Isabelle. "Le *Liber Nemroth* de astronomia. État de la question et nouveaux indices." *Revue d'Histoire des Textes* 13 (2018): 245–329. doi: 10.1484/J.RHT.5.114892.

Funk, Wolf-Peter. *Kephalaia I, 2. Hälfte, Lieferung 13/14*. Stuttgart: Kohlhammer, 1999.

Funk, Wolf-Peter. *Kephalaia I, 2. Hälfte, Lieferung 15/16*. Stuttgart: Kohlhammer, 2000.

Gail, Adalbert. "Planets and Pseudoplanets in Indian Literature and Art with Special Reference to Nepal." *East and West* 30, no. 1/4 (1980): 133–146. http://www.jstor.org/stable/29756562.

Gardner, Iain. *The Kephalaia of the Teacher: The Edited Coptic Manichaean Texts in Translation with Commentary*. Leiden: Brill, 1995.

Gardner, Iain, Jason BeDuhn, and Paul Dilley, eds. *Mani at the Court of the Persian King: Studies on the Chester Beatty* Kephalaia *Codex*. Leiden: Brill, 2015.

Geslani, Marko, Bill M. Mak, Michio Yano, and Kenneth Zysk. "Garga and Early Astral Science in India." *History of Science in South Asia* 5, no. 1 (2017): 151–191. doi: 10.18732/H2ND44.

Grenet, Frantz. "Some hitherto Unrecognized Mythological Figures on Sasanian Seals: Proposed Identifications." In *Commentationes iranicae. Vladimiro f. Aaron Livschits nonagenario donum natalicum / Sbornik statej k 90-letiyu Vladimira Aronoviča Livšica*, edited by Sergey Tokhtasev and Pavel Lurje, 202–210. St. Petersburg: Nestor-Historia, 2013.

Gulácsi, Zsuzsanna. *Mani's Pictures: The Didactic Images of the Manichaeans from Sasanian Mesopotamia to Uygur Central Asia and Tang-Ming China*. Leiden: Brill, 2015.

Hartner, Willy. "Djawzahar." In *Encyclopedia of Islam*. Second edition, edited by P. Bearman, Th. Bianquis, C.E. Bosworth, E. van Donzel, and W.P. Heinrichs. Accessed online February 4, 2021. doi: 10.1163/1573-3912_islam_SIM_2041.

Hartner, Willy. "The Pseudoplanetary Nodes of the Moon's Orbit in Hindu and Islamic Iconographies." *Ars Islamica* 5, no. 2 (1938): 112–154. https://www.jstor.org/stable/4520926.

Hayman, Peter. "The Dragon, the Axis Mundi, and Sefer Yeṣira § 59." In *"He unfurrowed his brow and laughed": Essays in Honour of Prof. Nicolas Wyatt*, edited by Wilfred Watson, 113–140. Münster: Ugarit-Verlag, 2007.

Henning, Walter Bruno. "An Astronomical Chapter in the Bundahishn." *Journal of the Royal Asiatic Society* 74, no. 3–4 (1942): 229–248.

Henning, Walter Bruno. "Ein manichäisches Henochbuch." *Sitzungsberichte der Preussischen Akademie der Wissenschaften, Philologisch-historische Klasse* 5 (1934): 27–35.

Henning, Walter Bruno. "A Sogdian Fragment of the Manichaean Cosmogony." *Bulletin of the School of Oriental and African Studies* 12, no. 2 (1948): 306–318.

Hintze, Almut. "The Cow That Came from the Moon: The Avestan Expression *māh-gaociϑra-*." *Bulletin of the Asia Institute* (*Iranian and Zoroastrian Studies in Honor of Prods Oktor Skjærvø*) 19 (2005 [2009]): 57–66.

Hutter, Manfred. *Manis kosmogonische Šābuhragān-Texte. Edition, Kommentar und literaturgeschichtliche Einordnung der manichäisch-mittelpersischen Handschriften M 98/99 I und M 7980–7984*. Wiesbaden: Harrassowitz, 1992.

Kennedy, Edward S., and David Pingree. *The Astrological History of Māshā'allāh*. Cambridge: Harvard University Press, 1971.

Klimkeit, Hans-Joachim. *Gnosis on the Silk Road: Gnostic Texts from Central Asia*. San Francisco: Harper, 1993.

Kósa, Gábor. "Atlas and Splenditenens in the Chinese Manichaean 'Cosmology Painting'." In *Gnostica et Manichaica. Festschrift für Aloïs van Tongerloo, anlässlich des 60. Geburtstages überreicht von Kollegen, Freunden und Schülern*, edited by Michael Knüppel und Luigi Cirillo, 63–88. Wiesbaden: Harrassowitz, 2012.

Kósa, Gábor. "Imprisoned evil forces in the textual and visual remains of Manichaeism." In *Fontes. Studies presented to Róbert Simon on the occasion of his 75th birthday. Tanulmányok a 75 éves Simon Róbert tiszteletére*, edited by Monika Pesthy-Simon, 64–87. Budapest: Corvina, 2014.

Kósa, Gábor. "The Manichaean Attitude to Natural Phenomena as Reflected in the Berlin Kephalaia." *Open Theology* 1 (2015): 255–268. doi: 10.1515/opth-2015-0011.

Kósa, Gábor. "Ships and Ferries in the Manichaean Cosmology Painting." In: *Danfeng canggui—Zhangxun bainian danchen jinian wenji* 丹枫苍檜—章巽百年诞辰纪念文集 [A Collection of Studies for the 100th Anniversary of Zhang Xun's Birthday], edited by Rui Chuanming 芮传明 and Zhang Jiaping 章嘉平, 41–67. Guangzhou: Guangdong Renmin Chubanshe, 2015.

Kósa, Gábor. "Translating the Eikōn. Some considerations on the relation of the Chinese Cosmology painting to the Eikōn." In *Vom Aramäischen zum Alttürkischen. Fragen zur Übersetzung von manichäischen Texten. Vorträge des Göttinger Symposiums vom 29./30. September 2011*, edited by Jens Peter Laut and Klaus Röhrborn, 49–84. Berlin: De Gruyter, 2014.

Kotyk, Jeffrey. "Astrological Iconography of Planetary Deities in Tang China: Near Eastern and Indian Icons in Chinese Buddhism." *Journal of Chinese Buddhist Studies* 30 (2017): 33–88.

Kuehn, Sara. *The Dragon in Medieval East Christian and Islamic Art*. Leiden: Brill, 2011.

Kuehn, Sara. "The Eclipse Demons Rāhu and Ketu in Islamic Astral Sciences," in *In Umbra: Demonologja kak semiotičeskaja sistema* (*Demonology as a Semiotic System*), edited by D. Antonov and O. Khristoforova, 211–244. Moscow: Indrik, 2016.

Lieu, Samuel, and Gunner B. Mikkelsen, *Tractatus Manichaicus Sinicus. Moni jiao can-*

jing 摩尼教殘經. *Pars Prima: Text, Translation and Indices*. In association with Lance Eccles, Enrico Morano, Nils A. Pedersen, Nicholas Sims-Williams, and Aloïs van Tongerloo. Turnhout: Brepols, 2017.

Ma, Xiaohe. "Wheel of Karma (Zodiac). A Study on the Cosmology Painting and Scripture of Buddha-Nature." In *Zur lichten Heimat. Studien zu Manichäismus, Iranistik und Zentralasienkunde im Gedenken an Werner Sundermann*, edited by Team Turfanforschung, 343–354. Wiesbaden: Harrassowitz, 2017.

MacKenzie, David N. "Bundahišn." In *Encyclopaedia Iranica*, online edition, New York, 1996–. Accessed February 4, 2021. https://iranicaonline.org/articles/bundahisn-primal-creation.

MacKenzie, David N. "Gōzihr." In *Encyclopaedia Iranica*, online edition, New York, 1996–. Accessed February 4, 2021. https://iranicaonline.org/articles/gozihr.

MacKenzie, David N. "Zoroastrian Astrology in the *Bundahišn*." *Bulletin of the School of Oriental and African Studies* 27, no. 3 (1964): 511–529.

Mak, Bill M. "Indian *Jyotiṣa* through the Lens of Chinese Buddist Canon." *Journal of Oriental Studies* 48, no. 1 (2015): 1–19.

Mak, Bill M. "The Transmission of Buddhist Astral Science from India to East Asia: The Central Asian Connection." *Historia scientiarum* 24, no. 2 (2015): 59–75.

Mak, Bill M. "The Transmission of the *Grahamātṛkādhāraṇī* and other Buddhist Planetary Astral Texts." *Pacific World* 20 (2018): 223–256.

Mak, Bill M. "Vedic Astral Lore and Planetary Science in the *Gārgīyajyotiṣa*." *History of Science in South Asia* 7 (2019): 52–71.

Mancuso, Piergabriele. *Shabbatai Donnolo's "Sefer Ḥakhmoni". Introduction, Critical Text, and annotated English Translation*. Leiden: Brill, 2010.

Markel, Stephen. "The Imagery and Iconographic Development of the Indian Planetary Deities Rāhu and Ketu." *South Asian Studies* 6 (1990): 9–26.

Markel, Stephen. *Origins of the Indian Planetary Deities*. Lewiston, NY: Mellen, 1995.

Mikkelsen, Gunner. *Dictionary of Manichaean Texts in Chinese*. Turnhout: Brepols, 2006.

Mitchiner, John. *The Yuga Purāṇa: Critical Edition with an English Translation and a detailed Introduction*. Calcutta: Asiatic Society, 1986.

Miyazaki, Tensho, Jundo Nagashima, Tatsushi Tamai, and Liqun Zhou. "The *Śārdūlakarṇāvadāna* from Central Asia". In *Buddhist Manuscripts from Central Asia. The St. Petersburg Sanskrit Fragments. Volume I*, edited by Seishi Karashima and Margarita I. Vorobyova-Desyatovskaya, 1–85. Tokyo: The Institute of Oriental Manuscripts of the Russian Academy of Sciences and The International Research Institute for Advanced Buddhology at Soka University, 2015.

Oort, Johannes van. "The Paraclete Mani as the Apostle of Jesus Christ and the Origins of the New Church." In *The Apostolic Age in Patristic Thought*, edited by Anthony Hilhorst, 139–157. Leiden: Brill, 2004.

Pakzad, Fazlollah. *Bundahišn. Zoroastrische Kosmogonie und Kosmologie. Band 1: Kritische Edition.* Tehran: Centre for the Great Islamic Encyclopaedia, 2005.

Panaino, Antonio. "Pahlavi Gwcyhl: Gōzihr o Gawčihr?" In *Scritti in onore di Giovanni M. D'Erme*, edited by Michele Bernardini and Natalia L. Tornesello, 795–826. Napoli: Università degli Studi di Napoli "L'Orientale", 2005.

Panaino, Antonio. *Tessere il cielo: considerazioni sulle tavole astronomiche gli oroscopi e la dottrina dei legamenti tra induismo, zoroastrismo, manicheismo e mandeismo.* Roma: Istituto italiano per l'Africa e l'Oriente, 1998.

Pettipiece, Timothy. *Pentadic Redaction in the Manichaean* Kephalaia. Leiden: Brill, 2009.

Pingree, David. "Astronomy and Astrology in India and Iran." *Isis* 54, no. 2 (1963): 229–246.

Pingree, David. "The Empires of Rudradāman and Yaśodharman: Evidence from Two Astrological Geographies." *Journal of the American Oriental Society* 79, no. 4 (1959): 267–270.

Pingree, David. *From Astral Omens to Astrology from Babylon to Bikaner.* Roma: Istituto per l'Africa e l'Oriente, 1998.

Pingree, David. "Indian Planetary Images and the Tradition of Astral Magic." *Journal of the Warburg and Courtauld Institutes* 52 (1989): 1–13.

Pingree, David. *Jyotiḥśāstra.* Wiesbaden: Harrassowitz, 1981.

Pirtea, Adrian. "From Lunar Nodes to Eclipse Dragons: The *Fundaments of the Chaldean Art* (CCAG V/2, 131–140) and the Reception of Arabo-Persian Astrology in Byzantium." In *Savoirs prédictifs et techniques divinatoires de l'Antiquité tardive à Byzance*, edited by Paul Magdalino and Andrei Timotin, 339–365. Seyssel: La Pomme d'or, 2019.

Pirtea, Adrian. "Is There an Eclipse Dragon in Manichaeism? Some Problems Concerning the Origin and Function of āṯālyā in Manichaean Sources." In *Zur lichten Heimat. Studien zu Manichäismus, Iranistik und Zentralasienkunde im Gedenken an Werner Sundermann*, edited by Team Turfanforschung, 535–554. Wiesbaden: Harrassowitz, 2017.

Ptolemy. *Tetrabiblos*. Edited and translated by Frank Robbins. Cambridge: Harvard University Press, 1940.

Raffaelli, Enrico. *L'oroscopo del mondo. Il tema di nascità del mondo e del primo uomo secondo l'astrologia zoroastriana.* Milano: Mimesis, 2001.

Reck, Christiane and Werner Sundermann. "Ein illustrierter mittelpersischer manichäischer Omen-Text aus Turfan." *Zentralasiatische Studien* 27 (1997 [1998]): 7–23.

Richter, Siegfried. "Ein manichäischer Sonnenhymnus." In *Studia Manichaica. IV. Internationaler Kongreß zum Manichäismus, Berlin, 14.–18. Juli 1997*, edited by Ronald E. Emmerick, Werner Sundermann and Peter Zieme, 481–493. Berlin: Akademie-Verlag, 2000.

Sharf, Andrew. *The Universe of Shabbetai Donnolo*. New York: Ktav, 1976.

Skjærvø, Prods Oktor, Djalal Khaleghi-Motlagh, and James R. Russell. "Aždahā." In *Encyclopaedia Iranica*, online edition, New York, 1996–. Accessed February 3, 2021. https://iranicaonline.org/articles/azdaha-dragon-various-kinds/.

Sørensen, Henrik. "Astrology and the Worship of the Planets in Esoteric Buddhism of the Tang." In *Esoteric Buddhism and the Tantras in East Asia*, edited by Charles Orzech, Henrik Sørensen, and Richard Payne, 230–244. Leiden: Brill, 2011.

Soudavar, Abolala. "The Significance of Av. čiθra, OP. čiça, MP. čihr, and NP. čehr, for the Iranian Cosmogony of Light." *Iranica Antiqua* 41 (2006): 151–185.

Stegemann, Viktor. "Zu Kapitel 69 der Kephalaia des Mani." *Zeitschrift für die neutestamentliche Wissenschaft und die Kunde der älteren Kirche* 37 (1938): 214–223.

Stuckrad, Kocku von. *Das Ringen um die Astrologie. Jüdische und christliche Beiträge zum antiken Zeitverständnis*. Berlin: De Gruyter, 2000.

Sundermann, Werner. *Mittelpersische und parthische kosmogonische und Parabeltexte der Manichäer*. Berlin: Akademie-Verlag, 1973.

Sundermann, Werner. "Der Paraklet in der ostmanichäischen Überlieferung." In *Manichaean Studies: Proceedings of the First International Conference on Manichaeism, August 5–9, 1987*, edited by Peter Bryder, 201–212. Lund: Plus Ultra, 1988.

Sundermann, Werner. *Der Sermon vom Licht-Nous. Eine Lehrschrift des östlichen Manichäismus. Edition der parthischen und soghdischen Version*. Berlin: Akademie-Verlag, 1992.

Sundermann, Werner. *Der Sermon von der Seele: Eine Lehrschrift des östlichen Manichäismus*. Turnhout: Brepols, 1997.

Sundermann, Werner. "Die vierzehn Wunden der lebendigen Seele." *Altorientalische Forschungen* 12, no. 2 (1985): 288–295.

Tubach, Jürgen. "Mani und das 366-tägige Sonnenjahr." *Le Muséon* 105, no. 1 (1992): 45–55.

Varāhamihira. *Bṛhatsaṃhitā*. Edited by H. Kern. Calcutta: Baptist Mission Press, 1865.

Villey, Emilie. "Les textes astronomiques syriaques (VIe et VIIe s.): établissement d'un corpus et de critères de datation. Édition, traduction et lexique." PhD diss., Université de Caen Normandie, 2012.

Yano, Michio. "The *Ch'i-yao jang-tsai-chüeh* and its Ephemerides." *Centaurus* 29, no. 1 (1986): 28–35.

Yoshida Yutaka 吉田豊. "Shinshutsu Manikyō kaiga no keijijō" 新出マニ教絵画の形而上. *Yamato Bunka* 大和文華 121 (2010): 1–34.

Yoshida Yutaka 吉田豊, and Shōichi Furukawa 古川攝一, eds. *Chūgoku Kōnan Manikyō Kaiga Kenkyū* 中国江南マニ教絵画研究. *Studies of the Chinese Manichaean Paintings of South Chinese Origin preserved in Japan*. Kyoto: Rinsenshoten, 2015.

CHAPTER 4

Deciphering Aristotle with Chinese Medical Cosmology: *Nanban Unkiron* and the Reception of Jesuit Cosmology in Early Modern Japan

Ryuji Hiraoka

1 Introduction

Japan first encountered Western cosmology through the Jesuit mission that began in 1549. The missionaries noticed that the Japanese showed a strong interest in European explanations of natural phenomena, and used these explanations to support their criticisms of Buddhism, their main religious rival in the country, and argue for God's existence on rational grounds.[1] The knowledge transmitted was the basic outline of the Aristotelian-Ptolemaic worldview, including the geocentric planetary system, the sphericity of the Earth, and Aristotelian meteorology. This was the first introduction of Western scientific knowledge to East Asia, preceding Matteo Ricci. Although the activities of these early Jesuits were put to an end in 1614 when the Tokugawa shogunate tightened its ban on Christianity, the Japanese treatises produced during this period survived the ban on Christianity itself and circulated widely in manuscript form throughout the following centuries.

Historians have portrayed this early transmission as an initial but insufficient step in the process by which Japan shifted from Chinese-influenced astronomy to a system influenced by the West. In this narrative, the astronomical and cosmological thinking of Japan's early modern period, which was initially dominated by Chinese scientific tradition, is gradually replaced and overturned by new ideas intermittently introduced from the West, especially since the eighteenth century.[2] One work in particular that advocated for this view was *A History of Japanese Astronomy*, the first comprehensive history of the field in English and still an influential work today, in which Nakayama Shigeru writes that early Jesuit cosmology "did not, however, replace traditional

1 See Hiraoka, "Jesuits, Creation and Cosmology."

2 See Nihon gakushiin, *Meiji zen Nihon tenmongaku-shi*, preface, 2; Nakayama, *A History of Japanese Astronomy*, 2–3; and Watanabe, *Kinsei Nihon tenmongaku-shi*, preface, 5.

 | DOI:10.1163/9789004511675_006

cosmology, with which it was largely incompatible. At best, Aristotelian theory was accepted as a facet of European learning and merely juxtaposed onto the Eastern theory."[3]

However, this view and narrative are oversimplified at best. Close examination of how early Jesuit treatises on cosmology were received and used by Japanese readers reveals a more complex pattern of interactions and negotiations between the Eastern and Western cosmologies that met in early modern Japan. This chapter considers the case of *Nanban unkiron* 南蠻運氣論 (*Yunqi* theory of the southern barbarians; hereafter NU), compiled in Nagasaki in the mid-seventeenth century, in this light. NU is unique in two important ways: its technical terminology is modified to conform to the Chinese medical cosmology of *yunqi* (Jp. *unki*) theory 運氣論 (the doctrine of five periods and six *qi*, or *wuyun liuqi* 五運六氣),[4] and it interpolates passages from Chinese medical texts such as *Huangdi neijing suwen* 黃帝内經素問 (Huangdi's inner classic, basic questions; hereafter *Suwen*). These interpolations have no counterpart in the Western sources of NU, resulting in a complex mixture of Chinese and Western elements in a single Japanese book. This immediately raises the question of how and why this cosmological overlap occurred.

The evidence regarding NU's authorship and origin, combined with the reception of the work in seventeenth-century Nagasaki and by later scholars, reveals not a simple replacement of the old cosmology by the new one, but an assimilation and absorption of Western knowledge into the existing cosmological framework. This shows how cosmologies—in the broader sense, covering a range of explanations about the structure and phenomena of the universe—can "overlap" when multiple theories from different sources coexist within a single scholarly community. The case of NU also sheds light on the nature of interaction between interpretive traditions, showing how one tradition's theories can incorporate concepts and assimilate texts from other traditions entirely.

2 Origin and Background of *Nanban Unkiron*

The origin of NU remains unclear due to a lack of bibliographic information in the document itself: it specifies no author or date, and some manuscripts even

3 Nakayama, *A History of Japanese Astronomy*, 114.

4 For a concise explanation of *yunqi* theory, see Yamada, *Ki no shizenshō*, 52–101, and Unschuld, *Huangdi neijing suwen*, 385–494.

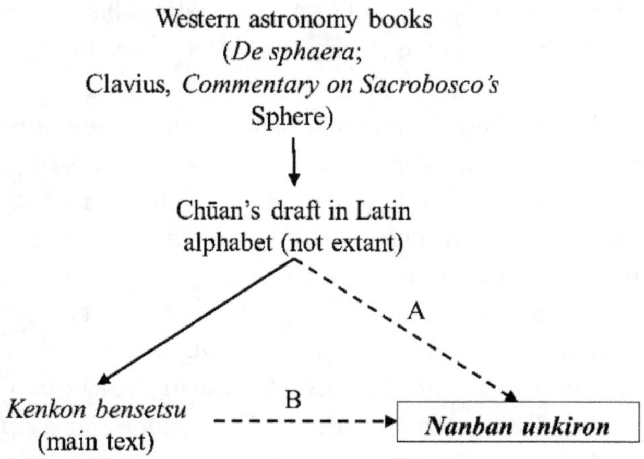

FIGURE 4.1 Possible derivations of *Nanban unkiron*

have different titles.[5] Previous studies have revealed, however, that the book's ancestry might be summarized as shown in figure 4.1.[6]

The direct source text for NU is assumed to be a draft by Sawano Chūan 澤野忠庵 (1580–1650). Born Cristóvão Ferreira, Chūan arrived in Japan as a Jesuit priest in 1609. Captured by the Shogunate in 1633, he apostatized, changed his name, and began to work for the Japanese authorities as an investigator and translator.[7] Sometime around 1645, Chūan was ordered to translate some Western books on astronomy (*tenmonsho* 天文書). While he could read and speak Japanese, he was not able to write well in Japanese, and he wrote his Japanese translation in the Latin script.[8] Historians have assumed that the books Chūan translated were (1) *De sphaera* (1593), a textbook in the Jesuit college in Japan, and (2) Christopher Clavius' *Commentary on Sacrobosco's Sphere*, which had a great influence on the Jesuit mission in early modern East Asia.[9]

There are two possibilities regarding the derivation of NU, represented as A and B in figure 4.1. Possibility A is that the book was transcribed directly into standard Japanese orthography from Chūan's draft, with editorial modifications based on Chinese *yunqi* theory as discussed in the next section. One

5 For the extant manuscripts of NU and their genealogy, see Hiraoka, *Nanbankei uchūron*, 157–191.

6 See Ōya, "Kenkon bensetsu no ichi ihon," 39; and Itō, *Bunmei ni okeru kagaku*, 226.

7 For Chūan's biography, see Cieslik, "The Case of Christovão Ferreira."

8 KB, 1–2.

9 Itō, *Bunmei ni okeru kagaku*, 231; Hiraoka, "Clavius and his Astronomical Data."

document suggests that this was done in mid-seventeenth-century Nagasaki, most likely by Shōgin 松吟 (1608–1691), head priest of Kōgenji temple in the city.[10]

It is known that Chūan's draft was separately transcribed into a more standard Japanese orthography, without changing the actual language or terminology used, around the mid-seventeenth century in Nagasaki by Mukai Genshō 向井玄松 (1609–1677), a Confucian physician, and Dutch interpreter Nishi Kichibyōe 西吉兵衛 (?–1666). This was done in order for it to be used as the main text of *Kenkon bensetsu* 乾坤辨説 (A discussion on heaven and earth with critical commentaries; hereafter KB). Possibility B, then, is that NU was edited not from Chūan's original but from the main text of KB, already transcribed into standard Japanese orthography by Mukai and Nishi.

The question of which possibility is more likely remains an open one, but in either case we can say with confidence that NU was established in mid-seventeenth-century Nagasaki, most likely with Shōgin as editor, derived directly or indirectly from Chūan's draft in Roman script.

Nagasaki was not only NU's birthplace but also the center of its subsequent circulation and reception. An early eighteenth-century document suggests that it was "copied by many people" in the city.[11] It should be noted that the contents of NU agree well with the research interests of the "Nagasaki school," a group of scholars of Western learning active in seventeenth-century Nagasaki. Figure 4.2 shows the master-pupil relationships within the school and the subjects transmitted.[12]

Astronomy (*tenmon* 天文 or *shōi* 象緯) was transmitted from master to pupil almost without exception. The next most commonly transmitted subject was geography (*chiri* 地理 or *yochi* 輿地), followed by *yunqi* (roughly equivalent to meteorology). The school's research interests were thus quite broad, ranging from the celestial to the terrestrial and taking in much of what lay between the two as well. This certainly matches the content of NU, which has a "Celestial Part" discussing Ptolemaic-Sacroboscan astronomy and a "Terrestrial Part" covering Aristotelian natural philosophy, including the concept of a spherical Earth, four-element theory, and explanations for meteorological phenomena.[13]

10 *Sokuryō higen*, 96; Ōya, "Kenkon bensetsu no ichi ihon," 36–37.

11 *Sokuryō higen*, 96; Ōya, "Kenkon bensetsu no ichi ihon," 36–37.

12 Wakaki et al., *Nagasaki senminden chūkai*, 45–51, 153–157. For Kobayashi Kentei, a representative scholar of the Nagasaki school, see Hiraoka, "Kobayashi Kentei den," and Hiraoka, "Kōtaiji no rakan sekizōgun to Kobayashi Kentei."

13 Note that the *Nigi ryakusetsu* 二儀略説 (Brief discussion of Heaven and Earth), a Japanese translation of *De sphaera* containing almost the same content as NU, was trans-

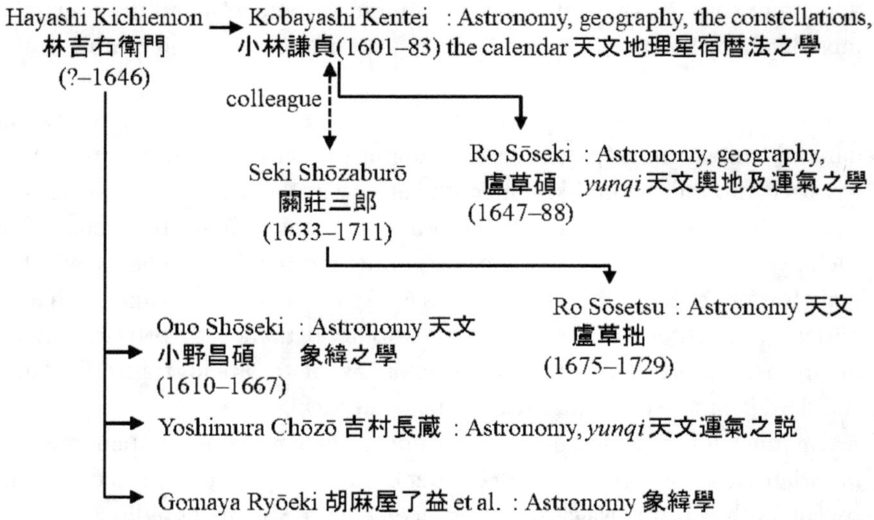

FIGURE 4.2 Master-pupil relationships and transmission of knowledge within the Nagasaki school

One of the most prominent figures among the next generation of Nagasaki scholars was Nishikawa Joken 西川如見 (1648–1724). According to Joken, the scholars of "Western-style *yunqi* in old Nagasaki," by which he most likely meant the scholars of the Nagasaki school, were strongly influenced by *Suwen rushi yunqi lunao* 素問入式運氣論奥 (Marvelous introductory remarks on the theory of *yunqi* according to the *Suwen*), a late eleventh-century text by Liu Wenshu 劉温舒.[14] As shown in the next section, the text of NU contains many quotations from the *Suwen*. Here, too, the intellectual trends of seventeenth-century Nagasaki, in which both new and old cosmologies coexisted, consistently match the content of NU.

Taken as a whole, this evidence strongly suggests that the Nagasaki school played an important role in the circulation and reception of NU in seventeenth-century Nagasaki. As a result, uncovering the philosophical characteristics of NU should help clarify the intellectual activities of the Nagasaki school, about which little is currently known.

mitted as the work of Kobayashi Kentei, a representative scholar of the Nagasaki school. See Hiraoka, *Nanbankei uchūron*, 103–128.

14 "Those who called themselves scholars of Western-style *yunqi* in old Nagasaki appeared to have never failed to learn Liu Wenshu's *yunqi* theory," *Tenmon giron*, 2:23a.

3 Deciphering Aristotle with Chinese *Yunqi* Theory

The use of the material from Chinese texts incorporated into NU's main text could be summarized as "deciphering Aristotle with Chinese *yunqi* theory." In the analysis below, I have juxtaposed the text of NU with the corresponding text in KB (fig. 4.1) to highlight the editorial changes made to NU. In explanatory notes to KB, Mukai Genshō emphasizes that he transliterated the original manuscript by Chūan "without changing a single word."[15] Therefore, when text or terminology that does not appear in KB is found in NU, we can assume that it was inserted or modified by the editor of NU.

Both KB and NU consist of around sixty chapters in total. Comparing their tables of contents, their overall structure shows a remarkable correspondence, although the order of the first and second volumes is switched, and each book has a few chapters the other does not. The text is also strikingly similar in every corresponding chapter, matching almost word-for-word in some.[16] This confirms that the two books are either both derived from a common parent, i.e. Chūan's draft (possibility A), or in a parent-child relationship (possibility B), as shown in figure 4.1.

On the other hand, there are characteristic differences between the KB and NU texts. Of particular note here are the following two points: (1) Both texts adopt systematically different words for specific terms. (2) Only NU quotes passages from Chinese treatises on *yunqi* theory throughout the text.[17]

Regarding (1), the texts use differing astronomical terminology throughout, as shown in table 4.1, virtually without exception. Another major difference is the terminology used for the four Aristotelian elements, as shown in table 4.2. KB refers to the elements as the "four great [elements]" (*shidai* 四大), and individual elements as the "fire great" (*kadai* 火大), "water great" (*suidai* 水大) and so on, presumably a borrowing from Buddhism's "five great [elements]" (*godai* 五大). However, the corresponding passages in NU use *yunqi* terminology[18] such as "minister fire" (*shōka* 相火), "water *qi*" (*suiki* 水氣) and "wind *qi*" (*fūki* 風氣).

---

15 KB, 3.

16 Compare, for example, Chapter 17 of KB's "Celestial Part" with Chapter 16 of NU's "Celestial Part." See KB, 77; and NU, 251.

17 Other differences noted by Ōya include: (3) Compared to KB, NU is written in a more concise style, closer to *Kanbun yomikudashi* style. (4) Only KB includes phonetically transcribed words from Western languages. (5) NU contains text that KB does not, but which is thought to have been in Chūan's draft. See Ōya, "Kenkon bensetsu no ichi ihon," 36–39.

18 See Unschuld, *Huangdi neijing suwen*, 417.

TABLE 4.1 Comparison of astronomical terminology

	KB	NU
Sun	日輪	陽日
Moon	月輪	月
Zodiac	逆道	逆旋
Pole(s)	軸	極, 辰

TABLE 4.2 Comparison of terminology for the four elements

	KB	NU
Earth	地大	地
Water	水大	水氣
Air	風大	風氣, 風中
Fire	火大	相火

These terms, of course, have specific meanings in their respective theories, and are used in this restricted sense in the Chinese texts that describe these theories. As a result, one set of terms cannot simply be replaced by the other in all contexts. An understanding of this fact is visible throughout KB, in which the "four great" and the related terms are used solely to refer to the Aristotelian elements, virtually without exception. This indicates that, as translator, Chūan took due care with specialist terminology.

On the other hand, the changes in terminology seen in NU evince no such care to ensure consistency of this sort between theory and terminology. It appears that the "four great" of Chūan's original manuscript were simply mechanically replaced in NU with *yunqi*-theory-based equivalents. This gives NU a unique, chimerical character as a text that seeks to explain the Aristotelian cosmology of the West, without changing its content or structure, using the terminology of *yunqi* theory.

What was the goal of NU's editor in carrying out these changes? An important clue to this question can be found in the quotations from Chinese medical works inserted here and there in NU.

Chapter 12 of the "Celestial Part," for example, follows Aristotle's "On the Heavens" in proposing that all heavy things move towards the bottom of the

TABLE 4.3 Comparison of *Kenkon bensetsu* and *Nanban unkiron* (1)

KB, 35–36	NU, 279
Chapter 12: Heavy Things Desire to Be at the Center of the Heavens	The Place of Heavy Things is at the Center of the Heavens
If so, for what reason is the element earth situated in the center of the heavens? As stated above, since the myriad things, according to their lightness and heaviness, have a nature that desires to move upward when light and downward when heavy, all heavy things desire the lowest [place]. As for where this lowest [place] is, it is the center of the heavens.	The myriad things, according to their lightness and heaviness, desire to move upward when light and downward when heavy. When we say that heavy things desire the lowest [place], this lowest [place] is the center of the heavens.
	This might be what is meant in the Suwen *by "[Earth is] below man and at the center of the Great Void."*
Therefore, the heavens are above in all places in relation to the element of earth. Since the element of earth is below in all places in relation to the heavens, the center of the heavens is the lowest place.	Since the heavens are above in all places in relation to the element of earth, the element of earth is below in all places in relation to the heavens. Therefore, the center of the heavens is the lowest place.

universe—"the lowest" (*shige* 至下) or "the center of the heavens" (*ten no seichū* 天ノ正中)—where Earth is located.[19] When we compare the corresponding passages in KB and NU (table 4.3), we find that only NU contains a section which notes that the *Suwen* says that Earth is "below man and at the center of the Great Void" (人之下太虚之中).[20] This quotation was obviously inserted to inform the reader that information matching the Western theories of NU can also be found in the *Suwen*. In other words, the editor seems to be "annotating" Aristotle with the Chinese classics.

In "On Earthquakes," Chapter 14 of the "Terrestrial Part," NU states that earthquakes are caused by vibration when air comes out of holes in the earth

19 Arist, *De caelo* II, 14, 296b10, 21–27. See also Hiraoka and Watanabe, "A Jesuit Cosmological Textbook," 147.

20 *Suwen*, 198.

TABLE 4.4 Comparison of *Kenkon bensetsu* and *Nanban unkiron* (II)

KB, 38	NU, 281
Chapter 14. On Earthquakes	On Earthquakes
	Now, if we examine the view of the Suwen, *it is stated that "Earth is at the center of the Great Void, and wind moves it," and "When wind dominates, it moves [Earth]."*
As written above, what is called an earthquake [begins with] the element of air blowing through holes in the element of earth so that it is hidden under the ground. However, because the proper place of element of air is above earth and water, air seeks to exit the ground to reach that [place].	Now, wind *qi* blows into the holes of the earth so that it is hidden under the ground. However, since the place of wind *qi* is above earth and water, it seeks to exit the ground to return to its original place.
	And then, at times, when spring *qi* is gained in the spring, and autumn *qi* in the autumn, wind seeks the same *qi*, so that the force of the wind increases more and more.
When there is no way out of the ground, it strongly desires to ascend, and the force yielded makes the body of the earth shake.	At that time, there is no way out of the ground. It strongly desires to ascend, and the force [of its attempts] to exit [the earth] again makes the earth shake.

(table 4.4). Here the quotations from the *Suwen* are inserted at the very beginning of the chapter: "Earth is at the center of the Great Void, and wind moves it" and "When wind dominates, it moves [Earth]."[21] Obviously, the editor hoped to make this introduction to Western earthquake theory more palatable by prefacing it with a quotation from a Chinese classic that closely resembled the subsequent discussion. Almost all of the Chinese material inserted into the NU text is of this sort, explicating and supplementing Western theories rather than criticizing them.

21 *Suwen*, 198–200.

Another book on Chinese *yunqi* theory quoted frequently in NU is the *Gezhi yulun* 格致餘論 (Further discourses on the properties of things) of Zhu Danxi 朱丹溪 (1282–1358), one of the four eminent physicians in the Jin and Yuan periods. This text on medicine is known to have been widely read and a number of Japanese editions and commentaries were published in seventeenth-century Japan.[22]

For example, when the main text of NU explains that moonlight is reflected sunlight and lunar eclipses happen when Earth is between the moon and the sun (Terrestrial Part, Chapter 8: "On the Sphericity of Earth"), this quotation is appended: "*Gezhi yulun* says that the moon (*yue* 月) is lacking (*que* 缺). It belongs to *Yin* and shines with the sunlight it receives."[23] Here, too, the quotation supplements the Western explanation of the lunar eclipse.

To give another example, in Chapter 1 of the "Celestial Part," the main text of NU explains that the heavens (*ten* 天, translating the concept of the ether in the Western original) are perpetual and immortal. After a relatively long quotation from *Gezhi yulun*, NU adds: "It is the heavens that have never waned from ancient times to the present day. Accordingly, it is said that they are exceedingly pure, durable, and unageing."[24] The reference to exceeding purity, durability, and unageingness is also evidently based on the *Gezhi yulun*, which says: "If they [the heavens] were not exceedingly pure, their durability could not have arisen in the earliest times and never grown old."[25]

For the modern historian, an approach combining two theories so distinct in nature and background might seem unlikely. A previous study suggested that scholars in Nagasaki, who feared being suspected of Christian sympathies, did this to "prolong the life of their learning" by "disguising it as Chinese thought" or "intensifying its Oriental flavor."[26] However, a closer examination of the circumstances of those involved with the creation of NU make this viewpoint untenable.

First, Chūan's original draft of what would become NU was created on the orders of Inoue Masashige 井上政重 (1585–1661), the Shogunate commissioner who supervised all of Chūan's activities. Clearly, therefore, the project was never at risk of censorship by the authorities—on the contrary, it was *commissioned* by the authorities from its earliest stages. It is also surely relevant that Inoue, despite being the Inquisitor-General against Christianity, had an

22 See Kumano, "Edo zenki ni okeru Chūgoku isho no juyō to ishazō."

23 NU, 272; *Gezhi yulun*, 15, 17.

24 NU, 237.

25 *Gezhi yulun*, 232–233.

26 Hirose, "Kinsei zenki no tenmon rekigaku," 459–460.

unusual passion for acquiring Western scientific and medical knowledge, even placing an order in 1647 with a Dutch merchant for a special telescope that would reveal "the four satellites of Jupiter, which cannot be seen with the naked eye."[27]

Furthermore, Shōgin, the most likely editor of NU, was a Buddhist priest with a close relationship with the Nagasaki Magistrate's office; his temple, Kōgenji, was founded with the Magistrate's assistance.[28] If he did indeed edit NU, he surely did so with the permission of Inoue or the Magistrate—possibly even on their orders. This is the exact opposite of the traditional view of the NU's editor as someone working under a cloud of fear, terrified of arrest and persecution as an underground Christian.

Re-examining the text with the assumption that its editor was secure in his position, there would be no reason to view the insertions of Chinese texts on *yunqi* theory as camouflage to avoid suspicion of covert Christianity. Rather, it seems more likely that, in the process of making the new Western knowledge his own and transmitting it to his readers, the Chinese theory was necessary for contextualizing and illuminating the Western theory. Given that the education of Japanese intellectuals at the time was founded on Chinese learning, particularly Confucianism, it is hardly surprising that scholars should seek to understand an unusual foreign *yunqi* theory through passages and concepts from the Chinese texts considered authoritative at the time. In Japan in particular, the study of Chinese medical texts from the Jin and Yuan periods that emphasized *yunqi* theory had been a lively field since the late medieval period. This scholarly activity reached its peak in the seventeenth century, and its core texts were the *Suwen rushi yunqi lunao* and *Gezhi yulun*. Multiple newly prepared and annotated versions of these texts were published in seventeenth-century Japan. Even among works discussing *yunqi* theory, they were the most widely read and the easiest to obtain.[29] For both the editor and readers of NU, it was surely the most appropriate cosmological framework for grasping this new Western theory in the context of their own intellectual world.

27 Michel, "Edo shoki no kōgaku seihin," 134.

28 Etchū et al., *Kōgenji-shi*, 60, 62.

29 Yamada, *Ki no shizenzō*, 40–42; Kumano, "Edo zenki ni okeru chūgoku isho no juyō to ishazō."

4 Nishikawa Joken's Heavenly Learning (天學)

The intellectual attitudes seen in NU were inherited and even developed further by Nishikawa Joken in the next generation.

There is no doubt that Joken consulted NU, since the figure "Foreign diagram of the mutual generation of four elements and eight *qi*" in his 1714 work *Ryōgi shūsetsu* 兩儀集說 (Collected theories on heaven and earth) is clearly taken from NU.[30] The account of the three regions of air (*sansai* 三際) in *Ryōgi shūsetsu* also coincides so markedly with the NU text that it is best viewed as a paraphrase.[31]

Joken also particularly emphasized the importance of the *Suwen rushi yunqi lunao* and *Gezhi yulun*: "You should never forget to learn Liu Wenshu's book on *yunqi* [i.e. *Suwen rushi yunqi lunao*] .... Among other medical books, *Gezhi yulun* is of great benefit to [the study of] astronomy and *yunqi*."[32] Joken sometimes quotes the same passages from the *Suwen* and *Gezhi yulun* as NU, confirming that both he and the NU editor regarded these texts as important for their discourses on heaven and earth.[33]

Furthermore, when Joken argues for his doctrine of a Chinese origin of Western knowledge, he once more invokes the authority of a passage from the *Suwen*:

> That Earth is spherical in shape and stays at the center of the heavens with no up or down had already been argued by the Chinese sages. It was not something first reported by [Western] Barbarians. It is only in recent years that these Barbarians and Red-Hairs, by sailing their ships and somehow drifting across the ocean, finally learned that Earth is spherical. The Chinese sages did not need to set sail; they learned this 4,000 years ago without rising from their seats [*Mengzi*, Lilou xia]. In *Huangdi neijing suwen*, Qi Bo 岐伯 says "Earth is below man and at the center of the Great Void." Huangdi asked "Is it supported?" Qi Bo replied "The Grand Qi holds it. Dryness dries Earth. Summer-heat steams it. Wind moves it. Dampness moistens it. Cold hardens it. Fire warms it," and so on. These were the founders of heavenly learning (*tengaku* 天學).[34]

---

30 See *Ryōgi Shūsetsu*, 7:37b; and NU, 265.

31 See *Ryōgi Shūsetsu*, 7:38a; and NU, 284.

32 *Tenmon giron*, 2:24a.

33 For example, see NU, 281 and *Tenmon giron*, 2:2b; NU, 237; and *Ryōgi shūsetsu*, 1:1a.

34 *Tenmon giron*, 1:8a–b; *Suwen*, 198–199. See also *Ryōgi shūsetsu*, introductory volume, 2b.

For Joken, the sphericity of Earth that Westerners argue for was already clearly shown in *Suwen* through the dialogue of Huangdi and Qi Bo, who were to be regarded as "the founders of heavenly learning."

Since such direct claim about the Chinese origins of Western learning cannot be found in NU, it is not clear if its editor shared Joken's views. However, even if the idea was original to Joken, it might have occurred to him via NU's editorial approach of annotating Western theory with Chinese text, requiring just one additional logical step. There is no doubt that Chinese *yunqi* theory was indispensable for both the NU editor and Joken in their discussion of heaven and earth, nor that they both belonged to the same tradition.

Of course, Joken's "heavenly learning" cannot be explained by Chinese *yunqi* theory alone. In his writings, Joken develops and describes several approaches not present in NU, such as the use of Chinese books on Western astronomy or geography produced by the Jesuits in China[35] and observation with a telescope.[36] Separating the elements in Joken's work that continue the previous intellectual tradition in Nagasaki from those that are original to him is a task for future investigations.

5 Reception of Western Theories without the West

Intriguingly, later generations of Japanese scholars who examined NU viewed it as a work of *yunqi* theory produced by a Japanese medical scholar, rather than a theory originating in the West. In the eighteenth century, when most of NU's surviving manuscripts were produced and circulated throughout Japan, some manuscripts were augmented with contemporary comments about the work's content and derivation.

One example is the so-called Yamauchi manuscript (entitled *Tenmon yōkai*, Outline of astronomy), which belonged to Tani Kakimori 谷垣守 (1698–1752), a Confucian scholar of the Tosa clan. According to an afterword added by Kakimori himself, the manuscript was made in Edo in 1736.[37] He evidently read the book with great care, adding his own annotations in red. For example, beside a passage that declares the number of fixed stars in the entire sky to be more than 1,020, Kakimori wrote, "This is the number of stars seen in Holland,"[38] indicat-

35 Ayuzawa, *Nishikawa Joken no sekai chiri kenkyū*, 25–26, 63–69.
36 *Ryōgi shūsetsu*, 2:53b–55a. *Idem*, 4:49a. *Tenmon giron*, 2:20b.
37 Hiraoka, *Nanbankei uchūron*, 162.
38 Quoted in Hiraoka, *Nanbankei uchūron*, 163.

ing that he correctly understood this information to have come from the West. Nevertheless, in his afterword, he assumes that the work is a variety of *yunqi* theory written by a medical scholar: "It seems to me that [this work] is a species of *yunqi* conjectural theory as taught by medical scholars. But its explanations of celestial phenomena are quite thorough, and it is worth referring to in conjunction with *Tianjing huowen*."[39] In all likelihood, based on comparison with and analogy to the works of *yunqi* theory around him at the time, Kakimori viewed NU as a particularly accomplished example of such a work.

Another example is the Hakuun manuscript (entitled *Nanban tenchiron*, Southern barbarian theory of heaven and earth), probably copied around the middle of the eighteenth century. The former owner and location of this manuscript are unknown, but its text is augmented with detailed annotations and several new diagrams, making it an important source of information on how NU was read by later generations. Particularly noteworthy is the following long annotation inserted at the beginning of the work: "Why was this work given the title '*Yunqi* theory of the Southern Barbarians'? I find this suspicious. Looking at its theories, they are arguments of Western calendrical science. When I asked a person knowledgeable about this matter, [he said that] the parts explaining *yunqi* were almost certainly Chinese *yunqi* theory, not a theory held by those from the South-West [Westerners?]. Given that this book is presumably the work of a medical scholar, this is just as expected."[40] In short, this unknown annotator found the mismatch between the "southern barbarians" mentioned in the work's title and its actual content suspicious enough to require explanation. After consulting with an expert, he came to the same conclusion as Tani Kakimori about the text being the work of a Japanese medical scholar. Unlike Kakimori, however, the annotator of the Hakuun manuscript was discerning enough to see "the arguments of Western calendrical science" as the essence of the work. The other annotations and newly added illustrations also reflect an accurate and critical understanding of the work's content, indicating that this annotator was highly knowledgeable about astronomy and calendars. This makes it all the more striking that even he concluded that the work was written by a Japanese medical scholar.

As these examples vividly show, for the intelligentsia of early modern Japan, the truth of NU's Western origins was extremely difficult—for some, even impossible—to discern.

---

39 Hiraoka, *Nanbankei uchūron*, 163. For *Tianjing huowen* and its diffusion in Edo Japan, see Hiraoka, "Printed Editions and Manuscripts of *Tianjing Huowen*."

40 Hiraoka, *Nanbankei uchūron*, 167.

6 Concluding Remarks

This chapter has examined the origins and later reception of the *Nanban unkiron*, a work of Western cosmology deriving from the early Jesuit mission to Japan, to reveal a process of encounter between early modern Japanese society and Western cosmology and their subsequent transformation.

From the earliest stages of NU's development, it strongly reflected East Asian interpretations. The changes and interpolations in NU relating to *yunqi* theory were not intended to camouflage its Christian origins, as traditionally believed. Rather, its editor viewed the Chinese theory as an essential framework for understanding the Western theory. Broadly speaking, the editor did not seek to analyze the content as a theory of new knowledge or weigh its advantages and disadvantages—on the contrary, great importance was placed on understanding and explaining the theory in a way that did not contradict existing knowledge. In seventeenth-century Nagasaki, where NU took shape, the landscape of cosmology saw Western and Chinese learning extending to the same horizon, connecting Aristotle and Chinese medical cosmology beyond time and space.

NU's reception by later scholars, too, can be viewed as a process of assimilation to and absorption by the dominant scholarly framework of *yunqi*-theory-based cosmology. Nishikawa Joken, a member of the next generation of Nagasaki scholars, argued that the theory of a spherical Earth was created not by the Westerners recently arrived in Japan but by the ancient Chinese sages who had developed *yunqi* theory. The eighteenth-century scholars across Japan involved in the manuscript's later reception identified it as work of a Japanese medical scholar, unable even to discern its origin in the West.

In short, throughout the entire encounter of Western cosmology with early modern Japanese society through NU, scholars sought to preserve as much consistency as possible with Chinese medical cosmology. New knowledge from the West was severed from its original context and reinterpreted and relocated within the existing local context. These nuanced interactions and negotiations between the Eastern and Western cosmologies that took place in early modern Japan have often been overlooked, but they must be considered as part of a comprehensive historiography of the period.

Acknowledgments

This is a revised and enlarged English version of the original Japanese article "Aristotle o unkiron de yomitoku." I thank the editors and anonymous readers of this volume for many important and critical comments reflected in this revision. My thanks also go to Matt Treyvaud for his assistance and valuable suggestions. The research that produced this work was funded by JSPS KAKENHI Grant Numbers 17K01179, 17H02392, 18H00641, and 20H05721.

Bibliography

Primary Sources

Gezhi yulun: Hasebe Eiichi 長谷部英一, Shinno Reiko 秦玲子, Uemura Asami 上村元顧, Matsushita Michinobu 松下道信 and Onda Hiromasa 恩田裕正. *Kakuchi yoron chūshaku* 格致餘論注釋. Tokyo: Iseisha, 2014.

Kenkon bensetsu (KB): Sawano Chūan 澤野忠庵 (Cristóvão Ferreira), Mukai Genshō 向井玄松 and Nishi Genpo 西玄甫. "Kenkon bensetsu" 乾坤辨說. In *Bunmei genryū sōsho* 文明源流叢書, vol. 2, edited by Kokusho Kankōkai 國書刊行会. Tokyo: Kokusho Kankōkai, 1914, 1–100.

Nanban unkiron (NU): Anonymous. "*Nanban unkiron*" 南蠻運氣論. In *Nanbankei uchūron no gententeki kenkyū* 南蠻系宇宙論の原典的研究 by Hiraoka Ryuji 平岡隆二, 157–191. Fukuoka: Hana Shoin, 2013.

Ryōgi shūsetsu: Nishikawa Tadasuke 西川忠亮, ed. *Ryōgi shūsetsu: Nishikawa Joken isho* 兩儀集說：西川如見遺書, vols. 15–18. Tokyo: Kyūrindo, 1907.

Sokuryō higen: Hiraoka Ryuji, and Hibi Kayoko 日比佳代子. "Shiryō shōkai: Hosoi Kōtaku-hen 'Sokuryō higen'" 史料紹介：細井廣沢編『測量秘言』. *Kagakushi kenkyū* Series II 43 (2004): 94–105.

Suwen: Paul U. Unschuld and Hermann Tessenow in collaboration with Zheng Jinsheng. *Huang Di Nei Jing Su Wen: An Annotated Translation of Huang Di's Inner Classic, Basic Questions*, vol. 2. Berkeley: University of California Press, 2011.

Tenmon giron: Nishikawa Tadasuke, ed. *Tenmon giron: Nishikawa Joken isho* 天文義論：西川如見遺書, vol. 2. Tokyo: Kyurindo, 1899.

Wakaki Taiichi 若木太一, Takahashi Masahiko 高橋昌彦, Kawahira Toshifumi 川平敏文, eds. *Nagasaki senminden chūkai—Kinsei Nagasaki no bun'en to gakugei* 長崎先民傳注解－近世長崎の文苑と學芸. Tokyo: Bensei Shuppan, 2016.

Secondary Sources

Ayuzawa Shintarō 鮎澤信太郎. *Nishikawa Joken no sekai chiri kenkyū* 西川如見の世界地理研究. Tokyo: Keisei Shuppanbu, 1944.

Cieslik, Hubert. "The Case of Christovão Ferreira." *Monumenta Nipponica* 29, no. 1 (1974): 1–54.

Etchū Tetsuya 越中哲也, Adachi Ken'ichi 足立巻一 and Etchū Isamu 越中勇. *Kōgenji-shi* 光源寺史. Nagasaki: Kōgenji, 1983.

Hiraoka Ryuji. "Aristotle o unkiron de yomitoku: *Nanban unkiron* to 17 seiki Nagasaki ni okeru seigaku rikai" アリストテレスを運氣論で読み解く：『南蠻運氣論』と17世紀長崎における西學理解. In *East-West Encounter in the Science of Heaven and Earth*, edited by Takeda Tokimasa and Bill Mak, 396–407. Kyoto: Institute for Research in Humanities, Kyoto University, 2019.

Hiraoka Ryuji. "Clavius and his Astronomical Data during the 'Christian Century' in Japan." *Historia scientiarum* 18, no. 3 (2009): 213–236.

Hiraoka Ryuji. "Jesuits, Cosmology and Creation in Japan's 'Christian Century' (1549–1650)." In *Visual and Textual Representations in Exchanges Between Europe and East Asia 16th–18th Centuries: History of Mathematical Sciences—Portugal and East Asia* V, edited by Luis Saraiva and Catherine Jami, 223–243. Singapore: World Scientific, 2018.

Hiraoka Ryuji. "Kobayashi Kentei den: Nagasaki no shiryō o chūshin ni" 小林謙貞傳ー長崎の史料を中心にー*Nagasakigaku: Nagasakishi Nagasakigaku kenkyūjo kiyō* 2 (2018): 19–33.

Hiraoka Ryuji. "Kōtaiji no rakan sekizōgun to Kobayashi Kentei: Teiennai no kō-gun o chūshin ni" 晧台寺の羅漢石像群と小林謙貞ー庭園内の甲群を中心に. *Nagasaki dansō* 100 (2015): 48–79.

Hiraoka Ryuji 平岡隆二. *Nanbankei uchūron no gententeki kenkyū* 南蠻系宇宙論の原典的研究. Fukuoka: Hana Shoin, 2013.

Hiraoka Ryuji. "Printed Editions and Manuscripts of *Tianjing Huowen*." *Historia scientiarum* 29, no. 1 (2019): 80–111.

Hiraoka Ryuji, and Watanabe Akihiko. "A Jesuit Cosmological Textbook in 'Christian Century' Japan: *De Sphaera* of Pedro Gomez (Part 2)." *SCIAMVS* 16 (2015): 125–223.

Hirose Hideo 廣瀬秀雄. "Kinsei zenki no tenmon rekigaku" 近世前期の天文暦學. In *Kinsei kagaku shisō II* 近世科學思想・下 (Scientific thought in the Edo period, vol. 2), edited by Hirose Hideo, Nakayama Shigeru 中山茂 and Ōtsuka Yoshinori 大塚敬節, 455–464. Tokyo: Iwanami Shoten, 1971.

Itō Shuntarō 伊東俊太郎. *Bunmei ni okeru kagaku* 文明における科學. Tokyo: Keisō Shobō, 1976.

Kumano Hiroko 熊野弘子. "Edo zenki ni okeru Chūgoku isho no juyō to ishazō: *Kakuchi yoron* o chūshin ni" 江戸前期における中國醫書の受容と醫者像—『格致餘論』を中心に. *Higashi Ajia Bunka Kōshōshi Kenkyū* 3 (2010): 161–178.

Michel, Wolfgang ヴォフルガング・ミヒェル. "Edo shoki no kōgaku seihin yunyū ni tsuite" 江戸初期の光學製品輸入について. *Yōgaku* 12 (2003): 119–164.

Nakayama, Shigeru. *A History of Japanese Astronomy: Chinese Background and Western Impact*. Cambridge: Harvard University Press, 1969.

Nihon Gakushiin 日本學士院, ed. *Meiji zen Nihon tenmongaku-shi* 明治前日本天文學史. Tokyo: Nihon Gakujutsu Shinkōkai, 1960.

Ōya Shin'ichi 大矢真一. "*Kenkon bensetsu* no ichi ihon: *Nanban unkiron*" 乾坤辨説の一異本： 南蠻運氣論. *Kagakushi kenkyū* 14 (1950): 35–39.

Unschuld, Paul U. *Huang Di Nei Jing Su Wen: Nature, Knowledge, Imagery in an Ancient Chinese Medical Text, with an Appendix, the Doctrine of the Five Periods and Six Qi in the Huang Di nei jing su wen*. Berkeley: University of California Press, 2003.

Watanabe Toshio 渡邊敏夫. *Kinsei Nihon tenmongaku-shi* 近世日本天文學史, vol. 1. Tokyo: Kōseisha Kōseikaku, 1986.

Yamada Keiji 山田慶兒. *Ki no shizenzō* 氣の自然像. Tokyo: Iwanami Shoten, 2002.

CHAPTER 5

Chinese-Muslims as Agents of Astral Knowledge in Late Imperial China

Dror Weil

Mei Wending 梅文鼎 (1633–1721),[1] the acclaimed Chinese astronomer of the seventeenth century, opens his essay on shadow measurement (published 1680) by ascribing its contents to his friend, Ma Decheng 馬德稱 who collected datasets from four Chinese mosques.[2] Ma Decheng, Mei explains, had family roots in the Western Regions (i.e., the Islamicate world). His ancestors were two Muslim translators, Ma-sha-yi-hei 馬沙亦黑 (<Shaykh Ma) and Ma-ha-ma 馬哈麻 (<Muḥammad), "who mastered the art of calendar-making and translated western works (*xishu* 西書) on the astral sciences at the court of the first Ming emperor."[3] In return, they were granted inheritable positions at the Ming Imperial Bureau of Astronomy. Mei's essay, as well as a number of his works, bring to light an unexplored aspect in the history of the astral sciences in China—the role of Chinese Muslims as brokers of astral knowledge between the late fourteenth and the mid-seventeenth centuries.

Grasping the structure and operations of the universe—including the underlying or overarching principles governing terrestrial and celestial phenomena, methods of measuring them, and the predictions of their effects on human society—has been a shared interest of pre-modern societies. Whereas native cosmological traditions developed in the course of their long histories into comprehensive, gradually expanding theories, encompassing various subject-matters and scholarly approaches, the limitations inherent in the importation of foreign knowledge entailed multiple processes of selection, extraction, translation and transformation. These processes were predicated by negotiations between the availability of knowledge and designated purposes specific to the recipient groups and individuals. Imported cosmological knowledge, as a result, was far more segmented, processed and discrete than

1 Special thanks are due to Lorraine Daston, Eric M. Gurevich and Yoni M. Siden for their insightful comments.

2 Mei, *Lisuan quanshu*, 19:16^{r}–21^{v}.

3 Mei, *Lisuan quanshu*, 19:16r.

 | DOI:10.1163/9789004511675_007

native cosmological traditions, dispersed between interest groups and learned communities.[4]

Cosmology and the astral sciences, the branch of cosmology that focuses on celestial phenomena, played different roles in the late imperial Chinese society. For the imperial dynasty, the ability to predict celestial phenomena was intertwined with the political doctrine of the Mandate of Heaven (*tianming* 天命) that contented that Heaven is the source of a dynasty or an individual ruler's legitimacy to govern. This legitimacy was conditioned on the ruler's correct performance of ritual that ensured that social order is in harmony with the cosmic order.[5] Warnings that Heaven had withdrawn its legitimacy would manifest as unpredictable terrestrial and celestial events. Chinese rulers, therefore, invested in capabilities to accurately predict events, such as occultations or movements of celestial bodies, to safeguard the legitimacy of their rule. For practitioners of Islam in China, it was the particular explanation of creation and the divine that cosmology offered that generated their interest. Added to that were practical needs of timing and dating the performance of Islamic religious rites, such as the five daily prayers and the month of fasting. These religious needs encouraged Chinese Muslims to delve into cosmological texts and apply methods to determine celestial movements. For late imperial China's literati, absorbed in classical erudition, cosmology offered the principles and patterns that govern the natural world, including celestial phenomena. Other literati, many of whom were involved in astral enterprises at court, were captivated by mathematical astronomy and its ability to explain natural phenomena. Thus, cosmology in late imperial China was broken into segments that catered to the interests and needs of particular groups.

Foreign cosmological knowledge provided convenient solutions to unresolved theoretical issues and intricate computations. Following the introduction of Buddhist cosmologies and Indian mathematical astronomy to China during the first millennium CE,[6] the utility of Western sciences (defined here as knowledge and practices that originated in lands to the west of China, Ch. *Xiyu* 西域, lit. "the western regions") gained currency in China. The increase in eastward movement of Arabic and Persian ideas, texts, artifacts, and peoples in the aftermath of the Islamic conquests imparted the reputation of Western sciences to ideas and texts that originated in the Islamicate world, making them attractive as complementary to, competing with, and even replacements

4 On theorization of scientific knowledge movement, see Huff, *Early Modern Science*; Montgomery, *Science in Translation*; Manning, "Introduction."

5 Sivin, *Granting the Seasons*, 36.

6 See Mak's chapter in this volume.

of earlier theories and practices among different parts of the late imperial Chinese society. In that process, Chinese Muslims were empowered as mediators in the importation of Arabo-Persian knowledge, based on their linguistic and cultural background, ability to read texts in Arabic and Persian, and simultaneous participation in activities at court, in local literati circles, and in mosque communities.

Shifting the analytic lens from canonical texts and established scholars to the tapestry of social relations of those who facilitated the making of astral knowledge in Nanjing and Beijing, this essay brings to light the unique intersection between membership in a religious community and participation in imperial science. In particular, it explores the instrumental roles that Chinese Muslims played in the study and development of astral knowledge in three different, yet connected, venues: imperial astronomy, astronomy in Muslim communities and mosques, and exchanges with non-Muslim local literati. Examining these three venues together illuminates the multiple manifestations of Arabo-Persian cosmology in China and the roles played by Chinese Muslims in brokering astral knowledge both prior to and following the arrival of European missionaries in late imperial China.

1 Establishing Arabo-Persian Astral Sciences at Court

The vast mobilization of experts, instruments, and texts under the Yuan dynasty introduced Arabo-Persian astral sciences to the Chinese court as useful alternatives to local methods of astronomical computation and astrological interpretation. Astral experts from various parts of central and western Asia carried out their activities at designated institutions at the Yuan capitals, Dadu and Shangdu, bringing with them Arabic and Persian texts and related instruments.[7] This institutional recognition of Arabo-Persian astral science granted by the Yuan court produced a space in which, for the next four centuries, astronomers and astrologers studied, put into use, and refined Arabo-Persian methods of predicting, measuring, and interpreting celestial phenomena.

With the fall of the Yuan capital in 1369, the Ming army took hold of the imperial library and transferred its rich contents to a different imperial library in the newly established Ming capital of Nanjing. Among these books, we are told by various records, were a few hundred Arabic and Persian texts, many of which were on the astral sciences.[8]

7 On Arabo-Persian astronomy under the Yuan, see Dalen, "Islamic and Chinese Astronomy," 327–356.

8 Wu, *Mingyi tianwenshu*, 21:296–298.

The first emperor of the Ming, Zhu Yuanzhang 朱元璋 (1328–1398) acknowledged the contributions of Arabo-Persian astral sciences to the Yuan court and sought ways to maintain the aspects that he found most beneficial to his rule. His main interest lay in Arabo-Persian astrological interpretation and astronomical computation. He is quoted praising the great precision and efficacy of celestial predictions carried out by Muslim astrologers (*xiyu yinyang jia* 西域陰陽家) and "their [computational] methods of planetary longitudes and latitudes that are not found in any of the Chinese books."[9] He was less interested, so it seems, in the broader Islamic cosmological theory.

In 1369, the first emperor recruited foreign experts, including ʿAlī Zheng 鄭阿里, ʿAbdallah 阿都剌, Darwīsh 迭里月實, Wu Ru 伍儒, and another eleven people to serve as astronomers at his court in Nanjing.[10] They were placed under the Directorate of Arabo-Persian Astronomy (Huihui sitian jian 回回司天監), a re-incarnation of a Yuan dynasty institution that was put in charge of using the Arabo-Persian astronomical system and computational methods. The Directorate of Arabo-Persian Astronomy worked in parallel to the Directorate of Astronomy (Sitian jian 司天監), which made use of native astronomical systems and computational methods.

When the emperor recognized that, in contrast to his Yuan predecessors, his court had only limited capacity to employ foreign experts who could make use of the Yuan's rich archive, he launched an imperial project in 1382 with the explicit aim to make Arabo-Persian mathematical astronomy and astrological interpretations accessible to the experts at his court. The project comprised two main components. First was the translation of relevant Arabic and Persian texts into Chinese. Second was the merger of the Directorate of Arabo-Persian Astronomy and the Directorate of Astronomy. Chinese Muslims played a central role in both parts: as Arabic and Persian translators and as skilled astronomers in imperial institutions of astronomy at the Ming and Qing courts.

Zhu Yuanzhang assigned the task of translating Arabo-Persian texts to a team of experts from the Hanlin Academy—a body of scholar-officials he formed to render scholarly services to his court.[11] The team of Hanlin academicians under the supervision of Wu Bozong 吴伯宗 (1334–1384) and Li Chong 李翀 could not read Arabic or Persian and recruited speakers of these languages to assist them in carrying out the translation. The recruited translators included people that worked at the Ming court, such as Ḥaydar 海達兒 from

9 Ma-ha-ma, *Tianwen xiangzong xizhan*, 21:3.

10 *Ming taizu shilu*, 35:637.

11 On the Hanlin academy's formation, see Mote, *Imperial China*, 598–600.

the Ming imperial observatory, as well as new recruits such as 'Aṭā' al-dīn 阿答尤丁, Shaykh Ma 馬沙亦黑, and Muḥammad 馬哈麻.[12]

Shaykh Ma and Muḥammad produced several Chinese translations, including that of Kushyār b. Labbān's (d. 1029) astrological treatise *al-Madkhal fī ṣināʿat aḥkām al-nujūm* (A primer in astrology), titled in Chinese *Tianwen shu* 天文書 (Treatise on astrology), and a Chinese work titled *Huihui lifa shili* 回回曆法釋例 (A detailed exposition on the Arabo-Persian astronomical system). This latter work explains methods to calculate planetary longitudes and latitudes. Genealogical research on the Ma family showed that Shaykh Ma, known also by his Chinese name Wu Liang 吳諒, and Muḥammad were two sons of a certain Ma De-lu-ding 馬德魯丁, who migrated to China from Samarqand or Arabia. They were among the foreign astronomers recruited by the Ming court in the late 1360s or 1370s. In addition to their assistance in translating Arabo-Persian texts, they served as astronomers at the Ming court in Nanjing, and, after 1421, in Beijing. They received inheritable positions at the Bureau of Astronomy that continued to be held by their descendants until the end of the Ming.[13]

The translation project was accompanied by reforms in the organization of the astronomical institutions at the Ming court to accommodate Arabo-Persian expertise. In that process, the Directorate of Astronomy was renamed Qintianjian 欽天監[14] and was vested with the responsibility to oversee all astronomical and astrological activities at the court in 1398. The Directorate of Arabo-Persian astronomy was abolished, and a Department of Arabo-Persian Calendar (Huihui li ke 回回曆科, also Huihui ke 回回科) was established under the Directorate of Astronomy to carry out Arabo-Persian computation and come up with predictions that could serve as counterparts to those made in other departments.[15]

12 On the translation of Arabic and Persian texts under the Ming, see Yabuuti and Dalen, "Islamic Astronomy in China," 11–43; Shi, "Islamic Astronomy," 41–61; Weil, " Fourteenth-Century Transformation," 345–370. On the Chinese translation of Kūshyār b. Labbān's *al-Madkhal fī ṣināʿat aḥkām al-nujūm* from the Arabic (or possibly the Persian) see Yano, *Kūšyār Ibn Labbān's Introduction to Astrology*.

13 On Ma-sha-yi-hei, see Haneda, "Ka-i yakugo no hensha Ma-sha-yi-hei," 437–446. Chen Jiujin 陳久金 studied Ma family's genealogy, titled *Juzhentang Mashi zongpu* 聚真堂馬氏宗譜, see "Ma-de-lu-ding fuzi he Huihui tianwenxue," 28–36. Shaykh Ma's biography is recorded also in Tang Jinhui, *Qingzhen shiyi buji*, 18:96.

14 For simplicity, I follow Hucker in translating both names as "Directorate of Astronomy." See Hucker, *Dictionary of Official Titles*, 169, 456.

15 Shi points out that Arabo-Persian methods were often employed in parallel to traditional Chinese methods to predict luni-solar eclipses—predictions that were central to court rituals ("Islamic Astronomy," 54–56).

To ensure the persistence of expertise in Arabo-Persian astronomy, Zhu Yuanzhang stipulated that employment in the department should be hereditary. He required, by law, that the descendants of the original recruits would be trained and continue to serve at the department.[16] This hereditary system continued throughout the Ming and until the abolishment of the department in 1658. As a result, descendants of the astronomers who were initially recruited by the first emperor of the Ming continued to be involved in imperial astronomy, constituting an important bridge between the court and the local Muslim communities.

Inheritable positions at the Directorate of Astronomy were not reserved only for Muslim astronomers. Yet, in the case of Muslim astronomers, the familial ties facilitated unwritten, oral transmissions of astronomical techniques across generations. Muslim astronomers in the Department of Arabo-Persian calendar, as the *Official History of the Ming* suggests, retained knowledge of "dust-and-board calculation" (*tupan busuan* 土盘布算)[17] and were able to "read texts of their native lands" (*reng yong qi benguo zhi shu* 仍用其本國之書).[18] This set of skills, we are told, allowed them to study the original texts and correct errors in translation. Mei Wending suggests that the unfolding of Arabo-Persian astral knowledge through translation into Chinese was only partial. Shaykh Ma, Muḥammad, and the other translators, Mei tells us, "had an excellent command of astronomical computations, but they did not fully record their techniques, leading to the loss of knowledge. For example, they did not record the factors of their tables (*libiao zhi gen* 立表之根), and so successive students lost the application of this method."[19] It was Muslim astronomers working at the Directorate who preserved and transmitted such knowledge orally, not the native astronomers.

Bei Lin 貝琳, the Vice Director of Astronomy during the Chenghua 成化 reign (1464–1487), expressed concern that Arabo-Persian astronomical computational methods transmitted orally among members of the Department of

16 On the hereditary system employed in the Directorate, see Chang, "Chinese Hereditary Mathematician Families." On the reasons behind it, see Weil, "Fourteenth-Century Transformation," 345–370.

17 The accurate meaning of the term "dust board calculation" is unclear. It might be related to the Arabic phrase used to designate Indian arithmetic: *takht wa-turāb* (writing board and dust). On this Arabic phrase and the application of Indian arithmetic in the Islamicate world, see Saidan, *Arithmetic of Al-Uqlīdisī*, 12–19. Naṣīr al-Dīn Ṭūsī's work on arithmetic is titled *Jawāmiʿ al-ḥisāb bil-takht wa-l-turāb* (Comprehensive arithmetic with writing board and dust).

18 *Mingshi*, 37:745.

19 Mei, *Lisuan quanshu*, 1:7r.

Arabo-Persian Astronomy but not published in writing would be lost. In particular, he was concerned that the computation based on the method devised by Shaykh Ma for the court astronomer Yuan Tong 元統[20] a few decades earlier had diminished in accuracy and required recalculation. He received the permission of the Chenghua Emperor, presented his amended text to the court in 1477, and later published it under the title *Qizheng tuibu* 七政推步 (Pacing the motions of the seven [celestial] governors).[21]

The continuous presence of Muslim astronomers at court reached its final stages in the mid-seventeenth century, when Jesuit missionaries gained prestige and power at the imperial court. Matteo Ricci, aware of the importance the Chinese court attached to the calendar, wished to demonstrate to the court the advancement of European astronomy. With the help of the Chinese collaborators Xu Guangqi 徐光啓 (1562–1633) and Li Zhizao 李之藻 (1565–1630), Matteo Ricci translated several works to Chinese, including the six books of Euclid's *Elements of Geometry*. Xu Guangqi, a Chinese literatus and high-ranking official in the Ministry of Rites, was instrumental in informing the court of the precision of the Jesuits' computation. In particular, he spotlighted the high level of accuracy they obtained in their prediction of a solar eclipse in December 1610 (the predictions calculated by astronomers at the Directorate of Astronomy were not as accurate as the Jesuits' and missed the eclipse by fifteen to forty-three minutes).[22] In 1629, Xu Guangqi proposed a calendar reform and the replacement of older astronomical systems, including the Arabo-Persian system, with an improved one that would be developed by the Jesuits.[23]

With the fall of the Ming and the establishment of the Qing court in 1644, the Jesuits enhanced their standing at court. The Qing reinstalled the Directorate of Astronomy and maintained the Department of Arabo-Persian Calendar under it. The German Jesuit Johann Adam Schall von Bell (1592–1666), who arrived in China 1621, was appointed in 1645 by the Shunzhi 順治 emperor (1644–1651) as the head of the Directorate of Astronomy. Schall von Bell sought to consolidate the use of European astronomy at court as the only Western science, at the expense of Arabo-Persian astronomy. Consequently, Schall von Bell disbanded the Department of Arabo-Persian Astronomy in 1658.

---

20 Yuan Tong was also one of the main compilers of the imperial calendar *Datong li* 大統曆 during the reign of the first emperor of the Ming.

21 Ruan, *Chouren zhuan*, 29:6–7; *Xianzong shilu juan* 171, 3093 (Chenghua 13/10/1). Bei Lin's translation, known as *Qizheng tuibu* 七政推步.

22 Elman, *On Their Own Terms*, 90.

23 Elman, *On Their Own Terms*, 94.

Wu Mingxuan 吳明炫 (also written 吳明烜), a Muslim astronomer at the Department of Arabo-Persian Astronomy and a member of Beijing Muslim community, was dismissed from his position after the department's disbandment. He petitioned the Emperor for the re-opening of the department. Wu's petition began by reminding the Emperor that his ancestor Shaykh Ma arrived in China from the Islamicate world (*Xiyu* 西域) together with another eighteen people, bringing with them Arabo-Persian astronomical knowledge (*lixue* 曆學) and translating astral texts for the Chinese court. He explained that his ancestors were put in charge of calculating the movements of the stars and interpreting their astrological effects, produced reports on the encroachments of the moon and the five planets, and developed predictions of solar and lunar eclipses as stipulated by imperial order. Since 1649, Wu reported, Schall von Bell had not allowed the Arabo-Persian Department to report to the throne on eclipses or encroachments, and his own reports were inaccurate on a number of occasions. Wu begged the emperor to reopen the department and its observatory in order to revive the lost Arabo-Persian astral knowledge (*juexue huochuan* 絕學獲傳).[24]

The political pressure exercised by the Jesuits and the consequent devaluation of Arabo-Persian astronomical prediction closed the Department of Arabo-Persian permanently. Wu Mingxuan was brought back in 1664 to serve as the Vice Director of Astronomy under Yang Guangxian 楊光先 (1597–1669), a fervent opposer of the Jesuits' activities at court, who enjoyed the Kangxi emperor's (r. 1661–1722) favor. Yang and Wu were once again removed from their office as the Kangxi emperor shifted his support to the Jesuits in 1668.[25] Wu Mingxuan's removal from office marked the closing note of the Chinese Muslims' official participation in the making of imperial astral science.

2 Chinese Communities and Mosques as Venues of Astral Knowledge Production

Interest in determining the positions and movements of celestial bodies was not limited to court astronomers and astrologers. China's Muslim mosque communities (that is, Muslim communities with mosques as their socio-religious center) produced and circulated texts on astral theories and computational methods, installed gnomons and erected moon-observation towers (*wangyue*

24 *Qing Shizu Zhanghuangdi shilu*, 109:835.

25 More on that issue, see Cullen and Jami, "Christmas 1668 and After."

ta 望月塔) in mosques' vicinities to measure time and lunation, and held heated debates on the application of mathematics for predicting new moons. Whereas the interest in Arabo-Persian astral sciences at court was mainly focused on improving the astronomical system and acquiring new methods to predict and interpret earthly effects of celestial phenomena, scholars in local Muslim communities were keen to link the astral sciences to a broader philosophy of nature, Islamic theology, and everyday religious practice. Scholars at the local level focused their attention on the exploration of cosmological principles that govern the structure and operation of the natural world as a whole and their implications for religious practice. This exploration made Chinese mosques and Muslim communities into spaces in which production, application, and development of astral knowledge took place.

The expertise of early Muslim migrants in the astral sciences and the historical service they rendered to the various Chinese courts has constituted a central facet in the collective memory of Islam in China. Stories about the arrival of foreigners from the Islamicate world with elevated expertise in astronomical computation and astrology are commonly celebrated tropes in mosque stele inscriptions, written records, and oral histories. A stele erected in 1602 in the central mosque of Jiaxing 嘉興 prefecture in Zhejiang 浙江 province in commemoration of the mosque's construction, for example, celebrates Zhu Yuanzhang's recruitment of the astronomer Zheng ʿAlī and eleven other Muslim astronomers to the imperial Directorate of Astronomy, as well as their contribution to the translation of Arabo-Persian astronomy.[26] Similarly, in the 1640s, one of the first Chinese Muslim theologians to publish on Islam in Chinese, Wang Daiyu 王岱輿 (1570–1660), opened his preface to a treatise on theology by glorifying his ancestors who were invited by the first emperor of the Ming to amend the flaws of the Chinese astronomical system, introduce their innovative techniques, and consequently granted office and residence in China for the next three centuries.[27] As seen in these two examples, the engagement with the astral sciences were an integral part of Chinese Muslims' sense of belonging at the same time that they brought out their unique religious and cultural identities.

An early instance of the intimate link between Muslim mosque communities and the imperial Directorate of Astronomy is found in the biography of Wu Ru 伍儒 (d. 1420). One of the founders of the Arabo-Persian Department of Astronomy at the Ming court, Wu Ru (courtesy name Dequan 德全) was

26 Yu and Lei, *Zhongguo huizu jinshilu*, 50. The original reads 我高皇帝龍興，徵本教鄭阿里等十一人，命儒臣譯其曆，特為置司天監.

27 Wang, *Zhengjiao zhenquan*, 16:30.

granted official residence in Nanjing in 1370 as well as inheritable positions at the Directorate of Astronomy. After retiring from office, Wu bequeathed his official residence to the Muslim community to serve as a mosque.[28] Wu's descendants continued to hold positions at the Directorate of Astronomy and constituted one of the leading families in the Nanjing Muslim community.[29]

Mosques, as spaces dedicated for religious activities, housed instruments and observation towers to determine the times of the five daily prayers, the beginning of the lunar months, and in particular the exact time of the beginning and end of the fasting days during the month of Ramaḍān. Towers labeled *wangyue lou* 望月樓 (observing of the moon minaret) became an integral part of Chinese mosque architecture during the Ming and Qing periods[30] and demonstrate the importance of lunar observation in the religious lives of Chinese Muslims.

Some Chinese mosques installed scientific instruments such as gnomons to determine time and date by measuring the sun's cast shadow, clearly demonstrating the intimate link between religious practice and scientific research. An example of a gnomon remains in the Huajue alley 化覺巷 mosque in Xi'an (established in the late fourteenth century on the ruins of a mosque from the eighth century). This gnomon, 117 cm in height, includes a square stone plate with edge length of 62 cm and thickness of 9 cm installed on a stone leg.[31]

The data of shadow measurement published by Mei Wending, the famous seventeenth-century astronomer quoted above, is an example of the application of mosque gnomons as late as the seventeenth century. In his essay titled *Sisheng biaoying licheng* 四省表影立成 (Tables of shadow measurement from four provinces, 1680), Mei included tables of shadow measurement gathered by his Muslim friend Ma Decheng from mosques in the provinces of Shaanxi 陝西, Henan 河南, Beizhi 北直, and Jiangnan 江南. Mei explains that two astronomical systems (*li* 曆) were in use in the western regions (i.e., the Islamicate world): a lunar system that was based on lunation stages and a solar system that was

28 The original reads 伍儒, 字德全, 樂善好施。洪武中賜籍金陵, 精曆學, 詔授漏刻科博士, 掌欽天監事, 子孫世其職，家傳不懈閱數百年，而猶有精其事者，賜官房一所, 後捨为寺. (Jiaqing 16) *Chongkan Jiangning fu zhi*, 36:2b. Another version of the bio appears in *Shangjiang Liangxian zhi*, 24:17b. The latter adds the address of the residence in the former Tianjing Road 古天津街, which might suggest that the mosque mentioned is the mosque Jingjuesi 淨覺寺 established in 1388.

29 On Wu's descendant, Wu Yuanzhi 伍元芝, see Gu, *Qingdai zhujuna jicheng*, 74:229–255.

30 Wu Jianwei, *Zhongguo qingzhensi zonglan*, 10.

31 For the full description of the sundial, see Bai, "Qingzhen dasi rigui chutan," 10–14. See also Chen, *Huihui tianwen xue shi yanjiu*, 262–271.

based on the twenty-four corrected *qi* (*dingqi* 定氣).[32] The first was put into use during the fasting period (that is, Ramaḍān), Mei explained, when Muslims observed the moon; the second was for timing the prayers "using shadow gnomons as evidence" (*gui ying weiping* 晷景為憑).[33] It is unclear, he wrote, why, of all the numerous mosques around China, these specific four places were chosen. The fact that these four mosques applied a single factor (*biaoying chidu gongzhi yishu* 表影尺度共祇一術) to account for the League Distance Correction (*licha* 里差, a correction to account for the difference in latitude and longitude of the four locations)[34] raised Ma Decheng's suspicion that there were inaccuracies in the data. He therefore asked for Mei's help in checking it.[35]

The centrality of astral knowledge to activities in Chinese mosques is further illustrated by an intriguing mosque stele that still stands at Xi'an's Huajue alley mosque and includes twenty-one diagrams of astral phenomena and measurements. The stele is 2.96 m high and 1.03 m wide, and seems to be inscribed by Jia Meng'ao 賈夢鰲, who served as a *ahong* 阿訇 (religious leader) at the mosque during the Xianfeng 咸豐 (1851–1861) period.[36] The diagrams display the positions of the twelve astrological houses and the twenty-eight lunar lodges, the normal and eccentric motions of the sun, moon, and the five planets, lunar and solar eclipses, as well as methods to compute corrections.[37] The diagrams include captions in Arabic and Persian, some of which are direct quotes from Naṣīr al-dīn Ṭūsī's *Memoir on Astronomy* (*al-Tadhkira fī ʿilm al-hayʾa*). Interestingly, astronomical theory is interwoven with religious concepts; illustrations of the sun, moon, and seven planetary spheres is complemented by identifying the seat of the fixed stars with *kursī* (Divine Footstool) and *ʿarsh* (Divine Throne) as the furthest sky.[38]

Debates among Chinese Muslims with regard to the proper method to time the beginning and conclusion of the fasting month of Ramaḍān are commemorated in records and steles. They bring to light the deliberation over the use of mathematical astronomy and observation in Chinese mosques. A historical record from Beijing's Niujie mosque recounts a heated debate between two religious leaders, Bai Yangheng 白養恆 and Ma Junxi 馬君錫, that took place in 1716.[39] Bai suggested the use of mathematical computation (*tuice zhi shu* 推

32 I follow Sivin in translating *dingqi* as "corrected *qi*" (*Granting the Seasons*, 395).

33 Mei, *Lisuan quanshu*, 19:16r.

34 On the *licha* correction, see Dalen, "Islamic and Chinese Astronomy," 331–333.

35 Mei, *Lisuan quanshu*, 19:16r.

36 Chen, *Huihui tianwen xue shi yanjiu*, 279.

37 Chen, *Huihui tianwen xue shi yanjiu*, 283.

38 Chen, *Huihui tianwen xue shi yanjiu*, 279.

39 *Beijing niujie zhi shu*, 40.

測之術); Ma Junxi stubbornly rejected the use of mathematics and emphasized the importance of naked-eye observation of the moon to accurately obey Sharia law. The community was divided between the two camps. In explaining the computational method his father advocated, Bai Yuanfu 白元輔, Bai Yangheng's son, asserted that the conjunction (*heshuo* 合朔) would take place during the double-hour *chou* 丑 (1–3 a.m.) on the first day of the lunar month. The moon and the sun would set apart by the double-hours *shen* 申 (3–5 p.m.) and *you* 酉 (5–7 p.m.) of the second day. Bai added that a comparison of the data from past years yielded that "when the previous month is long [30 days], a conjunction appears early, and the new moon is observed on the second day of the month; when the previous month is short [29 days], the conjunction is belated, and the new moon appears only on the third day."[40] Applying knowledge derived from past observations (*zaoyan zhi fa* 早驗之法), Bai Yuanfu insisted, would not lead to an error and hence did not violate the Sharia law. Mathematical astronomy (*tianwen suanfa zhi xue* 天文算法之學), according to Bai, had been the most important method in the western regions (i.e., the Islamicate world), yet it was neglected by his generation of Chinese Muslims.[41]

A similar debate is recorded in an inscription in Arabic engraved on a stele set up in the central mosque of Xi'an in 1732. The inscription provides an elaborate account of the strong opposition of religious leaders to the use of mathematics and prediction to determine the beginning and end of Ramaḍān. The inscription asserts that Sharia forbade the use of astronomical calculation (*ḥisāb al-nujūm*) or the science of the movements of the constellations and the moon (*ʿilm al-nujūm wa-sayr al-qamar*) to determine time, and that determining the time should be done exclusively by naked-eye observation of the moon. The inscription recounts that monthly observation of the moon in this mosque took place from the twenty-eighth day of the previous month and lasted until the moon was detected and the beginning of the new month was announced.[42]

While some Chinese Muslim religious leaders resisted the application of particular aspects of the astral sciences, and especially astrological interpretation and mathematical astronomy, Arabic and Persian texts on these themes could not evade the attention of curious members of the local Muslim community. Manuscripts on these themes were scattered in private libraries across China—some remained from the imperial libraries of the Yuan, while others were newly brought from the Islamicate world. Accounts of the various texts that circulated

40 *Beijing niujie zhi shu*, 40.

41 *Beijing niujie zhi shu*, 41.

42 Huart, "Inscriptions arabes et persanes," 295–312. See also Yu Zhen'gui, *Zhongguo huizu jinshilu*, 438–441.

during the Ming and Qing periods allow us to reconstruct the textual sources that were used for the study of the astral sciences during these periods.

A record of Ma Yonghe's 馬永和 (fl. early eighteenth century) serendipitous discovery of astrological texts in a coreligionist's library provides an example of the multiplicity of opinions among Chinese Muslims with regard to the astral sciences and the availability of texts—even texts that were rejected by religious authorities.[43] Ma, who later became a respected *ahong* (阿訇, religious leader) in Beijing, was at the time of the anecdote a young and curious student in Beijing's Niujie mosque. He was eager to know more about the legitimacy of practicing divination (*zhanbu zhi xue* 占卜之學) among Muslims and to learn methods to determine the length of the month of Ramaḍān. His mentor strongly opposed the use of astrology and mathematics, informing Ma that the use astrology was a strict religious taboo in the Islamicate world and that Sharia laws forbade the use of computation for predicting lunation and intercalation (*runyue zhi fa* 閏月之法) for religious practice. Frustrated by his mentor's dismissive responses, Ma happened on two tattered manuscripts in a Muslim friend's library a few years later. The manuscripts contained a Chinese text titled *Xiyu tianwen gao* 西域天文稿 (Essay on astrology in the western regions) and a Persian work entitled *Jahāndānish* (Knowledge of the universe).[44] Intrigued by the contents of these two texts, Ma Yonghe worked on emending and re-editing the manuscripts. He discovered that the two texts similarly expounded on the same Islamic cosmological model of nine heavens and seven continents (*jiutian qidi* 九天七地) and provided detailed descriptions of the principles that govern the motions of the sun, the moon, and the five planets. Further investigation led Ma to obtain a copy of Wu Bozong and Shaykh Ma's early Ming translation, *Qianfang mishu* 乾方秘書 (Secret book of the northwestern lands, also known as Treatise of astrology). The contents of this book, as Ma recorded, included elaborate discussions of astronomy, geography, astrological prognostication (*fengjiao zhanhou* 風角占候), and even predictions for military purposes. Ma Yonghe's growing engagement with Arabo-Persian astral sciences led him to encounter an Italian Jesuit he called Ma Weixian 馬偉賢 in 1721 in Beijing. The Jesuit fascinated Ma with his accounts of the richness of astronomical and astrological scholarship that he encountered during his travels through the Islamicate world and the prevalent use of astronomical computation, star charts, and methods for predicting eclipses and encroachments.

43 *Beijing niujie zhi shu*, 21–26.

44 *Beijing niujie zhi shu*, 22.

A more systematic record of the Arabic and Persian astral texts that circulated in China during the late seventeenth and early eighteenth centuries is found in Liu Zhi's 劉智 works (1660–1730). A prolific Chinese Muslim scholar, Liu Zhi traveled extensively throughout China in search of Arabic and Persian manuscripts forgotten in private libraries or newly brought to China by foreign visitors and Islamic missionaries.[45] Some of the texts he discovered served as sources for his trilogy on Islamic cosmology and praxis. The first work in this trilogy, *Tianfang xingli* 天方性理 (On the principles of nature in Islam) was published in 1704 and is dedicated to cosmology. The second, *Tianfang dianli zeyao jie* 天方典禮擇要解 (Commentaries on selected Islamic rituals) was published in 1710 and focuses on the Islamic ritual and social interaction. The third, *Tianfang zhisheng shilu nianpu* 天方至聖實錄年譜 (The veritable records and chronological tables of the life of Islam's most revered) was completed in 1724 and provides an account of the life of the Prophet Muhammad. Whereas the third work is based on a single Persian text, the first two provide extracts and summaries from numerous Arabic and Persian texts and include rather detailed bibliographies of their sources.

Liu Zhi's sources include texts on astrological interpretation and astronomical computation. In his bibliographies, Liu gives the titles of his sources in Chinese characters representing original Arabic or Persian sounds, followed by a short explanation in Chinese.[46] In some cases, the reconstruction of the original titles is uncertain or refers to an unknown text. Among the identified titles are *A-sa-er ou-liu-wei* 阿撒爾歐六巍—a transliteration of the Persian *Āthār-i ʿulūwī* (Super-terrestrial phenomena). This title seems to refer to the Arabic (or maybe Persian) translation of Aristotle's treatise *Meteorology*, which addresses weather phenomena and some astronomical phenomena.[47] The accompanying explanation to the entry suggests that the work refers to the "interpretation of profound sky phenomena" (*xuanqiong xiang jie* 玄穹象解). Another entry, *Zhe-han da-ni-shi* 哲罕打尼識, refers to the Persian astronomical treatise *Jahāndānish* (Knowledge of the universe) mentioned above. This work provides explanations and diagrams of the planetary system and the movement of the celestial bodies and was originally composed in Arabic by Muḥammad

45 On the circulation of Arabic and Persian texts in late imperial China, see Weil, "Islamicated China," 36–60, and Weil, "Libraries," 90–92.

46 On the reconstruction of the full bibliographical lists, see Leslie and Wassel, "Arabic and Persian Sources," 78–104, and Weil, "Vicissitudes," Appendix B.

47 On the Arabic translations of Aristotle's *Meteorology*, see Sezgin, *Geschichte Des Arabischen Schrifttums*, 7:212–215; Peters, *Aristoteles Arabus*, 39–40. Leslie and Wassel read the transliterated title as the Arabic *al-Āthār al-ʿulūwīyya* and identify it as a work of Avicenna by the same name ("Arabic and Persian Sources," 93).

b. Masʿūd al-Masʿūdī al-Ghaznawī (fl. ca. 1274) under the title *al-Kifāyāh fī ʿilm hayʾat al-ʿālam* (comp. 1245). Liu Zhi described the work as an "exposition on the universe" (*huanyu shujie* 寰宇述解).[48] Among the unidentified titles are *E-he-ke-mu ke-wa-qi-bi* 額合克目克瓦乞蔔 (from the Persian *Aḥkām-i kawākib*, Precepts of heavenly bodies), described as a work that focuses on the particular natures of the celestial paths (*tianjing qingxing* 天經情性); *E-fu-a-lu e-fu-la-qi* 額福阿祿額福剌乞 (from *Afāʿl-i aflāk,* The work of the stars), described as a work on the principles of the operation of heavenly bodies (*tiande yuanji* 天德元機); *Er-lin ya-fa-ge* 爾林亞法格 (from *ʿIlm al-āfāq*, The science of the horizons), referred to as a description of the universe (*huanyu shu* 寰宇述); *Ye-wa-ji-te* 葉瓦基特 (which seems to stand for the Arabic *Yawāqīt*, Rubies), described as a work on lunation records (*yueling ji* 月令紀);[49] and an unidentified compendium of astronomy (*Lixue daquan* 歷學大全) entitled *Er-shu-du ke-bi-er* 二數度克比爾.

Liu Zhi's *Tianfang xingli* includes various explanations of the structure and operation of the cosmos. The diagrams included in this work illustrate the nine heavens—that is, the seven planets plus *kursī* (Divine Footstool) and *ʿarsh* (Divine Throne)—their rotations, and their distances from Earth. References to mathematical computations are embedded in Liu's descriptions of the ratio of a circle's circumference to its diameter (*jingyi wei san* 徑一圍三), the movements of the celestial bodies with respect to *chidao* 赤道 (lit. the red path, i.e., the equator) and *huangdao* 黃道 (lit. the yellow path, i.e., the ecliptic). Additionally, Liu dedicated sections of the work to explain the different natures of the planets using the four elements as an analytical framework. In the chapter on the seven continents, Liu included a diagram of the earthly sphere with markings of latitude, longitude, and the two poles. His explanations connect each of the seven continents to their respective seven planets and elaborate on interdependency between the two sets.

Other Chinese-Muslim authors also published on aspects of the astral sciences. Ma Zhu 馬注 (1640–1711) dedicated a number of chapters in his *Qingzhen zhinan* 清真指南 (Compass to Islam, 1683) to intercalation, eclipses, and the Jesuit globe. Ma Dexin 馬德新 (1794–1874), a prolific scholar of Islamic theology who widely travelled in the Middle East and who introduced new visions of Islam to his Chinese coreligionists, published works on cosmology and the

48 On the history of Jahāndānish in China, see Weil "Islamicated China," 36; Weil, "Collation and Articulation."

49 The term *yueling* 月令 refers to the historical custom of the seasonal announcement of rituals and duties in early China. On the term and its history, see Cullen, *Astronomy and Mathematics*; Henderson, *Development and Decline*, 20–24.

Islamic astronomical system titled *Huanyu shuyao* 寰宇述要 (Exposition on cosmology, 1862) and *Tianfang liyuan* 天方曆源 (The source of the Islamic calendar, 1875). His works merged older material, such as what he read in Liu Zhi's works, with new post-Copernican astronomical knowledge.[50] Whereas the earlier work includes the algorithmic instructions required to compile the Islamic calendar, the latter includes explanations and diagrams of astronomical phenomena and computation methods.

By circulating texts on the astral sciences, engaging in debates over methods to determine and time of astral phenomena, and setting up instruments to observe and record such phenomena, China's Muslims communities—with mosques at their core—constituted a space in which astral knowledge was produced, studied, and developed.

3 Chinese Muslims' Exchanges with Non-Muslim Local Literati

With more than one hundred thousand Chinese Muslims living in the cultural metropolis of Nanjing and a similar figure residing in Beijing, Muslims were not alien to local intellectual communities. Active at court and in local Muslim centers, Chinese Muslims in these regions played important roles in circulating astral knowledge between the court, Muslim communities, and non-Muslim local literati. Such involvement with non-Muslim literati who were interested in the astral sciences took the forms of circulating texts, sharing astral datasets, and expounding the Islamic cosmological models.

The translation of Arabo-Persian texts on astrology and astronomy at court, and publications on Islamic cosmology locally, attracted new, non-Muslim audiences to the study of Arabo-Persian astral sciences. Court astronomers such as Yuan Tong and Bei Lin (mentioned above) were examples for non-Muslim astronomers who could become acquainted with Arabo-Persian astronomy and astrology through their work at the imperial Directorate of Astronomy. Whereas Yuan Tong embedded his knowledge in his making of the Ming Datong 大統 calendar,[51] Liu Bei joined with Chinese Muslim astronomers at the Directorate to update computations and publish a new edition of the Islamic calendar.

These Chinese works, as well as the translations of Arabic and Persian texts that were made at court, produced interest in the study of the Arabo-Persian

---

50 Chen, *Huihui tianwen xue shi yanjiu*, 307–308. On these works, see also Petersen, *Interpreting Islam in China*.

51 Shi, "Yuantong 'weidu taiyang tongjing'," 241–249.

astronomical system, or as it became known, *Huihui lifa* 回回曆法 (also *Huihui lixue* 回回曆學), that went beyond the walls of imperial institutions. The encyclopedic *Biographies of Men of Science* (*Chouren zhuan* 疇人傳, 1799) provides accounts of non-Muslim literati who mastered *Huihui lixue* outside the court and published their studies during sixteenth and early seventeenth centuries. Among them are celebrated astronomy scholars such as Tang Shunzhi 唐順之 (1507–1560), Zhou Shuxue 周述學, Chen Rang 陳壤, Lei Zong 雷宗, and Yuan Huang 袁黃 (1533–1606).[52] Whereas some Muslim astronomers at the Directorate of Astronomy could still use the writings of their native lands (*reng yong qi benguo zhi shu* 仍用其本國之書), the *Official History of the Ming* (1732) tells us that the investigation of *Huihui lifa* 回回曆法 in these circles exclusively focused on texts in Chinese, due to lack of access to original texts and knowledge. Hence, they produced their own scholarly discourse (*zicheng yijia yan* 自成一家言) on Arabo-Persian astronomy.[53]

Friendships and informal associations between Chinese Muslims and their non-Muslim literati neighbors constituted another venue where informal exchange of astral knowledge took place. Mei Wending, for example, recounted in one of his essays that visits to Chinese Muslim astronomers (*xiyu guansheng* 西域官生) were a common practice among those interested in the study of the astral sciences, and that he himself paid multiple visits to astronomers working at the court and local men of science.[54]

Mei Wending's friendship with Ma Decheng (courtesy name Ruji 儒驥), a Chinese Muslim from Nanjing who had access to gnomon data in mosques across China, is an example of such a case. In his published essay on shadow measurement, Mei Wending recalled his visit to his Muslim friend Ma Decheng in 1680, who provided him with gnomon data that he obtained from four Chinese mosques and asked his opinion on the uniform correction factors used by the data compilers. Mei mentioned that he knew of the long tradition of Islamic astronomy and its merit, but he also applauded his friend's critical investigation of his ancestral tradition. In his deliberation on the issue of the uniform correction factors, Mei asserted that it was reasonable to believe that mathematical astronomy in the Islamicate world had developed a method to correct the difference in positions. He postulated, however, a number of

52 Ruan, *Chouren zhuan*, 30:292–293, 298–299.

53 *Mingshi*, 37:745. On this issue, see Weil, "Fourteenth-Century Transformation," 262–274.

54 This quote is included in the biography of Mei Wending composed by Mao Jike 毛際可 (1633–1708): "He paid visits to former observatory officials, disciples of science and Muslim officials." 若舊臺官、疇人子弟及西域官生，皆折節造訪. See Mei, *Wu'an lisuan shumu*, 19:2r.

reasons why such techniques were not implemented in that instance: the method existed in the Islamicate world, but the relevant texts had not been translated into Chinese; texts on the method had been translated into Chinese but not put into practice; or, alternatively, the method had been translated into Chinese and put into practice but vanished over time. Mei concludes his preface to the essay with the hope that Ma Decheng and his Chinese Muslim peers would find the method and allow the wider community to correct the computational errors.[55] In another essay, Mei Wending suggests that Ma Decheng was the one to introduce him to Jan Mikołaj Smogulecki (known in China as Mu Nige 穆尼閣, 1610–1656), a Polish Jesuit living in Nanjing and interested in astronomy who became one of Mei Wending's informants on European astronomy.[56]

Finally, private libraries of members of China's Muslim community constituted repositories of long-lost works and became venues for Muslim and non-Muslim scholars to search for astronomical texts. The Chinese Muslim scholar Liu Zhi recorded at length his extensive journeys across China to find relevant texts on cosmology. In a preface to one of his works, Liu recounted that when he sought relevant texts for his study on "the interaction of Heaven, Earth and human society," he happened on the several tens of volumes (*ce* 冊) of Arabic and Persian manuscripts (*xiguo yuanben* 西國原本, lit. "original texts from western lands") housed at the library of the Wu family 吳氏 in Beijing. These texts, Liu explained, were works on the astral sciences and geography (*tianwen dili zhi xue* 天文地理之學) that originated in the imperial libraries of the Yuan court and were dispersed by bandits after the fall of the dynasty.[57] Similarly, Mei Wending recorded in one of the bibliographical essays that he found a copy of Wu Xinmin's 吳信民 mathematical compendium *Jiuzhang bilei*

55 Mei, *Lisuan quanshu*, 19:17r–17v. A version of the preface was published by the philologist Hang Shijun 杭世駿 (1696–1773) as follows: "Mosques in the four provinces of Shaanxi, Henan, Beizhi and Jiangnan transmitted their shadow measurements, among them were also place of erroneous [measurements]. Ma Ruji of the Western Regions inquired about these. Thereafter he edited and added the techniques in order to fill in the gaps and compiled 'The Tables of Shadow Measurement of Four Provinces,' in one fascicle." 陝西河南北直江南四省禮拜寺有其表景之傳而其中亦有傳訛之處，西域馬儒驥以此致詢遂為訂定並附用法, 以補其闕作四省表景立成一卷. Hang, *Daogu tang wenji*, 30:4.

56 In *Wu'an lisuan shumu*, under the short entry on Jan Mikołaj Smogulecki's book, *Tianxue fangtong* 天學防通, Mei recorded: "In year 1675, I was introduced to him [Nikolaus Smogulecki 穆尼閣] during a conversation with my elder Ma Decheng" 歲乙卯，晤馬德稱諸君始知之，則其歸已久. Mei, *Wu'an lisuan shumu*, 34r.

57 Liu, *Tianfang Zhisheng shilu*, 14:41.

九章比類 (*Jiuzhang suanfa bilei daquan* 九章算法比類大全, 1450) in the library of the Chinese Muslim Wu Erzhang 伍爾章 (courtesy name Zuntao 遵韜) and asked to borrow it.[58]

4 Conclusion

The official recognition granted by the Yuan dynasty to Arabo-Persian astral sciences ushered in four centuries of Chinese engagement with texts, methods, and traditions for various aspects of Arabo-Persian astral sciences. This essay has explored the astronomical activities of Chinese Muslims at court, astronomical enterprises among Muslims and in mosques, and interactions between Chinese Muslims and their non-Muslim literati neighbors. These three venues represent sites in which astral knowledge was imported, developed, and refined.

The Ming and Qing courts established designated institutions to accommodate Arabo-Persian computational methods and astrological predictions. They assigned scholars to select and translate Arabic and Persian texts for the benefit of non-speakers. Chinese Muslims played in important roles as translators and practitioners versed in Arabo-Persian computational techniques. Through inheritable positions at the Directorate of Astronomy, Chinese Muslims maintained their involvement in imperial science for more than four centuries and constituted a direct bridge between China's Muslim communities and the court.

To a modern reader, it may seem remarkable that ethno-religious communities and their worship halls were engaged in the study of the sciences. In China's Muslim communities, however, astral sciences played important roles in religious life and in the collective memory. Cosmological theories and observations of celestial phenomena were integral aspects of Islamic thought and ritual. Through circulation of texts and ideas, use of scientific instruments and buildings, and heated epistemic debates, Chinese Muslim communities produced a space in which astral knowledge was studied and implemented.

Interactions between Muslim and non-Muslim literati with a shared interest in the astral sciences is a rather unknown aspect of the history of Arabo-Persian astral sciences in China. With their direct links to mosques and the astronomical bureaus, Chinese Muslims served as sources of astral information for local savants. Local Muslim and non-Muslim literati engaged in conversations and

58 Mei, *Wu'an lisuan shuji*, 44ʳ.

exchanges of information and books that opened up a space for learning and developing astral concepts.

Despite the shared interest in celestial phenomena, the various spaces in which Chinese Muslim scholars operated differed in their engagement with Arabo-Persian astral sciences. The court was mainly interested in importing the computational methods for predicting celestial motions and occultations and astrological interpretations, read against methods and data from other disciplines. Religious scholars and agents in mosques focused on the aspects of cosmology that were relevant to Islamic theology and practice. Literati found interest in overarching cosmological theory as well as particular methods of computation. These three independent, yet related, spaces facilitated the development of Arabo-Persian astral sciences and were complementary in their roles as transmitters of Arabo-Persian astral knowledge to late imperial China.

Bibliography

Bai Shangshu 白尚恕. "Qingzhen dasi rigui chutan" 清真大寺日晷初探. *Xibei daxue xuebao* 17, no. 2 (1987): 10–14.

Beijing niujie zhi shu 'Gangzhi' 北京牛街志書 '岡志.' Edited by Liu Dongsheng 劉東生 and Liu Shenglin 劉盛林. Beijing: Beijing Chubanshe, 1991.

Chang, Ping-Ying. "Chinese Hereditary Mathematician Families of the Astronomical Bureau, 1620–1850." PhD diss., City University of New York, 2015.

Chen Jiujin 陳久金. *Huihui tianwen xue shi yanjiu* 回回天文學史研究. Nanning Shi: Guangxi Kexue Jishu Chubanshe, 1996.

Chen Jiujin 陳久金. "Ma-de-lu-ding fuzi he Huihui tianwenxue" 馬德魯丁父子和回回天文學. *Ziran kexue shi yanjiu* 8, no. 1 (1989): 28–36.

Chongkan jiangning fu zhi 重刊江寧府志. Nanjing shi: Jiangsu Guangling Guji Ke yinshe, 1986.

Cullen, Christopher. *Astronomy and Mathematics in Ancient China: the Zhou Bi Suan Jing*. Cambridge: Cambridge University Press, 2007.

Cullen, Christopher and Catherine Jami. "Christmas 1668 and After: How Jesuit Astronomy Was Restored to Power in Beijing." *Journal for the History of Astronomy* 51, no. 1 (2020): 3–50.

Dalen, Benno van. "Islamic and Chinese Astronomy Under the Mongols: A Little-Known Case of Transmission." In *From China to Paris: 2000 Years Transmission of Mathematical Ideas*, edited by Yvonne Dold-Samplonius et al., 327–356. Stuttgart: Franz Steiner Verlag, 2002.

Elman, Benjamin A. *On Their Own Terms: Science in China 1550–1900*. Cambridge: Harvard University Press, 2005.

Haneda Tōru 羽田亨. "Ka-i yakugo no hensha Ma-sha-yi-hei" 華夷訳語の編者馬沙亦黒. *Tōyō gakuhō* 7, no. 3 (1917): 437–446.

Hang Shijun 杭世駿. *Daogu tang wenji* 道古堂文集. Shanghai: Shanghai Guji Chubanshe, 2009.

Henderson, John B. *The Development and Decline of Chinese Cosmology*. New York: Columbia University Press, 1984.

Huart, Clément. "Inscriptions arabes et persanes des mosquées chinoises de K'ai-fong-fou et de Si-ngan-fou." T'oung Pao 6, no. 3 (1905): 295–312.

Hucker, Charles O. *A Dictionary of Official Titles in Imperial China*. Taipei: Southern Material Center, 1985.

Huff, Toby E. *The Rise of Early Modern Science: Islam, China and the West*. Cambridge: Cambridge University Press, 2003.

Leslie, Donald Daniel, and Mohamed Wassel. "Arabic and Persian Sources Used by Liu Chih." *Central Asiatic Journal* 26, no. 1–2 (1982): 78–104.

Liu Zhi 劉智. *Tianfang Zhisheng shilu* 天方至聖實錄. In *Qingzhen Dadian* 清真大典, edited by Huang Xiefan 周燮藩 et al., 14:1–365. Hefei shi: Huangshan shushe, 2005.

Ma-ha-ma 馬哈麻. *Tianwen xiangzong xizhan* 天文象宗西. In *Qingzhen Dadian* 清真大典, edited by Huang Xiefan 周燮藩 et al., 21:1–58. Hefei shi: Huangshan shushe, 2005.

Manning, Patrick. "Introduction: Knowledge in Translation." In *Knowledge in Translation: Global Patterns of Scientific Exchange, 1000–1800 CE*, edited by Patrick Manning and Abigail Owen. 1–16. Pittsburg: Pittsburg University Press, 2018.

Mei Wending 梅文鼎. *Lisuan quanshu* 曆算全書. Shanghai: Shanghai Guji Chubanshe, 1987. Siku Quanshu edition.

Mei Wending 梅文鼎. *Sisheng biaoying licheng* 四省表影立成. In *Lisuan quanshu* 曆算全書, 19:16r–21v. Shanghai: Shanghai Guji Chubanshe, 1987. Siku Quanshu edition.

Mei Wending 梅文鼎. *Wu'an lisuan shumu* 勿菴歷算書目. In *Zhibuzu zhai congshu* 知不足齋叢書, edited by Pao Tingbo 鮑廷博 et al. Shanghai: Gushu Liutongchu, 1921.

Montgomery, Scott L. *Science in Translation: Movements of Knowledge Through Cultures and Time*. Chicago: University of Chicago Press, 2000.

Mote, Fredrick W. *Imperial China, 900–1800*. Cambridge: Harvard University Press, 1999.

Ming taizu shilu 明太祖實錄. Taibei: Academia Sinica, 1962–1968.

Petersen, Kristian. *Interpreting Islam in China pilgrimage, scripture, and language in the Han Kitab*. Oxford: Oxford University Press, 2018.

Peters, Francis E. *Aristoteles Arabus: The Oriental Translations and Commentaries on the Aristotelian Corpus*. Leiden: E.J. Brill, 1968.

Gu Tinglong 顧廷龍. *Qingdai zhujuna jicheng* 清代硃卷集成. Taipei: Chengwen chubanshe, 1992.

Qing Shizu Zhanghuangdi shilu 清世祖章皇帝實錄. Beijing: Zhonghua Shuju, 2008.

Ruan Yuan 阮元. *Chouren zhuan* 疇人傳. Taipei: Yiwen Yinshuguan, 1977.

Saidan, A.S. *Abū al-Ḥasan Aḥmad b. Ibrāhīm al-Uqlīdisī. The Arithmetic of Al-Uqlīdisī: The Story of Hindu-Arabic Arithmetic as Told in Kitāb al-Fuṣūl fī al-Ḥisāb al-Hindī*. Dordrecht: D. Reidel Publishing Company, 1978.

Sezgin, Fuat. *Geschichte Des Arabischen Schrifttums*. Leiden: Brill, 1967.

Shangjiang liangxian zhi tong zhi 上江兩縣志. Taipei: Chengwen chubanshe, 1970. Tongzhi 13 edition.

Shi, Yunli. " Islamic Astronomy in the Service of Yuan and Ming Monarchs." *International Journal for the History of the Exact and Natural Sciences in Islamic Civilisation* 13 (2014): 41–61.

Shi Yunli. "Yuantong 'weidu taiyang tongjing': Huihui lifa shishang de yifen zhongyao wenxian" 元統 '緯度太陽通徑': 回回曆法史上的一份重要文獻. *Gyujanggak* 규장각 33 (2008): 241–249.

Sivin, Nathan. *Granting the Seasons: the Chinese Astronomical Reform of 1280*. New York: Springer, 2009.

Tang Jinhui 唐晉徽. *Qingzhen shiyi buji* 清真釋疑補輯. In *Qingzhen Dadian* 清真大典 edited by Huang Xiefan 周燮藩 et al. 18:37–115. Hefei shi: Huangshan shushe, 2005.

Wang Daiyu 王岱輿. *Zhengjiao zhenquan* 正教真詮. In *Qingzhen Dadian* 清真大典, edited by Huang Xiefan 周燮藩 et al., 16:23–166. Hefei shi: Huangshan shushe, 2005.

Weil, Dror. "Collation and Articulation of Arabo-Persian Texts in Early Modern China." In *Routledge Handbook on Science in the Islamicate World*, edited by Sonja Brentjes. New York: Routledge, forthcoming.

Weil, Dror. "The Fourteenth-Century Transformation in China's Reception of Arabo-Persian Astronomy." In *Knowledge in Translation: Global Patterns of Scientific Exchange*, 1000–1800 CE, edited by Patrick Manning and Abigail Owen, 345–370. Pittsburg: Pittsburg University Press, 2018.

Weil, Dror. "Islamicated China: China's Participation in the Islamicate Book Culture During the Seventeenth and Eighteenth Centuries." *Histories of Books in the Islamicate World—Intellectual History of the Islamicate World* 4 (2016): 36–60.

Weil, Dror. "Libraries of Arabic and Persian texts in late imperial China." In *Encyclopaedia of Islam—Three*, edited by Kate Fleet et al. 90–92. Leiden: Brill, 2020.

Weil, Dror. "The Vicissitudes Of Late Imperial China's Accommodation of Arabo-Persian Knowledge of the Natural World, 16th–18th Centuries." PhD diss., Princeton University, 2016.

Wu Bozong 吳伯宗. *Mingyi tianwenshu*. 明譯天文書 In *Qingzhen Dadian* 清真大典 edited by Huang Xiefan 周燮藩 et al., 21:295–380. Hefei shi: Huangshan Shushe, 2005.

Wu Jianwei 吳建偉, *Zhongguo qingzhensi zonglan* 中國清真寺綜覽. Yinchuan: Ningxia Renmin Chubanshe, 1995.

Xianzong shilu 憲宗實錄. Taipei: Academia Sinica, 1962–1968.

Yabuuti, Kiyosi, and Benno van Dalen. "Islamic Astronomy in China During the Yuan and Ming Dynasties." *Historia Scientiarum* (1997), 11–43.

Yano, Michio. *Kūšyār Ibn Labbān's Introduction to Astrology*. Tokyo: Institute for the Study of Languages and Cultures of Asia and Africa, 1997.

Yu Zhengui 余振貴, and Lei Xiaojing 雷晓静. *Zhongguo huizu jinshilu* 中國回族金石錄 Yinchuan: Ningxia Renmen Chubanshe, 2001.

Mingshi 明史 (Ming history). Compiled by Zhang Tingyu 張廷玉 et al. Beijing: Zhonghua shuju, 1974.

CHAPTER 6

Geometrizing Chinese Astronomy? The View from a Diagram in the *Kashf al-ḥaqāʾiq* by al-Nīsābūrī (d. ca. 1330)

Yoichi Isahaya

1 Introduction

This chapter deals with an overlap between the cosmologies of eastern and western Eurasia through a Persian astronomical work in which a part of a Chinese calendrical system is represented in terms of western Eurasian geometrical cosmology through a diagram. The work in question is *Kashf al-ḥaqāʾiq-i zīj-i īlkhānī* (Uncovering the truths of the *Zīj-i īlkhānī*) by Niẓām al-Dīn al-Nīsābūrī (d. ca. 1330), who was active mainly in the Iranian plateau during the thirteenth and fourteenth centuries, when Iran was under the domination of the Mongol empire (1206–1368)—that is, the period of the Īlkhānate (ca. 1256–1357). In a section of his commentary on the lunar anomaly in a Chinese calendar, al-Nīsābūrī appropriated eastern Eurasian numerical procedures by means of a Ptolemaic geometrical understanding typical of the western Eurasian astral tradition. In this chapter, through textual analysis of the relevant section of the *Kashf al-ḥaqāʾiq* (including a critical edition and an English translation), I will attempt to shed light on the manipulation used by Muslim scholars to represent a Chinese computation method on their own cosmology.

The *Kashf al-ḥaqāʾiq-i zīj-i īlkhānī* is perhaps the earliest material to reveal a Ptolemaic understanding of Chinese astronomy, and in some ways it may be considered unique. As the title shows, this work is a commentary upon the *Zīj-i īlkhānī* (Īlkhānid astronomical handbook) by Naṣīr al-Dīn al-Ṭūsī (1201–1274), the first Islamicate astronomical handbook (*zīj*) to include explanation of a Chinese calendar. The *zīj* represents a genre of astronomical literature in the Islamicate world that originated in the Pahlavi and Indic astral traditions but later incorporated Greek scientific thought, especially Ptolemaic astronomy. The *zīj* consists of a series of tables representing all the functions for calculating planetary longitudes and other quantities of interest. The tables are also accompanied by texts that explain the use of the tables.[1]

1 For the *zīj*, for example, see King and Samsó, "ZĪDJ," and Mercier, "From Tantra to Zīj."

 | DOI:10.1163/9789004511675_008

The *Zīj-i īlkhānī* was the nominal major production resulting from collaboration among a group of astronomers working at the Maragha observatory.[2] The observatory was well-known as the foremost intellectual center on a Eurasian scale, with not only Muslim intellectuals but also Chinese and Syriac scholars active there. It was founded under the initiative of al-Ṭūsī, who served as a trusted councilor to Hülegü (1218–1265), the founder of the Īlkhānate. The adjective "nominal" applies on the grounds that this astronomical handbook was not necessarily based on observations carried out at the Maragha observatory. According to al-Ṭūsī's remarks in the introduction of the *Zīj-i īlkhānī*, Hülegü's principal aim in constructing the observatory and compiling the *zīj* was to envisage heavenly portents.[3] In other words, for Hülegü, the observatory and astronomical handbook were to function as tools for his dynasty's legitimatization and policy-making. He could not wait a very long time for such a practical purpose. Under these circumstances, al-Ṭūsī relied heavily on his predecessors' observations and the astronomical handbooks based on those observations for the hasty completion of his own. In fact, we know that the parameters underlying the solar, lunar, and planetary tables (the core parts of astronomical handbooks) were mainly taken from the astronomical handbooks of Ibn al-Aʿlam (d. ca. 985) and Ibn Yūnus (d. 1009), and the trigonometric tables were copied from Ibn Yūnus and Abū Rayḥān al-Bīrūnī (973–ca. 1050).[4] Still, according to Mohammad Mozaffari and Georg Zotti, "the *Zīj-i īlkhānī* is not based completely on the earlier *zīj*es and at least some independent observations were made in the observatory to measure the longitude of the solar apogee."[5] Ṭūsī, according to his own words, compiled the *Zīj-i īlkhānī* by comparing "what we got [from predecessors' observations] with what we knew from our observation."[6]

The fact that the *Zīj-i īlkhānī*, "completed tentatively" in 1272, was not based on new observations in Maragha was well known among scholars already in the Īlkhānid period.[7] Among scholars who aimed at revising the uncompleted *zīj*, ʿAlā Munajjim al-Bukhārī can be regarded as the pioneer, compiling a Persian *zīj* entitled *ʿUmdat al-īlkhānīya* (The cream of the [*Zīj-i*] *īlkhānī*) (1287/88).

2 Saliba, "Horoscopes and Planetary Theory," 363.

3 Ṭūsī, *Zīj-i īlkhānī*, 3$^{r–v}$.

4 King and Samsó, "ZĪDJ," 499. Ṭūsī's reliance is further confirmed by the scrutiny of the mean motion tables of the *Zīj-i īlkhānī* in Sawādī and Nīk-Fahm, "Ḥarakat-i wasaṭ-i kawākib dar *Zīj-i īlkhānī*."

5 Mozaffari and Zotti, "Observational Instruments," 56.

6 Ṭūsī, *Zīj-i īlkhānī*, 3^{v}.

7 Sayılı, *Observatory in Islam*, 214.

Nīsābūrī also recounted in the introduction of the *Kashf al-ḥaqā'iq* that al-Ṭūsī passed away before the close proofreading of his *zīj*:[8]

> When looking at the already-made *zīj*es, one did not find any [*zīj*] as useful and full of benefit as the *Zīj-i īlkhānī*, which is attributed to the righteous philosopher and the absolute intellectual ... Naṣīr al-Ḥaqq wa-l-Milla wa-l-Dīn Muḥammad b. Muḥammad al-Ṭūsī—may God glorify his soul, and may his intimacy [with God] increase in the gardens of Paradise. That is because, in this *zīj*, paying attention to the brevity of words, he did so in referring to clear definitions. He incorporated whatever was known by himself through blessed observation—which will be mentioned—into the abstract of what came to be known through his predecessors and contemporaries, as "there are all kinds of game in the belly of the wild ass."[9] [But] it is not far from [the fact] that that book, as is known, was not read for him in his lifetime, and he did not inquire into some of the problems and difficulties at that time. For that reason, if a slip of the pen occurred in some places, he did not reexamine them.[10]

Whatever the facts were regarding the compiling of the *Zīj-i īlkhānī*, this description makes clear that the scholars of the time realized its incompleteness, so much so that a scholar close to the Īlkhānid court admitted it.

As we have seen, the *Zīj-i īlkhānī*, despite its significant influence, was not necessarily based on actual observations carried out at the Maragha observatory, and it still adhered to the Ptolemaic tradition that was greatly challenged by Maragha astronomers.[11] What is new in the *Zīj-i īlkhānī* rather appears in the calendrical materials.[12] Among these materials, the "Cathay calendar" (*tārīkh-i Qitā*) is most remarkable in being the first description of any sort of Chinese calendar in a *zīj*.[13] In the *Zīj-i īlkhānī*, a long chapter divided into twelve sections is devoted to the description of the Cathay calendar.[14] The Cathay calendar is known to have been produced by al-Ṭūsī through a dialogue with a

8 Sawādī and Nīk-Fahm, "Ḥarakat-i wasaṭ-i kawākib dar *Zīj-i īlkhānī*," 371.

9 A famous Arabic proverb that denotes someone or something that combines all good qualities and advantages and makes everything else dispensable (Cowan, *Arabic-English Dictionary*, 821).

10 Nīsābūrī, *Kashf al-ḥaqā'iq-i zīj-i īlkhānī*, RR, 3ʳ; IF, 2ᵛ–3ʳ.

11 For example, see Ragep, "Copernicus and his Islamic Predecessors."

12 King and Samsó, "ZĪDJ," 499.

13 For studies on the Chinese calendar in the Islamicate astronomical handbooks, see Isahaya, "*Tārīkh-i Qitā*," 150–152.

14 For the contents, see Isahaya, "*Tārīkh-i Qitā*," 164–165.

"sage of Cathay," Fu Mengzhi 傅孟質.[15] Close investigations into the contents of the Cathay calendar have ascertained that the calendar can be considered largely a sort of amalgam of two Chinese astronomical systems: the *Chongxiu daming li* 重修大明曆 (Revised great enlightenment astronomical system) and the *Futian li* 符天曆 (Astronomical system tallying with heaven).[16] However, as shown in the analysis below, some elements are not identical to either of these astronomical systems. The incorporation of a Chinese calendar into an Islamicate astronomical handbook created a moment of overlap between two Eurasian cosmologies.

Of the various cosmologies found in western Eurasia, Aristotelian physics was most influential among Greek-writing intellectuals from the Hellenistic period onward. In line with Jamil Ragep's explanation, this cosmology was based on a finite and spherical universe centered on an immobile and spherical Earth. Moreover, the celestial world was differentiated from the sublunar one. The former consisted of spherical orbs and bodies that carried the "planets," including the sun and the moon. The celestial realm, comprising the planets and stars, was thought to be composed of a perfect substance called aether, within which celestial bodies rotate in uniform circular motion.[17] Scholars strove to represent all celestial motions as uniform and circular in their geometrical models in accordance with the strict principles of this ideal. The tradition was systematized by Ptolemy (ca. 83–ca. 168) in an influential book later called the *Almagest* (after ca. 150). The Ptolemaic system, along with the aforementioned Aristotelian cosmology, was integrated into the intellectual arena of the Islamicate world through the so-called ʿAbbāsid translation movement, a sociopolitical movement encompassing a wide range of circles consisting of the rulers, bureaucrats, and intellectuals of the ʿAbbāsid dynasty (750–1258) in the eighth to tenth century CE.[18]

In contrast, eastern Eurasia—especially the Chinese-writing world—did not represent celestial phenomena with geometrical models as its west Eurasian counterpart did. For the motion of a given celestial body, Ptolemaic astronomy set out to use observations to find the parameters (radii, speeds

15 For this "astronomical dialogue," see Isahaya, "Fu Mengzhi."

16 For the *Futian li*, see, for example, Isahaya and Lin, "Entangled Representation of Heaven," 163–165.

17 Ragep, "Islamic Culture," 43.

18 Gutas, *Greek Thought, Arabic Culture*. Kevin van Bladel's recent essay challenged the current understanding of the translation movement—especially its early phase—in relation to eastern factors such as the Tang dynasty (618–690, 705–907 CE) and the Indic astral tradition (van Bladel, "Eighth-Century Indian Astronomy").

of rotation, etc.) of the components of a geometrical model. When parameters that matched the recorded observations had been found, trigonometrical calculation was used to generate tables enabling users to calculate the apparent position of the body in question on the celestial sphere at any given moment, without the need to use much more than simple arithmetical processes. In the case of its east Eurasian counterpart, however, such tables were constructed solely on the basis of the numerical regularities in the series of past observations, without any reference to an underlying geometrical model as in Ptolemaic astronomy.[19] Regarding the east Eurasian astral tradition, the term "numerical cosmology" is borrowed from Christopher Cullen.[20]

The core of the Chinese astral tradition, *li* 曆, was based on the numerical cosmology of the eastern Eurasian astral tradition. *Li* is conventionally translated as "calendar," yet holds multiple connotations that cannot be expressed by that single word. Following Nathan Sivin's usage, therefore, I call this set of components an astronomical system.[21] Several scholars specializing in the Chinese astral tradition have pointed out commonalities between the *zīj* and the *li*, both of which include not only the motion of all luminaries but also other important celestial phenomena such as solar and lunar eclipses, solstices, equinoxes, the exact timing of noon and midnight, and planetary conjunctions. Sivin states that the *li* astronomical system shares a basic similarity to the *zīj* and western treatises from Ptolemy to Georg von Peuerbach (1423–1461).[22] Furthermore, Jean-Claude Martzloff, who assigns the term "astronomical canon" to the *li*, also attaches the term to the *zīj* for the following two reasons. First, a *zīj* translated into Chinese in the Ming period (1368–1644) bears the title *Huihui li* 回回曆 (Islamic astronomical system). Second, the *zīj* corresponds with the Greek "canon" (κανών), and the word *qānūn*—the Arabic equivalent

---

19 Cullen, *Heavenly Numbers*, 179–222.

20 Cullen, *Heavenly Numbers*, 2. The cosmological notion of a "spherical heaven" based on west-Eurasian geometrical cosmology was also conceived in China, under the term *huntian* 渾天, already in the early imperial period—the period of the Qin and Han dynasties (*Heavenly Numbers*, 218–222). However, while in western Eurasia the celestial sphere was a geometrical fiction, since the celestial bodies really moved on separate complex systems of spheres and circles, in eastern Eurasia all the celestial bodies were understood to really move on, or very close to, the inner surface of a single celestial sphere (*Heavenly Numbers*, 223–292).

21 Sivin refers to the astronomical system in terms of the following four aspects: (1) the art of computing the motion or location of certain celestial phenomena, (2) the sequence of computations necessary to make a complete ephemeris, (3) the embodiment of these first two in a computational treatise, and (4) almanacs based on the treatise (*Granting the Seasons*, 38–39).

22 Sivin, *Granting the Seasons*, 38.

of the Greek canon—appears in the title of al-Bīrūnī's *zīj*, *al-Qānūn al-masʿūdī* (Masʿūdic canon).[23]

Despite their similarities, the difference between the cosmologies of the two genres is remarkable. The *zīj* is absolutely based on the geometrical cosmology of the western Eurasian astral tradition. In the case of the *li*, such a geometrical representation did not appear in official systems prior to the Ming period. The *li* system embodied the numerical cosmology of the eastern Eurasian astral tradition, which consists of constants and algorithms with a specific nomenclature but not diagrams. This difference is also reflected in the *Zīj-i īlkhānī* in the sense that there is neither a geometrical diagram nor its explanation in the chapter on the Chinese calendar. However, in the *Kashf al-ḥaqā'iq*, al-Nīsābūrī seems to take a different approach to the Chinese calendar, especially in a section concerning the lunar anomaly, in which he uses a geometrical diagram to explain a mathematical procedure. In other words, a fourteenth-century scholar in Iran interpreted one aspect of Chinese astronomy through Ptolemaic geometrical representation, of which Islamicate astronomers were masters. This material gives us valuable insight into the way an Islamicate intellectual appropriated Chinese astronomy into his own astral tradition. In the next section, al-Nīsābūrī and his works, including the *Kashf al-ḥaqā'iq*, are discussed as a prelude to investigating his geometrical representation of Chinese astronomy.

2 Nīsābūrī and His *Kashf al-ḥaqā'iq*

The profile and achievements of al-Ḥasan b. Muḥammad b. al-Ḥusayn Niẓām al-Dīn al-Aʿraj al-Nīsābūrī have already been scrutinized by Robert Morrison in his book *Islam and Science*. Among the "foreign" or "ancient" sciences entering Arabic culture mainly through the aforementioned ʿAbbāsid translation movement, mathematics—especially geometry and arithmetic—were earlier justified within the Islamic religious norms. Furthermore, more advanced branches of mathematics had practical applications; for example, algebra was utilized for calculating inheritances, plane trigonometry was applied to calculating prayer times, and spherical trigonometry was important to determine the direction of Mecca, which was also of significance for daily prayers. Astronomy included in the mathematics of the Islamicate intellectual tradition was incorporated into religious tradition, which resulted in facilitating the development of this

23 Martzloff, *Le calendrier chinois*, 371–372.

discipline.[24] Such a rapprochement between Islam and science provides us with a clue to understanding the career of al-Nīsābūrī, who composed several astronomical texts in the fourteenth-century Iranian plateau. His best-known text was a *Qurʾān* commentary entitled *Gharāʾib al-qurʾān wa-raghāʾib al-furqān* (The curiosities of the *Qurʾān* and the desiderata of the demonstration), in which he demonstrated the importance of science for religious scholars.

Following Ibn al-Akfānī's (d. 1348) encyclopedic treatise, the *Irshād al-qāsid ilā asnāʾ al-maqāsid* (The guiding of the searcher to the most sublime purposes), Morrison explains that within the framework of the ethical classifications of Islamic law, the use of astronomy for the calculation of prayer times was not only justified but "incumbent" (*wājib*), and applying astronomy to discerning the existence of the Creator and His power was "recommended" (*mandūb*). Even believing in the stars' influence on terrestrial events was "permitted" (*mubāḥ*) unless one began to think that the stars influence terrestrial events independently, which was "forbidden" (*maḥẓūr*). In his *Sharḥ taḥrīr al-majisṭī* (Commentary on the revision of the *Almagest*, completed by mid-1303)—a commentary upon al-Ṭūsī's *Taḥrīr al-majisṭī* (Revision of the *Almagest* by Ptolemy)—al-Nīsābūrī elaborated on the recommended (*mandūb*) application that formed a theological motivation for the study of astronomy, with an emphasis on astronomy's ability to mediate between physics and the divine science. Among astronomical practices, he attached high importance to observation, which not only served practical purposes but also provided reminders of God's omnipotent role in creation.[25]

In 1308/09, after the compilation of the aforementioned *Sharḥ taḥrīr al-majisṭī*, al-Nīsābūrī completed the main text under discussion in this chapter, the *Kashf al-ḥaqāʾiq*, which was also a commentary upon al-Ṭūsī's work, the *Zīj-i īlkhānī*, the nominal *chef-d'oeuvre* of the Maragha observatory. One of the main applications of a *zīj* like the *Zīj-i īlkhānī* was astrological predictions. In consideration of the fact that al-Nīsābūrī dedicated the *Kashf al-ḥaqāʾiq* to Saʿd al-Dīn al-Sāwajī (d. 1311)—co-vizier together with Rashīd al-Dīn al-Hamadānī (1247–1318) until the former's execution in February 1311—in Persian, the work could well have served his patron's personal interest in astrology.[26] On the other

24 Morrison, *Islam and Science*, 13–14.

25 Morrison, *Islam and Science*, 28–29.

26 The dedication passage to al-Sāwajī appearing in RR—the autograph from 1310—is omitted in IF, which was copied earlier in 1308/09 (Nīsābūrī, *Kashf al-ḥaqāʾiq-i zīj-i īlkhānī*, RR, 4ʳ). Regarding sources mentioning Saʿd al-Dīn al-Sāwajī, see Pfeiffer, "Confessional Ambiguity," 148n66. At that time, even in the Persianate world, Arabic was the first language of astronomy in the sense that the discipline was part of religious scholarship like the case of

hand, through the scrutiny of this work, Morrison reveals al-Nīsābūrī's attitude toward astrology, which was, as already mentioned, "permitted" except in the case of believing the stars to be independent intermediary causes between celestial and terrestrial realms. According to Morrison, the *Kashf al-ḥaqā'iq* clarified al-Nīsābūrī's position on the validity of astrology to provide humans with conjectural knowledge of how God controls the earth through celestial motions, with the caveat that any conclusion should rely on repeated observations.[27]

Nīsābūrī's attitude, attaching high importance to observational astronomy, was crystallized in his representative work in the field of *'ilm al-hay'a* (cosmography), the *Tawḍīḥ al-tadhkira* (Elucidation of the *Tadhkira*, ca. 1311), a commentary on Ṭūsī's *al-Tadhkira fī 'ilm al-hay'a* (Memento on the science of the configuration of the universe).[28] For al-Nīsābūrī, *'ilm al-hay'a* should be based on physics and geometry, and less on metaphysics. In this regard, his thought on astronomy was also associated with Islamic jurisprudence (*fiqh*) by the use of the term *uṣūl* (principals), which often appeared in the context of jurisprudence.[29] Using the term in his astronomical treatises, al-Nīsābūrī tried to show that observation and science could contribute to legal reasoning.[30]

In this way, Morrison contextualizes the *Kashf al-ḥaqā'iq* within al-Nīsābūrī's intellectual career, which revolves around his views on astrology. Instead of such a horizontal approach, encompassing all al-Nīsābūrī's works, I take a vertical approach to deal with the *Kashf al-ḥaqā'iq* in the tradition of the *zīj* compilations, whereby another remarkable feature of the *Kashf al-ḥaqā'iq* is singled out from among other *zīj*es. Table 6.1 lists *zīj*es in which the description of Chinese calendar appears, from the *Zīj-i īlkhānī* to the *Zīj-i jadīd-i sulṭānī* by Ulugh Bīg (1394–1449), which completely replaced the former.[31]

A remarkable feature of the *Kashf al-ḥaqā'iq* among the listed *zīj*es is the geometrical explanation of its contents. In the *Kashf al-ḥaqā'iq*, al-Nīsābūrī inserted many sentences into an introductory part of the *Zīj-i īlkhānī* to provide a geometrical explanation of planetary motions on the basis of the *Ele-*

Sharḥ taḥrīr al-majisṭī, written in Arabic and dedicated to al-Sāwajī, as well as the *Kashf al-ḥaqā'iq* (Morrison, *Islam and Science*, 64–65). In the introduction of the *Kashf al-ḥaqā'iq*, al-Nīsābūrī explained his language choice in that writing in Persian made the content and its explanation accessible to a wider audience (Nīsābūrī, *Kashf al-ḥaqā'iq-i zīj-i īlkhānī*, RR, 4^v; IF, 3^v); cf. Morrison, *Islam and Science*, 66.

27 Morrison, *Islam and Science*, 63–70.

28 For the *'ilm al-hay'a*, for example, see, Ragep, *Jaghmīnī's Mulakhkhaṣ*.

29 Morrison, *Islam and Science*, 79.

30 Morrison, *Islam and Science*, 112–113.

31 King and Samsó, "ZĪDJ," 498.

TABLE 6.1 List of *zīj*es including the Chinese calendar

	Author	Title	Year of compilation
1	Naṣīr al-Dīn al-Ṭūsī	*Zīj-i īlkhānī*	1272
2	Muḥyī al-Dīn al-Maghribī	*Adwār al-anwār*	1276
3	Jamāl al-Dīn b. Maḥfūẓ al-Baghdādī	unknown	1286
4	ʿAlā al-Munajjim al-Bukhārī	*ʿUmdat al-īlkhānīya*	1287/88
5	Sayf al-Munajjim al-Bāyzdiwī	*Zīj-i ashrafī*	ca. 1303
6	**Niẓām al-Dīn al-Nīsābūrī**	***Kashf al-ḥaqā'iq-i zīj-i īlkhānī***	**1308/09**
7	Nāṣir b. Haydar al-Shīrāzī	*Zīj-i nāṣirī*	ca. 1310
8	Shams al-Munajjim al-Wābkanawī	*Zīj al-muḥaqqaq al-sulṭānī*	ca. 1320
9	Ghiyāth al-Dīn Jamshīd al-Kāshī	*Zīj-i khāqānī*	1413
10	Ulugh Bīg	*Zīj-i jadīd-i sulṭānī*	1440

ments and *Almagest* together with diagrams.[32] Such geometrical explanations appear far less in the other *zīj*es listed above—the only comparable *zīj* is the *Zīj-i khāqānī* by Ghiyāth al-Dīn Jamshīd al-Kāshī (1380–1492), a distinguished mathematician of the Samarqand school.[33] Overall, Persian *zīj*es after the *Zīj-i īlkhānī* tended to be weighted more toward the practical aspect of the contents, such as the tables themselves and instruction for their use, rather than geometrical proofs with diagrams, to which a larger portion of the contents had been devoted in earlier *zīj*es like that of al-Ḥabash al-Ḥāsib (ca. 850), Kūshyār's *al-Zīj al-jāmiʿ* (Comprehensive astronomical handbook, ca. 1000), and al-Bīrūnī's *al-Qānūn al-masʿūdī* (1030).[34] In this sense, al-Nīsābūrī's *Kashf al-ḥaqā'iq* reversed the contemporary trend, probably due to his great interest in theoretical astronomy. To the contrary, as Morrison points out, al-Nīsābūrī pays less attention to the sections on astrological prediction.[35]

32 Nīsābūrī, *Kashf al-ḥaqā'iq-i zīj-i īlkhānī*, RR, 7ᵛ–19ʳ; IF, 6ᵛ–19ʳ; cf. Morrison, *Islam and Science*, 65.

33 For the contents of the *Zīj-i khāqānī*, see Kennedy, *Contents and Significance*, and for mathematicians and astronomers at the Samarqand observatory, for example, see Fazlioglu, "The Samarqand Mathematical-Astronomical School."

34 For the contents of *al-Qānūn al-masʿūdī*, see Kennedy, "al-Bīrūnī's Masudic Canon." I was instructed regarding earlier *zīj*es, especially in terms of their geometrical proofs, by Dr. Tarō Mimura.

35 Morrison, *Islam and Science*, 65.

Importantly in our context, his interest in theoretical astronomy, especially geometrical verification, seems to extend to the description of the Chinese calendar, in which no geometrical explanation is found in the original *Zīj-i īlkhānī*. In the *Kashf al-ḥaqā'iq*, al-Nīsābūrī assigns much space to comments on the contents of the Cathay calendar, section by section, though he does not deal with the tables as much. In the next section, I investigate in particular his comments on the seventh section concerning the lunar anomaly, because this is where he makes use of a diagram to explain the mathematical procedures of the Cathay calendar.

3 The Section on the Lunar Anomaly

The section on the lunar anomaly of the *Kashf al-ḥaqā'iq* is divided into two parts: text (*matn*) and commentary (*sharḥ*). In the text part, the original sentences of the *Zīj-i īlkhānī* are "quoted." It is likely, however, that al-Nīsābūrī did not quote the sentences as they were but rather changed some of them—perhaps to clarify the meanings—since the quoted sentences, although very close to the "embedded" version of the *Zīj-i īlkhānī*, are not completely identical to it.[36]

The commentary is also subdivided into two parts: example (*mithāl*) and verification (*taḥqīq*). In the example part, we are provided with an example of how to calculate the lunar argument for the beginning of the year 677 Yazdegerdi (1308 CE). The verification following that example is where the geometrical explanation for Chinese astronomy appears.

The seventh section states how to compute the lunar argument, needed especially for the lunar equation of the center in consideration of the lunar anomaly. Astral sciences in both eastern and western Eurasia, in computing the inequality of the lunar motion, took into account the anomalistic month—that is, the period of the lunar motion from its perigee/apogee back to perigee/apogee. Since there was no explicit concept of perigee or apogee in the astronomical vocabulary of eastern Eurasian astral science, the anomalistic month was seen as a cycle called the *junjūn* (*zhuanzhong* 轉終).[37]

36 For the categorization of manuscripts in the *Zīj-i īlkhānī*, see Isahaya, "*Tārīkh-i Qitā*," 156–162.

37 Sivin, *Granting the Seasons*, 101–102.

In this section, the initial value of the lunar argument (*aṣl-i ḥiṣṣa-yi māh*) means the interval between the beginning of the solar year—that is called the *wūshī* (*yushui* 雨水), "rain waters," or the beginning of Pisces (330°) in the Cathay calendar—and the beginning of the anomalistic month preceding the former at the beginning of the first year of the Superior Epoch (*shangyuan* 上元)—that is 1264 CE. The value called the *jūnjūn kā* (*zhuanzhong ying* 轉終應?) in Chinese is given as 78.3948 days. This value changes annually by 7.0338 days, called the *jūnjā* (*zhuancha* 轉差), which means the excess of a solar year over thirteen anomalistic months.

To compute the interval between the beginnings of the solar year and of the anomalistic month preceding the former in another year, therefore, the initial value of the lunar argument must be added to the product of the multiplication of the interval between the epoch year (1264 CE) and a required year by 7.0338 (the *jūnjā*) if the required year falls after the epoch year. The interval between the beginnings of the solar year and of the anomalistic month preceding the former in a required year is called "the first number to be kept" (*maḥfūẓ-i awwal*). Then, the interval between the beginnings of the first month and the solar year—which was already computed in the previous section—is subtracted from "the first number to be kept" to yield the interval between the beginnings of the first month and of the anomalistic month preceding the former, which is called "the second number to be kept" (*maḥfūẓ-i duwwum*). If subtraction is not possible, the period of the anomalistic month—the value of which is 27.5556 days—must be added prior to the subtraction. From "the second number to be kept," the period of the anomalistic month is subtracted as many times as possible. That is because the interval between the beginnings of the first month and of the anomalistic month preceding the former must be shorter than the period of an anomalistic month. The result is one-ninth of the lunar argument in the required year, thus it must be multiplied by 9 to obtain the actual value of that lunar argument.[38] As a reference, table 6.2 lists the lunar argument in the *Zīj-i īlkhānī*.

In the seventh section of the Chinese calendar in the *Kashf al-ḥaqā'iq*, the original text of the *Zīj-i īlkhānī* is followed by a commentary which, as already mentioned, consists of the example and verification parts. The example part shows the mathematical procedure used to calculate the lunar argument for the year of 677 Yazdegerdi (1308 CE). The operation is represented as table 6.3.

In this case—that is, the beginning of the year 677 Yazdegerdi (1308 CE)—through the operation, 82.0764 days (that is, the 83rd day) are obtained as the

38 Isahaya, "*Tārīkh-i Qitā*," 200.

TABLE 6.2 Table of the lunar equation

Positive		Lunar equation	Negative		Positive		Lunar equation	Negative	
Lunar argument			Lunar argument		Lunar argument			Lunar equation	
0	124	0, 0, 0	124	248	32	92	0, 49, 4	156	216
1	123	0, 2, 3	125	247	33	91	0, 50, 3	157	215
2	122	0, 4, 4	126	246	34	90	0, 51, 0	158	214
3	121	0, 6, 3	127	245	35	89	0, 51, 55	159	213
4	120	0, 8, 0	128	244	36	88	0, 52, 48	160	212
5	119	0, 9, 55	129	243	37	87	0, 53, 39	161	211
6	118	0, 11, 48	130	242	38	86	0, 54, 28	162	210
7	117	0, 13, 39	131	241	39	85	0, 55, 15	163	209
8	116	0, 15, 28	132	240	40	84	0, 56, 0	164	208
9	115	0, 17, 15	133	239	41	83	0, 56, 43	165	207
10	114	0, 19, 0	134	238	42	82	0, 57, 24	166	206
11	113	0, 20, 43	135	237	43	81	0, 58, 3	167	205
12	112	0, 22, 24	136	236	44	80	0, 58, 40	168	204
13	111	0, 24, 3	137	235	45	79	0, 59, 15	169	203
14	110	0, 25, 40	138	234	46	78	0, 59, 48	170	202
15	109	0, 27, 15	139	233	47	77	1, 0, 19	171	201
16	108	0, 28, 48	140	232	48	76	1, 0, 48	172	200
17	107	0, 30, 19	141	231	49	75	1, 1, 15	173	199
18	106	0, 31, 48	142	230	50	74	1, 1, 40	174	198
19	105	0, 33, 15	143	229	51	73	1, 2, 3	175	197
20	104	0, 34, 40	144	228	52	72	1, 2, 24	176	196
21	103	0, 36, 3	145	227	53	71	1, 2, 43	177	195
22	102	0, 37, 24	146	226	54	70	1, 3, 0	178	194
23	101	0, 38, 43	147	225	55	69	1, 3, 15	179	193
24	100	0, 40, 0	148	224	56	68	1, 3, 28	180	192
25	99	0, 41, 15	149	223	57	67	1, 3, 39	181	191
26	98	0, 42, 28	150	222	58	66	1, 3, 48	182	190
27	97	0, 43, 39	151	221	59	65	1, 3, 55	183	189
28	96	0, 44, 48	152	220	60	64	1, 4, 0	184	188
29	95	0, 45, 55	153	219	61	63	1, 4, 3	185	187
30	94	0, 47, 0	154	218	62	62	1, 4, 4	186	186
31	93	0, 48, 3	155	217					

lunar argument, so that, in referring to table 6.2,[39] we obtain 0,56,43 *funks* as the value of the lunar equation.[40]

39 This table is extracted from the relevant part of the original table in the *Zīj-i īlkhānī* (Isahaya, "*Tārīkh-i Qitā*," 183).

40 In the Cathay calendar, a day is divided into 10,000 *funks* (Isahaya, "*Tārīkh-i Qitā*," 188).

TABLE 6.3 Tabular representation of the "example" part

Terms	Chinese equivalents	Values
The required year		1308 CE
The initial value of the lunar argument in the epoch year	*zhuanzhong ying* 轉終應	78.3948 days
The interval between the epoch and required years		44 years
The excess of a solar year over thirteen "anomalistic months"[41]	*zhuancha* 轉差	7.0338 days
"The first number to be kept"		387.882 days
The interval between the beginnings of the first month and of the solar year in 1308		20.5396 days
"The second number to be kept"		367.3424 days
An anomalistic month	*zhuanzhong* 轉終	27.5556 days
One-ninth of the lunar argument		9.1196 days
The lunar argument at the beginning of 1308		82.0764 days

After showing an example, al-Nīsābūrī begins the verification of the procedure, which is the focal point of this paper due to the fact that he makes use of a diagram to explain the procedure.

In the verification section, al-Nīsābūrī represents the procedure for the lunar anomaly with a simple epicycle model by means of a diagram (figs. 6.1–3). In other words, al-Nīsābūrī projects the numerical values of the Chinese procedure onto a diagram based on the west Eurasian astral tradition, as described in table 6.4.[42]

For "the second number to be kept" in this calculation, the result would be *LZ*, not *LT*, by the subtraction of *MKZ* from *LZKHM*. This point is clarified through a marginal note in IF as follows: "Because, for arc *LT* and arc *LTZ*, the

41 The value of an "anomalistic month" (27.5546 days) expressed with the phrase "the mean lunar cycle" (*dawr-i qamarī-yi awsaṭ*) in this part is in fact different from the later value appearing as the cycle of the lunar argument (*dawr-i ḥiṣṣa-yi māh*) (27.5556 days). See Isahaya, "The *Tārīkh-i Qitā* in the *Zīj-i Īlkhānī*," 200–201.

42 For an explanation of the epicycle model for the lunar anomaly in the Ptolemaic system, see, for example, Pedersen, *Survey of the Almagest*, 166–167; Ôhashi, "Mathematical Structure," 86.

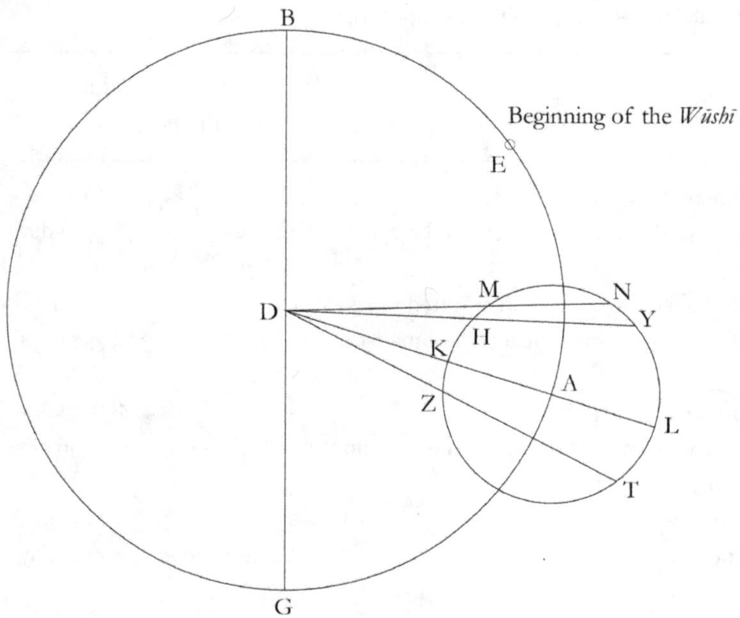

FIGURE 6.1 Diagram for the lunar anomaly

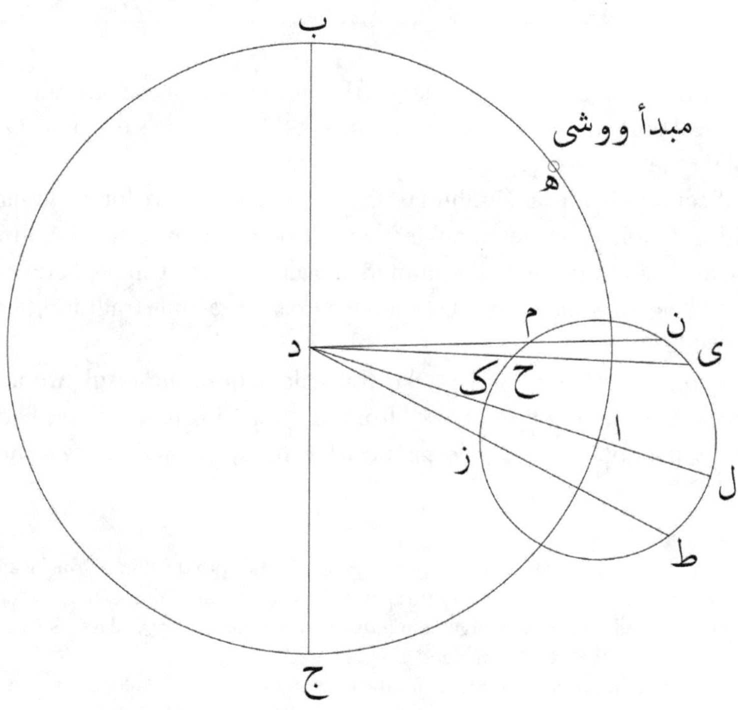

FIGURE 6.2 Diagram for the lunar anomaly

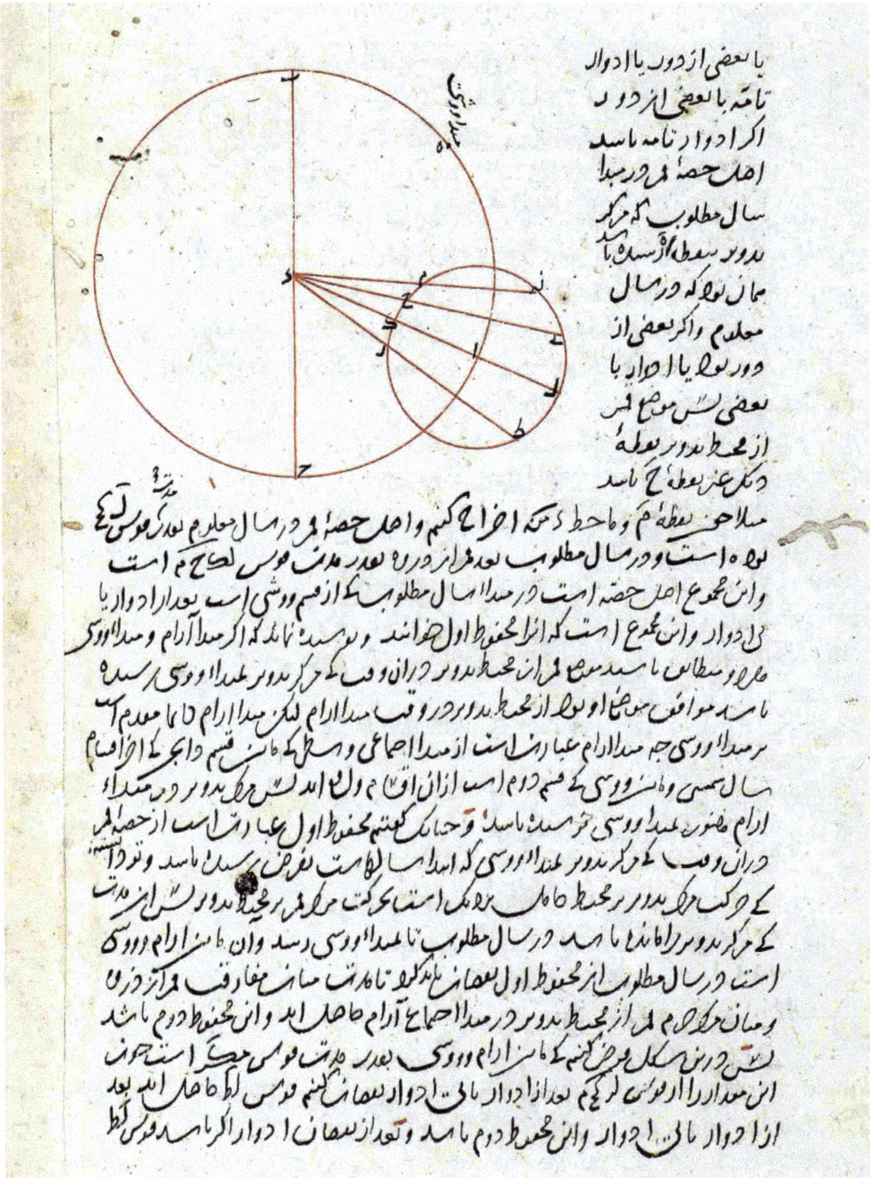

FIGURE 6.3 A diagram in the *Kashf al-ḥaqāʾiq* (RR, 33v)

TABLE 6.4 Tabular representation of the "verification" part

Terms	Corresponding letters	Values
The beginning of the solar year in the required year	Point *E*	
The interval between the beginnings of the first month and of the solar year in the epoch year.	Arc *AE*	14.4676 days
The apogee of the anomalistic month	Point *L*	
The interval between the beginnings of the solar year and of the anomalistic month preceding the former in the epoch year	Magnitude *LKH*	78.3948 days
The excess of a solar year over thirteen anomalistic months	Somewhere beyond point *H*	7.0338 days
(Case 1) A lunar position at the beginning of the solar year in a required year	Point *M*	
"The first number to be kept" in the required year	Magnitude *LZKHM*	
The interval between the beginnings of the first month and of the solar year in the required year	Magnitude *MKZ*	
"The second number to be kept" in the required year	Arc *LT*	
(Case 2) "The first number to be kept" in another required year	Arc *LT*	
The position of the anomalistic month at the beginning of the first month in the required year.	Point *Y*	
"The second number to be kept" in the required year	Arc *LTKY*	

angle created by these becomes the same (*cha qaws* LṬ *wa qaws* LṬZ *rā yik ḥukm dar iḥdāth-i zāwiyat bar mī-gard*)."[43] That is, for both *LZ* and *LT*, the angle from point *D* (angle *LDZ* and angle *LDT*) are the same. So, "the second number to be kept" is described as *LT* in this part.

43 Nīsābūrī, *Kashf al-ḥaqā'iq-i zīj-i īlkhānī*, IF, 35ᵛ.

4 A Ptolemaic Understanding of Chinese Astronomy

By examining the verification by al-Nīsābūrī, we now know how he represented the Chinese method of computing the lunar anomaly by means of a geometrical epicycle model peculiar to the west Eurasian astral tradition. Among the contents of the Cathay calendar in the *Zīj-i īlkhānī*, this section is by no means the only one which allows such a geometric representation. The sixth to ninth sections, concerning the computation of the new year and the solar and lunar equation of the center, could also be explained in such a way. However, al-Nīsābūrī used a geometrical explanation with a diagram only for the seventh section, and more specifically, only for the lunar anomaly in this section. A specific question arises here as to what prompted al-Nīsābūrī to make use of geometrical representation only in the seventh section.

Importantly, the part on the lunar anomaly in the seventh section of the original *Zīj-i īlkhānī* is unique in that it is the only section where the basic constant is not derived from the original Chinese astronomical system, the *Chongxiu daming li*.

As seen from table 6.5, the basic constants of the epoch year outside the seventh section in the *Zīj-i īlkhānī* are almost in accordance with those of the *Chongxiu daming li*, one of the two main sources of the Cathay calendar. There is a significant difference, however, in the two values of "the interval between the beginnings of the *wūshī* and the anomalistic month preceding the former," the computation of which is explained in the seventh section. The given values are 23.2836 and 9.49, with a difference of 13.7936 days that appears to be around half the value of the anomalistic month (13.7773 days in the case of a 27.5546 day month and 13.7778 days in the case of a 27.5556 day month—as already mentioned, there are two values of an anomalistic month in the Cathay calendar). In many Chinese astronomical systems, including the *Chongxiu daming li*, despite there being no equivalent notion of the moon's perigee or apogee, an anomalistic month was counted as a cycle from the perigee, when the moon is closest to the earth and its apparent speed is fastest.[44] In the Ptolemaic system, the starting point for calculation was its apogee. Therefore, the difference between the *Zīj-i īlkhānī* and the *Chongxiu daming li* must be ascribed to the difference of the standard points of the calculation, that is, the apogee in the *Zīj-i īlkhānī* (23.2836 days) and the perigee in the *Chongxiu daming li* (9.49 days). The difference between the apogee and the perigee must be half the value of the anomalistic month.[45]

---

44 Isahaya, "*Tārīkh-i Qitā*," 200.

45 Dr. Yukio Ôhashi was of great help in the discussion of this paragraph.

TABLE 6.5 Comparison of constants between the *Zīj-i īlkhānī* and the *Chongxiu daming li*

Values in the epoch year (1264 CE)	*Zīj-i īlkhānī*	*Chongxiu daming li*
Grand epoch[a]	88,639,679 years from 1203 CE	88,639,656 years from 1180 CE
The length of the solar year[b]	365.2436 days	365.2435946 days
The length of the lunar month[c]	29.5306 days	29.5305927 days
Interval between the winter solstice and the beginning of the *lījun*.[d]	11.7660 days	11.7662 days
Interval between the beginnings of the first month and the *wūshī*.[e]	14.4676 days	14.4724 days
Interval between the beginnings of the *wūshī* and the anomalistic month preceding the former[f]	**23.2836 days**	**9.49 days**

a Isahaya, "The *Tārīkh-i Qitā* in the *Zīj-i Īlkhānī*," 192–193.
b Isahaya, "The *Tārīkh-i Qitā* in the *Zīj-i Īlkhānī*," 169n1.
c Isahaya, "The *Tārīkh-i Qitā* in the *Zīj-i Īlkhānī*," 176n3.
d Isahaya, "The *Tārīkh-i Qitā* in the *Zīj-i Īlkhānī*," 193–195.
e Isahaya, "The *Tārīkh-i Qitā* in the *Zīj-i Īlkhānī*," 197–199.
f Isahaya, "The *Tārīkh-i Qitā* in the *Zīj-i Īlkhānī*," 201–202.

We are also able to compute the interval between the beginnings of the solar year and of the anomalistic month preceding the former in the epoch year from the main planetary tables of the *Zīj-i īlkhānī*. At noon on the beginning of the solar year, the *wūshī* (February 13, 1264 CE), the lunar mean anomaly has the value of 30, 22°.[46] That corresponds to 83.7% of its cycle from the apogee, which can be represented as 23.1 days. This value is very close to that of the Cathay calendar in the *Zīj-i īlkhānī* (23.2836 days).[47] Based on this fact, it seems likely that Naṣīr al-Dīn al-Ṭūsī shifted the initial value of the lunar argument in accordance with his own astral tradition, in which the apogee was the standard point for calculation. The position of the apogee at that time would be calculated on the basis of the planetary table of his *Zīj-i īlkhānī*. Then, this shift was actually visualized as a diagram by al-Nīsābūrī, who had an immense interest

46 The computation was conducted by means of the Devplo program, written by Dr. Raymond Mercier (http://www.raymondm.co.uk/).

47 I received great assistance regarding the computation from Dr. Benno van Dalen (Isahaya, "*Tārīkh-i Qitā*," 202).

in providing geometrical verification for the operations in the *Zīj-i īlkhānī*. That could be interpreted as a moment of overlap between the two cosmologies of Eurasia.

Such an overlapping did not necessarily lead to a positive evaluation of the Chinese calendar by Muslim scholars like al-Nīsābūrī or even al-Ṭūsī. In the Chinese numerical cosmology, every constant was combined with specific nomenclatures that resulted in complicated mathematical procedures. For Muslim scholars, in contrast, the mathematical procedures of the epicycle model were visualized quite simply in geometrical representations that must be considered fundamental to western Eurasian astral science. Based on this difference, a relatively low opinion of Chinese astronomy appears in the *Tānksūqnāma-yi īlkhān dar funūn-i ʿulūm-i khiṭāʾī* (The treasure book of the Īlkhān on the Cathay arts and sciences, ca. 1313 CE), which was the main product of the so-called Īlkhānid translation project, an enterprise of Chinese-to-Persian translation described by Rashīd al-Dīn al-Hamadānī (1249–1318), the famed Persian statesman and historian.[48] In the introduction of the *Tānksūqnāma*, Rashīd al-Dīn gives rich information on Chinese society and culture. In one part, after mentioning the astronomical dialogue between al-Ṭūsī and Fu Mengzhi, he continues as follows:

> [Fu Mengzhi's astral knowledge] harmed the reputation of the Cathay scholars since in this land (Iran) it is [now] assumed that their astral science is at such a [low] level and that they are unfamiliar with *ʿilm al-hayʾa*, the *Almagest*, and all that relates to it. What the person (Fu Mengzhi) reported to the late Naṣīr al-Dīn [al-Ṭūsī] was at such [a low] level that his (Ṭūsī's) knowledge already encompassed it, and the book that this person read was an abridgement that beginners study [in Iran].[49]

The Chinese calendar transmitted by Fu Mengzhi on the basis of the numerical cosmology was evaluated as "what beginners study" by Muslim intellectuals. What is worthy of attention in our context is that such an evaluation is narrated in light of the geometrical cosmology represented in the works of the *ʿilm al-hayʾa* and the *Almagest*. The visualization of the Chinese astronomical system connected two Eurasian astral traditions, but, in this case, such overlapping cosmologies led to the underestimation of the eastern Eurasian astral tradition by western Eurasian astronomers.

48 For the work and translation project, see Isahaya, "Sino-Iranica in Pax Mongolica."

49 Rashīd al-Dīn, *Tānksūqnāma*, Ms. Istanbul, 8^v–9^r; Facsimile, 16–17; Isahaya, "Fu Mengzhi," 247.

Appendix A: Translation

Here, I provide the translation of the seventh section. I made use of the following two manuscripts of the *Kashf al-ḥaqāʾiq*:

RR MS Persian 1203, Raza Library, Rampur (autograph dated to April 19, 1310), 32^{v}–34^{r}.

IF MS 3421, Fatih Millet Kütüphanesi, Istanbul, copied from an original in 1308/09, 34^{r}–35^{v}.

In the translation, the folio numbers of RR are denoted with curly { } brackets. Additions to the text are marked with square [ ] brackets, and explanatory words are inserted within round ( ) brackets. For letter-names in the Persian text, I use the transliteration listed in table 6.6.

[*The Part of the Text* (Matn)]

Text: For the lunar argument, the initial value of the lunar argument (*aṣl-i ḥiṣṣa-yi māh*) at the beginning of one year must be known. That is 78 days and 3,948 *funks*—78^{d} 1, 5, 48^{f}—in the first year of the cycle of the Superior Epoch (*dawr-i shānk wan*).[50] That value is called *jūnjūn kā* in the language of Cathay. Then, we obtain the difference between that year and the required year. We multiply the difference by 7 days and 338 *funks*—7^{d} 0, 5, 38^{f}. That value is called *jūnjā* in the language of Cathay. In my opinion, that is the excess of a solar year over thirteen times of the mean lunar cycle (*dawr-i qamarī-yi awsaṭ*). If the required year is after [the year of] the *jūnjūn kā*, we add the result to that (*jūnjūn kā*). If [the required year is] before that, we subtract [the result] from *jūnjūn kā*. If it cannot be subtracted, we add some times the period of one cycle of the lunar argument (*dawr-i ḥiṣṣa-yi māh*) to the initial value of the lunar argument—that is, *jūnjūn kā*—so that the result could be subtracted from it. What is obtained after addition to, or what remains after subtraction from, the initial value of the lunar argument in that year, we call "the first number to be kept." Then, we subtract the interval between the beginning of the first month (*ārām*)[51] and that of the *wūshī*[52] from that amount. If it cannot be subtracted, we add one cycle of the lunar cycle (*dawr-i qamarī*)—that is 27^{d} 1, 32, 36^{f}—once or more to the days, and then subtract. We call what is obtained "the second number to be kept." Then, from "the second number to be kept," we subtract the period of the

50 The first year of the cycle is 1264 CE. For the cycle in the Cathay calendar, see Isahaya, "*Tārīkh-i Qitā*," 191–193.

51 For the names of the months of the Cathay calendar, see Isahaya, "*Tārīkh-i Qitā*," 196.

52 For the *wūshī*, see Isahaya, "*Tārīkh-i Qitā*," 190–191.

TABLE 6.6 Persian-Latin letter-names

Persian	Latin	Persian	Latin
ا	A	ط	T
ب	B	ی	Y
ج	G	ک	K
د	D	ل	L
ه	E	م	M
ز	Z	ن	N
ح	H		

cycle of the lunar argument, the value of which is 27 days and 5,556 *funk*s—27^{d} 1, 32, 36^{f}—and called *junjūn* by them, as many times as [the remainder] becomes smaller than one cycle. What remains is a ninth of "the initial value of the lunar argument." We multiply that by 9 to obtain the lunar argument of the beginning in the required year.

[*The Part of an Example* (Mithāl)]

Commentary: For this procedure, first, we refer to an example. After that, we engage in the investigation of that.

Its example: We want to know the lunar argument in the year 677 Yazdegerdi [1308 CE]—which is the year of the *wūshin* (45th year of the sexagenerary cycle) in the language of the Cathay people and the year of the monkey in the language of the Turks. Since a year for which the initial value of the lunar argument is known {33^{r}} is the first year of the cycle of the Superior Epoch corresponding with 633 Yazdegerdi (1264 CE), then the difference between the known year and required year is 44. We multiple that by the *jūnjā*—7^{d} 0, 5, 38^{f}. 308 days come as the result with 14,872 *funk*s. In place of 10,000 *funk*s, we add one to the days, so that it becomes 309 days. As for *funk*s, 4,872 comes to—1, 21, 12^{f}. We add this result to the initial value of the lunar argument in the known year, 78^{d} 1, 5, 48^{f}, because the known year is prior to [the required year, and the result] becomes 387 days and 2, 27, 0^{f}. This is "the first number to be kept." Then, from this amount, "the first number to be kept," we subtract the difference between the beginning of the first month and the beginning of the *wūshī* in this year, which is 20 days and 1, 29, 56^{f}, as we have shown before in an example. 367 days and 0, 57, 4^{f} remains. This is "the second number to be kept." Since from this amount thirteen times of the *junjūn*, that is, 358^{d} 0, 37, 8^{f}, is subtracted, 9^{d} and 0, 19, 56^{f} remains. This is a ninth of the lunar argument. We multiple this by

9. 82 days and 0, 12, 44^{f} come as the result. This is the lunar argument at the beginning of the required year.

An example in the case of the required year being prior to the known year must be also deduced from this.

[*The Part of Verification* (Taḥqīq)]

Then, the verification of this procedure (*ʿamal*) aims at a ninth of the lunar argument equal to the period of the lunar motion from the apogee of an epicycle (*dhurwa-yi tadwīr*) to the place where the center of the body (*markaz-i jarm*) is on the circumference of the epicycle. We, for explanation, draw a circle ABG around the center *D*. We assume point *A* on its circumference as the center of the epicycle in a required year and point *E* as the beginning of the *wūshī* also in this year. We produce lines DZT, DHY, DKAL.

Then, what can be known from the previous section is the period between AE. Then, it is required to obtain the period between a true conjunction (*ijtimāʿ-i ḥaqīqī*) and the beginning of the *wūshī*. In this section, it is required to obtain a ninth [of the lunar argument] equal with the period of the distance between the apogee and the center of the lunar body on the circumference of the epicycle in an assumed year in accordance with the position of the epicycle's center on the outer circumference (*muḥīṭ-i khārij*). To explain this, we say that the period of the motion of the center of the lunar body on the circumference of the epicycle is closely related to the period of the motion of the epicycle's center on the circumference of the circle ABG.

Then, in a known year—for example, the first year of the Superior Epoch on the condition that the beginning of the [solar] year is at the first point of division of the *wūshī*—we assume the distance of the moon from point *L*, the apogee, as the magnitude of LKH and the center of the lunar epicycle as point *A* in each year of the solar years, the beginning of which is also at the beginning of the division of the *wūshī*. The center of the lunar body passes point *H* by the magnitude of the excess of a solar year over the thirteen lunar cycles—that is approximately 7^{d} 0, 5, 38^{f}. Since the number of the solar years which is between the required year and the known year is multiplied by this excess, the total of this is the result of the difference of those years. That is either complete revolutions, {33^{v}} part of a revolution, or revolutions with a part of a revolution. If it is complete revolutions, the initial value of the lunar argument at the beginning of the required year, when the center of the epicycle comes to point *E*, is same as [that of] the known year. If it is part of a revolution or revolutions with a part, then the position of the moon on the circumference of the epicycle is a point other than point *H*.

For example, if it is point *M*, we produce line DMN. The initial value of the lunar argument in the known year is the magnitude of the period of arc LKH. In the required year, the distance of the moon from the apogee is the magnitude of the period of arc *LKHM*. This total is the initial value of the argument at the beginning of the required year which is from [the beginning] of the division of the *wūshī*—after revolutions, or before a revolution. This total is that called "the first number to be kept."

There is no doubt that, if the beginning of the first month and the beginning of the *wūshī* correspond to each other, the position of the moon on the circumference of the epicycle at that time when the center of the epicycle comes to the beginning of the *wūshī* corresponds with its position on the circumference of the epicycle at the time of the beginning of the first month. However, the beginning of the first month is always prior to the beginning of the *wūshī*, since the beginning of the first month means the beginning of a mean conjunction (*ijtimāʿī wasaṭī*) which occurs between the division of the *dāychin*—the last division of a solar year—and the *wūshī*—the second division. Thus, the center of the epicycle at the beginning of the first month does not come to the beginning of the *wūshī*.

As we mentioned, "the first number to be kept" means the lunar argument at the time when the center of the epicycle is assumed to come to the beginning of the *wūshī*—the beginning of [the solar] years. You have known that the motion of the epicycle center on the circumference of the deferent (*ḥāmil*) is other than the motion of the lunar center on the circumference of the epicycle. Thus, the period in which the center of the epicycle goes in a required year is until it comes to the beginning of the *wūshī*. That is the difference between the first month and the *wūshī* in the required year. It must be subtracted from "the first number to be kept," so that the result comes as the period of the distance between the lunar apogee and the center of the lunar body on the circumference of the epicycle at the beginning of the conjunction of the first month. This is "the second number to be kept." Then, in this diagram, we assume that [the difference] between the first month and the *wūshī* is the magnitude of the period of arc *MKZ*. Since we subtract this magnitude from arc *LZKHM*—after revolving or not, arc *LT* comes as the result—after revolving or not. That is "the second number to be kept." After the subtraction of revolutions, if there are any, arc *LT* {34^r} remains as one-ninth of the lunar argument.

For example, this is the case that "the first number to be kept" is the period of arc *LT*, and the period between the first month and the *wūshī* is larger than this. Then, the position of the moon on the circumference of the epicycle at the beginning of the first month stays behind point *L*, the apogee. For example, it is on point *Y*. Then, at that time, the period of one cycle, 27^d 1, 32, 36^f, must be added to the period of arc *LT*. After that, the period between the first month and the *wūshī* is subtracted from the amount until the period of arc *LTKY* remains. That is one-ninth of the lunar argument. [That] must not exceed two cycles as is apparent from reflection. Since one-ninth of the argument is multiplied by 9, the result is the lunar argument of the beginning of the first month in the required year. Investigation to clarify the reason for multiplying by 9 will be mentioned in the ninth section. He is the One who guides to what is correct.

Appendix B: Critical Edition

As the second appendix to this chapter, I provide the edited text of the section on the lunar anomaly from the *Kashf al-ḥaqāʾiq* translated above. The edition is based on the two abovementioned manuscripts, RR and IF. As discussed, RR is the extant autograph dated to Dhū al-Qaʿda 10, 709 (April 19, 1310). Since the autograph reflects the author's intention, this manuscript is chosen as the base text for the edition. Its folio numbers appear in the edition with curly brackets { }. RR was collated with IF, the scribe of which records that it was copied from an original in 1308/09—that is, earlier than the extant autograph! Value in IF is also found in a marginal note that clarifies the meaning of an original sentence as shown above. In addition, IF tends to transcribe the Chinese technical terms more accurately than the autograph manuscript, sometimes even with diacritic marks, as al-Nīsābūrī was obviously not familiar with such foreign terms.

The editorial principle of punctuating the text to enhance facility of reading has been used, following English conventions, in line with the above translation. Regarding diacritical marks, the conventions of classical Persian grammar are followed in the main text, while variants in the manuscripts are noted in the critical apparatus.

In the critical notes, I use the following abbreviations:[53]

line	a gloss, addition or correction found between lines of text
marg.	a gloss, addition or correction found in the margin
(–)	omitted.

53 In the editorial procedure, I follow the method explained in Sidoli and Isahaya, *Thābit ibn Qurra's Restoration*, 30–31.

متن: امّا حصّهٔ ماه را باید که اصل[54] حصّهٔ ماه در مبدأ یک سال معلوم باشد، و آن در سال اوّل از دور «شانک‌ون»[55] هفتاد و هشت روز و سه هزار و نهصد و چهل و هشت فنک بوده- است که ارقام آن این است عحاهمح و به لغت قتا آن را «جونجونکا»[56] خوانند. پس تفاوت میان آن سال و سال مطلوب بگیریم. و آن را در هفت روز و سیصد و سی و هشت فنک که ارقام آن این است ز٠هلح و آن را به لغت قتا «جونجا»[57] خوانند—و ظن من آن است که آن فضل سال شمسی بر سیزده دور قمری اوسط است—ضرب کنیم. آنچه حاصل آید، اگر سال مطلوب بعد از «جونجونکا» باشد، حاصل را بر آن افزاییم. و اگر پیش از آن بود، از «جونجونکا» بکاهیم. و اگر نتوان کاست، چندان بار مدّت یک دور حصهٔ ماه بر اصل حصهٔ ماه—یعنی «جونجونکا»— افزاییم که حاصل از او بتوان کاست. آنچه بعد از زیادت حاصل شود یا از نقصان[58] باقی بماند اصل حصهٔ ماه در آن سال،[59] آن را «محفوظ اوّل» خوانیم. پس مابین اوّل آرام و اوّل ووشی را از آن مبلغ نقصان کنیم. اگر نقصان نتوان کرد، یک دور قمری—و آن کزالب‌لو باشد—یک بار یا بیشتر بر روزها افزاییم، و نقصان کنیم. آنچه حاصل آید، آن را «محفوظ دوّم» خوانند. پس مدّت یک دور حصهٔ ماه که مقدار آن بیست و هفت روز و پنج هزار و پانصد و پنجاه و شش فنک باشد و ارقامش این است کزالب‌لو و آن را «جُونجُون»[60] خوانند، از «محفوظ دوّم» نقصان می کنیم تا کمتر از یک دور بماند. آنچه بماند، تُسع اصل حصّهٔ ماه باشد. آن را در نه ضرب کنیم تا حصّهٔ ماه در اوّل سال مطلوب حاصل شود.

شرح: از بهر این عمل اولاً مثالی ایراد کنیم. و بعد از آن به تحقیق آن مشغول شویم.

مثالش: خواستیم که حصّهٔ قمر در سال شصت و هفتاد و هفت یزدجردی که سال ووشن باشد

---

54 IF *marg.* [اصل

55 RR شانک‌ون] سانک‌ون .*marg* در سال موش موافق خلج یزدجردی صح

56 RR جونجونکا] جومحونکا

57 RR جونجا] حونحا

58 براصل حصهٔ ماه یعنی جونجونکا افزاییم که حاصل از او بتوان کاست آنچه بعد از زیادت حاصل شود یا از نقصان] IF *marg.*

59 IF اصل حصهٔ ماه در آن سال] اصل حصهٔ ماه باشد در آن سال

60 RR جُونجُون] ححون

به لغت اهل قتا و سال بیچین[61] به لغت ترکان معلوم کنیم. چون سالی که اصل حصّهٔ قمر در آن سال معلوم بوده است، {33r} سال اوّل است از دور «شانک‌ون»[62] مطابق به سیصد و سی و سیم یزدجردی، پس تفاوت میان سال معلوم و سال مطلوب چهل و چهال باشد. آن را در «جونجا»[63] که ز۰ه‌لح است، ضرب کنیم. از ایّام سیصد و هشت حاصل آید، و از فنکات چهارده هزار و هشتصد و هفتاد و دو. از بهر ده هزار فنک، یکی بر ایّام افزودیم تا سیصد و نه گشت. و فنکات با چهار هزار و هفتصد و هفتاد و دو آمد که مدفوع آن چنین باشد اکایب. و چون این حاصل را بر اصل حصّهٔ قمر در سال معلوم که عح‌اه‌مح است افزاییم، به سبب آنکه سال معلوم مقدم است بر سال مطلوب، ایّام سیصد و هشتاد و هفت شود، و فنکات ب کز۰ فنکاً گردد. و این «محفوظ اوّل» است. پس مابین اوّل آرام و اوّل ووشی را درین سال که کَ یوماً اکط‌نو فنکاً است، چنانکه پیش ازین در مثال نموده‌ایم، ازین مبلغ که «محفوظ اوّل» است، نقصان کردیم. از ایّام سیصد و شست و هفت روز و از فنکات ۰نزد باقی ماند. و این «محفوظ دوّم» است. و چون ازین مبلغ تضاعیف «جنجون»[64] به سیزده بار—و آن سنح۰لزح است—نقصان کنند، طَ ایّام ۰یط‌نو فنکاً باقی ماند. و این تُسع حصّهٔ قمر است. این را در نه ضرب کنیم.[65] از ایّام هفتاد و دو حاصل آید، و از فنکات ۰یب‌مد. و این خصهٔ قمر باشد در اوّل سال مطلوب.

و مثال اینکه سال مطلوب سابق بود بر سال معلوم هم، برین قیاس باید کرد.

و امّا به تحقیق این عمل آن است که مراد به حصّهٔ قمر تُسعهٔ امثال مدّت حرکت قمر است از ذروهٔ تدویر تا آنجا که مرکز جرم او بود از محیط تدویر. و ما از بهر توضیح، دایرهٔ[66] ابج بر مرکز دَ رسم کنیم. و نقطهٔ آ را از محیط او مرکز تدویر فرض کنیم در سال مطلوب، نقطهٔ هَ مبدأ ووشی هم در این سال. و خطوط دزط دحی دکال اخراج کنیم.

61 بیچین] ـحن RR & IF

62 شانک‌ون] سانک‌ون RR

63 جونجا] سونحا RR؛ جونحا IF

64 جنجون] حجون RR

65 کنیم] IF *line*

66 دایرهٔ] دایرهٔ هم ~~در آن سال~~ IF

پس آنچه از فصل متقدم معلوم تواند شد، مدّت مابین اه است. و من بعد مطلوب آن است که مدّت مابین اجتماع حقیقی و مبدأ ووشی حاصل آید. و درین فصل مطلوب آن است که تُسعة امثال مدّت[67] مابین مفارقت قمر از ذروه و میان مرکز جرم قمر از محیط تدویر در سال مفروض—و به سبب موضع مرکز تدویر از محیط خارج—حاصل آید. و از بهر بیان این، می گوییم مدّت حرکت مرکز جرم قمر بر محیط تدویر با مدّت حرکت مرکز تدویر از[68] محیط دایره ابج متقارب اند.

و چون در سالی[69] معلوم مثلاً چون سال اوّل از دور «شانک‌ون» به شرط آنکه مبدأ آن سال از ابتداء قسم ووشی باشد، بُعد قمر را از نقطهٔ ل که ذروهٔ است به قدر لکح فرض کنیم، و مرکز تدویر قمر را بر نقطهٔ آ در هر سالی از سالهای شمسی که مبدأ آن سالها هم از اوّل قسم ووشی باشد. مرکز جرم قمر بعد از ادوار نقطهٔ ح تجاوز کرده باشد به قدر فضل[70] سال شمسی بر سیزده دور قمری—و آن ز٠ه‌لح است تقریباً. چون[71] عدد سالهای شمسی که مابین سال مطلوب و سال معلوم بود، درین فضل ضرب کنند، مجموع این تفاوت در آن سالها حاصل آید. و آن یا ادوار تامّه باشد {33v} یا بعضی از دوریا[72] ادوار[73] تامّه یا بعضی از دور. اگر ادوار تامّه باشد، اصل خصهٔ قمر در مبدأ سال مطلوب که مرکز تدویر[74] به نقطهٔ ه رسیده‌باشد، همان بود که در سال معلوم. و اگر بعضی از دور بود، یا ادوار با بعضی، پس موضع قمر از محیط تدویر به نقطهٔ دیگر غیر نقطهٔ ح باشد.

مثلاً چون نقطهٔ م، و ما خط دمن اخراج کنیم. و اصل خصهٔ قمر در سال معلوم به قدر مدّت[75] قوس لکح بوده‌است. و در سال مطلوب بُعد قمر از ذروه به قدر مدّت قوس لکحم است. و این مجموع اصل حصّه است در مبدأ سال مطلوب که از قسم ووشی است بعد از ادوار یا بی‌ادوار. و این مجموع است[76] که آن را «محفوظ اوّل» خوانند.

---

67 مدت] IF *marg.*

68 از] IF *line*

69 سالی] سال IF

70 فضل] IF *marg.*

71 چون] پس چون IF

72 دوریا] IF *marg.*

73 ادوار] ادوار ~~مجموع این تفاوت~~ IF

74 تدویر] تدویر ~~بذروه~~ IF

75 مدت] RR *line*

76 بعد از ادوار یا بی‌ادوار و این مجموع است] IF *marg.*

و پوشیده نماند که اگر مبدأ آرام و مبدأ ووشی هر دو متطابق باشند، موضع قمر از محیط تدویر در آن وقت که مرکز تدویر به مبدأ ووشی رسیده‌باشد، موافق موضع او بود از محیط تدویر در وقت مبدأ آرام. لیکن مبدأ آرام دایماً مقدّم است بر مبدأ ووشی، چه مبدأ آرام عبارت است از مبدأ اجتماعی وسطی که مابین قسم دایچن[77] که آخر اقسام سال شمسی و مابین ووشی که قسم دوّم است از آن اقسام واقع آید. پس مرکز تدویر در مبدأ آرام هنوز به مبدأ ووشی نرسیده‌باشد.

و چنانکه گفتیم، «محفوظ اوّل» عبارت است از خصهٔ قمر در آن وقت که مرکز تدویر به مبدأ ووشی که ابتداء سالها است، به فرض رسیده‌باشد. و تو دانسته‌ای[78] که حرکت مرکز تدویر بر محیط حامل بر دیگر است به حرکت مرکز قمر بر محیط تدویر. پس آن مدّت که مرکز تدویر را مانده‌باشد در سال مطلوب تا به مبدأ ووشی رسد—و آن مابین آرام و ووشی است در سال مطلوب—از «محفوظ اوّل» نقصان باید کرد تا مدّت میان مفارقت قمر از ذروه و میان مرکز جرم قمر از محیط تدویر در مبدأ اجتماع آرام حاصل آید. و این «محفوظ دوّم» باشد. پس درین شکل فرض کنیم که مابین آرام و ووشی به قدر مدّت قوس مکز است. چون این مقدار را از قوس لزکم بعد از ادوار یا بی‌ادوار نقصان کنیم، قوس لط[79] حاصل آید بعد از ادوار یا بی‌ادوار—و این «محفوظ دوّم» باشد. و بعد از نقصان ادوار—اگر باشد—قوس لط {34r} باقی ماند که تُسع خصهٔ قمر.

و بود که «محفوظ اوّل» بود در مدّت قوس لط[80] باشد مثلاً، و مدّت مابین آرام و ووشی بیشتر ازین باشد. پس موضع قمر از محیط تدویر در مبدأ آرام متخلّف بود از نقطهٔ ل که ذروه است، مثلاً بر نقطهٔ ی باشد. پس آنگاه مدّت یک دور—و آن کزالب لو است—بر مدّت قوس لط زیادت باید کرد. و بعد از آن مدّت مابین آرام و ووشی را از مبلغ نقصان کرد تا مدّت قوس لطکی بماند. و آن تُسع حصّهٔ قمر باشد. و نباشد که دو دور زیادت باید کرد، چنانکه به تأمل ظاهر گردد. و چون تُسع حصّهٔ را در نه ضرب کنند، حصّهٔ ماه حاصل آید، مبدأ آرام را در سال مطلوب. و به تحقیق آنکه چرا در نه ضرب باید کرد، در فصل نهم گفته شود. و هو الموفّق للصواب.

---

77 دایچین] داجن RR، دایجین IF

78 تو دانسته‌ای] تو دانستهٔ RR & IF

79 لط] لط *marg.* حاشیه: چه قوس لط و قوس لطز را یک حکم در احداث زاویه بر می گرد IF

80 باقی ماند که تسع حصّهٔ قمر و بود که محفوظ اول بود در مدت قوس لط] (–) IF

Acknowledgments

I greatly benefited from the insightful comments of Dr. Yukio Ôhashi and Dr. Takanori Suzuki on an earlier draft, although I alone remain responsible for any deficiency in this chapter. I would also like to express my deep appreciation to Dr. Abu Sad Islahi, a library officer at the Raza Library in Rampur, for his great assistance in accessing the autograph manuscript of the *Kashf al-ḥaqā'iq*, as well as for permitting the publishing of the folios. Later, I submitted a draft of this chapter as the paper for my seminar at the Slavic-Eurasian Research Center, Hokkaido University, to which I currently belong. Valuable remarks by Dr. Tarô Mimura—the commentator of the seminar—as well as those by the faculty members had considerable importance for the completion of this chapter. The last indispensable piece for the completion of the chapter was provided by Dr. Christopher Cullen. I would also like to convey my gratitude to Dr. Leigh Chipman for her English editing. This work is supported by JSPS KAKENHI Grant Number JP18J40179 and JP19H00546. The work was completed after the abrupt passing of Dr. Ôhashi, whose assistance enabled me to write the core part of this chapter. I devote this work to his memory with profound gratitude.

Bibliography

Cowan, J., ed. *Arabic-English Dictionary: The Hans Wehr Dictionary of Modern Written Arabic*, 4th edition. Wiesbaden: Harrassowitz, 1979.

Cullen, C. *Heavenly Numbers: Astronomy and Authority in Early Imperial China*. Oxford: Oxford University Press, 2017.

Fazlioglu, I. "The Samarqand Mathematical-Astronomical School: A Basis for Ottoman Philosophy and Science." *Journal for the History of Arabic Science* 14 (2008): 3–68.

Gutas, D. *Greek Thought, Arabic Culture: The Graeco-Arabic Translation Movement in Baghdad and Early 'Abbāsid Society (2nd–4th/8th–10th Centuries)*. London and New York: Routledge, 1998.

Isahaya, Y. "Fu Mengzhi: 'The Sage of Cathay' in Mongol Iran and the Two-Way Exchange of Astral Sciences along the Silk Roads." In *Along the Mongol Silk Roads: Merchants, Generals, Religious Experts*, edited by M. Biran et al., 238–254. Berkeley: University of California Press, 2020.

Isahaya, Y. "Sino-Iranica in Pax Mongolica: The Elusive Participation of Syriac-Rite Christians in the Ilkhanid Translation Project." In *Marco Polo and the Silk Road (10th–14th Centuries)*, edited by Rong X. and Dang B., 341–362. Beijing: Peking University Press, 2019.

Isahaya, Y. "The *Tārīkh-i Qitā* in the *Zīj-i Īlkhānī*: the Chinese Calendar in Persian." *SCIAMVS* 14 (2013): 149–258.

Isahaya Y., and J.F. Lin, "Entangled Representation of Heaven: A Chinese Divination Text from a Tenth-Century Dunhuang Fragment (P. 4071)." *Historia Scientiarum* 26/3 (2017): 153–171.

Kennedy, E. "al-Bīrūnī's Masudic Canon." *al-Abhath* 24 (1971): 59–81.

Kennedy, E. *On the Contents and Significance of the Khāqānī Zīj by Jamshīd Ghiyāth al-Dīn al-Kāshī*. Frankfurt: Institute for the History of Arabic-Islamic Science, 1998.

King, D., and J. Samsó. "ZĪDJ." In *The Encyclopaedia of Islam*, new edition, 11: 496–508. Leiden: Brill, 2002.

Martzloff, J.C. *Le calendrier chinois: structure et calcus (104 av. J.–C.–1644)*. Paris: Honoré Champion, 2009.

Mercier. R. "From Tantra to Zīj." In *Sic itur ad astra: Studien zur Geschichte der Mathematik und Naturwissenschaften*, edited by M. Folkerts and R. Lorch, 451–460. Wiesbaden: Harrassowitz, 2000.

Morrison, R. *Islam and Science: The Intellectual Career of Nīẓām al-Dīn al-Nīsābūrī*. London: Routledge, 2007.

Mozaffari, M., and G. Zotti. "The Observational Instruments at the Maragha Observatory after AD 1300." *Suhayl* 12 (2013): 45–179.

Nīsābūrī, Niẓām al-Dīn. *Kashf al-ḥaqā'iq-i zīj-i īlkhānī*. In two manuscripts, MS Persian 1203, Raza Library, Rampur (abbreviated RR); MS 3421, Fatih Millet Kütüphanesi, Istanbul (abbreviated IF).

Ôhashi, Y. "Mathematical Structure of the Eccentric and Epicycle Models in Ancient Greece and India." In *History of Mathematical Science II*, edited by B. Yadav and S. Singh, 83–102. Cambridge: Cambridge Scientific Publishers, 2011.

Pedersen, O. *A Survey of the Almagest: With Annotation and Revised Commentary by Alexander Jones*. New York: Springer, 2011.

Pfeiffer, J. "Confessional Ambiguity vs. Confessional Polarization: Politics and the Negotiation of Religious Boundaries in the Ilkhanate." In *Politics, Patronage and the Transmission of Knowledge in 13th–15th Century Tabriz*, edited by J. Pfeiffer, 129–168. Leiden: Brill, 2014.

Ragep, J. "Copernicus and his Islamic Predecessors: Some Historical Remarks." *History of Science* 45/1 (2007): 65–81.

Ragep, J. "Islamic Culture and the Natural Sciences." In *The Cambridge History of Science*, vol. 2, *Medieval Science*, edited by D. Lindberg and M. Shank, 27–61. Cambridge: Cambridge University Press, 2013.

Ragep, S. *Jaghmīnī's Mulakhkhaṣ: An Islamic Introduction to Ptolemaic Astronomy*. Cham: Springer, 2016.

Rashīd al-Dīn, *Tānksūqnāma-yi īlkhān dar funūn-i ʿulūm-i khiṭā'ī*. In two versions, MS Aya Sofya 3596, Süleymaniye Kütüphanesi, Istanbul; facsimile (intro. M. Minūwī). Teheran: Intishārāt-i dānishkada-yi adabiyyāt wa ʿulūm-i insānī, 1972.

Saliba, G. "Horoscopes and Planetary Theory: Ilkhanid Patronage of Astronomers." In *Beyond the Legacy of Genghis Khan*, edited by L. Komaroff, 357–368. Leiden: Brill, 2006.

Sawādī F. and Nīk-Fahm, S. "Ḥarakat-i wasaṭ-i kawākib dar *Zīj-i īlkhānī* wa naqd-hā-yi wārid bar ān" (The mean motion of the planets in *Zīj-i Īlkhānī* and the criticisms regarding it). In *Ustād-i bashar: pazhūhish-hā dar zindigī, rūzgār, falsafa wa 'ilm-i Khwāja Naṣīr al-Dīn Ṭūsī*, edited by M. Ma'ṣūmī-Hamadānī, 365–472. Tehran: Mīrāth-i maktūb, 2012.

Sayılı, A. *The Observatory in Islam and Its Place in the General History of the Observatory*. Ankara: Türk Tarih Kurumu Basımevi, 1960.

Sidoli, N., and Y. Isahaya. *Thābit ibn Qurra's Restoration of Euclid's Data: Text, Translation, Commentary*. Cham: Springer, 2018.

Sivin, N. *Granting the Seasons: The Chinese Astronomical Reform of 1280, with a Study of Its Many Dimensions and an Annotated Translation of Its Record*. New York: Springer, 2009.

Toomer, G. *Ptolemy's Almagest*. London: Duckworth, 1984.

Ṭūsī, Naṣīr al-Dīn. *Zīj-i īlkhānī*. Ms. London, British Library Or. 7464.

Van Bladel, K. "Eighth-Century Indian Astronomy in the Two Cities of Peace." In *Islamic Cultures, Islamic Contexts: Essays in Honor of Professor Patricia Crone*, edited by B. Sadeghi et al., 257–294. Leiden: Brill, 2015.

CHAPTER 7

From Heaven to Earth: Circles and the Construction of the Imperial Order in Late Ming and Early Qing China

Catherine Jami

The late Ming and early Qing period (late sixteenth to early eighteenth century) saw an unprecedented circulation of goods, persons, and knowledge between the two ends of the Eurasian continent. Most of the Europeans who actually settled in China at that time were Catholic missionaries. Among the various religious orders present there, it was the Jesuits who had the greatest influence on Chinese elites, as they introduced some of the elements of their own academic training, including the sciences, through the medium of printed books written in classical Chinese, in addition to personal contacts.

The Jesuits' scientific activity in China can be divided into three periods. During the first decades of their mission, their main interlocutors were literati, with whom they produced a number of translations of books from European languages and some books composed by Jesuits for a specifically Chinese audience. When they worked on the calendar reform starting in 1629, and then implemented it at the Astronomical Bureau (Qintianjian 欽天監) after the advent of the Qing (1644–1911), their mathematical output took a more technical turn. Finally, when they served the Kangxi 康熙 emperor (r. 1662–1722) personally as court savants, they contributed to the production of maps and an imperial textbook of mathematics. To these three periods corresponded three stages in the way they taught the sciences in China. At each stage, new elements were introduced, resulting not only in an accumulation but also in a reshaping of what was presented as "Western knowledge" (*xixue* 西學) relevant to the sciences.

In imperial China, the calendar was more than an administrative matter: in the worldview that prevailed among the elite, the emperor acted as an intermediary between the cosmos and the human realm, and he was charged with ensuring that the latter conformed to the order and pace of the former. Imperial ability to promulgate an accurate calendar every year was regarded as a token of dynastic legitimacy. The Jesuits' work on calendar reform thus turned them into crucial technical actors in the upholding of the cosmological imperial order. Moreover, the reform they proposed was not a mere change in astronomical

 | DOI:10.1163/9789004511675_009

constants or in the algorithms hitherto used to compute the calendar every year. Instead, they introduced a geometrical model as a fundamental explanatory mechanism for observed solar, lunar, and planetary motion. Thus, in the eyes of some scholars of the time, whereas in the Chinese tradition astronomical systems were normative, prescribing "what must be so" (*suodangran* 所當然), the Jesuits explained "why it must be so" (*suoyiran* 所以然), providing imperial astronomy with a new kind of foundation.[1]

Although modern scholarship has tended to separate the history of science from the history of religion, sometimes presenting the Jesuits' use of science as a bait on the hook of their religion, the Jesuits themselves understood what they brought to China as forming a single body of knowledge. It was commonly held in Europe at the time that the sciences, as tools for better understanding nature, were a way that led one closer to God, who had created the cosmological order that underlay nature.[2] In particular, this belief resulted in the establishment of the teaching of mathematics in the Society of Jesus during the second half of the sixteenth century. The founder of this teaching, Christoph Clavius (1538–1612), was also the author of the Gregorian calendar (1582). Both as a teacher and as an astronomer, he provided a role model for Jesuits in China, all the more so since several of those who worked in China in the early seventeenth century had been his students at the Roman College.

In the early modern European understanding of the word, mathematics encompassed several fields that later came to be regarded as separate disciplines. This was a legacy from Greek antiquity: throughout the Middle Ages, mathematics was represented in the curriculum by the quadrivium, which comprised arithmetic, geometry, astronomy, and harmonics. Clavius proposed a modified definition, which took into account fields that had gained importance in his time. For him, mathematics was divided into pure mathematics, which consisted of arithmetic and geometry, and mixed mathematics, which included six branches: natural astrology (in modern terms, astronomy), perspective, geodesy (the representation and measurement of the earth), music, practical arithmetic, and mechanics.

The astronomical reform on which the Jesuits worked in Beijing relied on the Tychonic system that placed the Earth at the center of the universe and posited that the sun and moon rotated around the Earth while the five planets rotated around the sun. The mathematical object that lay at the heart of

---

1 Jami, *The Emperor's New Mathematics*, 16, 328.

2 Cullen and Jami, "Christmas 1668 and After," 4.

this new system was the circle, which, in the scientific tradition inherited by the Jesuits, was regarded as central to cosmology: the motion of heavenly bodies, regarded as perfect, was assumed to result from a combination of uniform circular motions. Therefore, the definition and status of the circle in the mathematical knowledge that underlay the Jesuits' astronomy in China are worthy of attention. They presented the circle as an object defined within the Euclidian tradition of geometry, and which was furthermore measured by means of trigonometry, two fields which they introduced into China early on. The spherical shape of the Earth, also introduced into China by the Jesuits during the first decades of their mission, long remained controversial there.[3] This shape and the surveying techniques used to map the Earth in general and later the Qing empire in particular were also premised on circles and therefore on the mastery of both Euclidian geometry and trigonometry.

In the present contribution, I will focus on the circle, examining how it was defined in three printed geometrical treatises in Chinese, each corresponding to one of the three periods of Jesuit activity in China mentioned above. Also relying on the sources of these treatises, written in Latin, Chinese, French, and Manchu, I will show how these definitions, as well as the content, structure, and style of the treatises in which they are found, changed as mathematics became first a tool for astronomers and then an imperially sponsored field of learning closely connected to cartographic techniques.

Unlike some geometric objects such as points and angles, which were first introduced into China by the Jesuits, the circle was present in Chinese mathematical texts as represented by the founding classic of the first century CE, *Nine chapters on mathematical procedures* (*Jiuzhang suanshu* 九章算術). This being said, it is well known that Chinese mathematical texts prior to the late Ming period do not by and large define the objects on which they work.[4] However, the most widely read mathematical treatise of the late imperial period, Cheng Dawei's 程大位 (1533–1606) *Suanfa tongzong* 算法統宗 (Unified lineage of mathematical methods, 1592), provided "Notes on the characters used" (*Yongzi fanli* 用字凡例) at the beginning of its first chapter. They include the following glosses:

---

3 Chu, "Trust, Instruments."

4 The *Mojing* 墨經 (fifth century BCE) gives the following definition: 圜，一中同長也。, which A.C. Graham rendered as: "*Yüan* (circular) is having the same lengths from one centre" (*Later Mohist Logic*, 307, 309). There is no evidence, however, that this or any other definition of the circle formed part of the early Chinese mathematical corpus.

周外圍也 。

Circumference: external enclosure.

…

徑周中之弦。

Diameter: the hypotenuse/chord through the middle of a circumference.[5]

The first term mentioned here does not refer to a two-dimensional area but rather to a line that can be measured. There is no gloss of *yuan* 圓, the character commonly used to refer to circles in the *Suanfa tongzong*. This term occurs in chapter 3, where one finds figures illustrating it in most of its occurrences. Like most other shapes given there, *yuan* 圓 is followed by the suffix *tian* 田 (field) in the text to refer to a circular field, the area of which is being discussed.[6] This usage is representative of the Chinese mathematical tradition.

1 The First Jesuit Treatise on Geometry and Its Definition of the Circle

Matteo Ricci (1552–1610), the founder of the Jesuit mission in China, had been one of Clavius's students at the Roman College. He translated a number of Clavius's works into Chinese. The most famous one, *Jihe yuanben* 幾何原本, is a translation of the first six books of Euclid's *Elements of Geometry*, done from Clavius's Latin edition and completed in 1607 by Ricci and Xu Guangqi 徐光啓 (1562–1633), a high official and Catholic convert. This first documented introduction of geometry in the Euclidian tradition had a lasting impact on mathematics in China and beyond. The fact that the modern term for geometry in Chinese, Japanese, and Korean (Ch. *jihexue* 幾何學) was derived from the title of this translation bears witness to this. In their respective prefaces, the two translators both stressed the importance of mathematics as providing a foundation for Western knowledge.[7]

5 ZKJDT, 2:1230; *xian* 弦 (the hypotenuse of a right triangle or the cord of an arc) has been explained shortly before, both in the context of *gougu* 勾股 (base and altitude, the equivalent of a right triangle; *gou* and *gu* refer respectively to the shorter and to the longer sides of the right angle in a right triangle) and in that of *hu* 弧 (arc).

6 ZKJDT, 2:1265–1266.

7 Engelfriet, *Euclid in China*, 454–460, 291–297.

The *Jihe yuanben* contains the earliest known definition of the circle in the Chinese mathematical literature. Found in Book 1, this definition reads as follows:

> 第十五界
> 圜者。一形於平地居一界之間。自界至中心作直線俱等[8]

> Definition 15
> A circle is a figure on even ground, occupying the space [within] a boundary; the straight lines drawn from the boundary to the middle are all equal.

Let us compare this to Clavius's definition in Latin:

> XV
> Circulus est, figura plana sub vna linea comprehensa, quæ peripheria appellatur; ad quam ab vno puncto eorum, quæ intra figuram sunt posita, cadentes omnes rectæ lineæ inter se sunt æquales.[9]

> 15
> A circle is a plane figure contained within a single line, which is called circumference; towards which all the straight lines from one of the points inside the figure are equal.

While the Chinese text conveys the same meaning as the Latin one, it is not in strict conformity with it according to the Euclidian style, in which technical and ordinary words are by no means interchangeable. In this respect, we can point to three discrepancies. Firstly, whereas the notion of plane surface (Lat. *plana superficies*, Ch. *pingmian* 平面) is found in definition 7, the translators have chosen to characterize the circle as being on even ground (*pingdi* 平地). Secondly, in the same sentence, they have preferred the use of "boundary" (*jie* 界) to that of "line" (*xian* 線).[10] Thirdly, in the second sentence, they have used the non-technical term "middle" (*zhongxin* 中心) instead of a clause that would have rendered "one of the points of the figure." The point in ques-

8 *Jihe yuanben* 1607, 5b. I reproduce the punctuation of this edition, in which there are no punctuation marks at the end of paragraphs.

9 Clavius, *Euclidis Elementorum libri XV*, 12.

10 The fact that *jie* 界 is mostly used in *Jihe yuanben* as an abbreviation of *jieshuo* 界說 (definition) must not have made the reader's task easier.

tion is defined as the center of the circle (*yuanxin* 圜心) in the next definition. This being said, it is interesting to note that the Chinese definition uses the character *yuan* 圜 (also read *huan*, a variant of 圓 attested since early texts) rather than the character *yuan* 圓 found in the *Suanfa tongzong* and elsewhere to name the circle. As we shall see below, this choice most likely resulted from the translators' wish to emphasize the fact that the object defined here was not merely the intuitively obvious circular shape represented in these texts. This contributes to placing the definition in the Euclidian realm, where geometric objects are defined abstractly by their properties, without any reference to the material world.

The definition given in the Chinese text quoted above is followed by three sentences of clarification, which are taken from Clavius's lengthy commentary:

> 若甲乙丙為圜。丁為中心。則自甲至丁、與乙至丁、丙至丁。其線俱等[11]
>
> If ABC is a circle, and D is the middle, then the lines from A to D, B to D, and C to D are all equal.
>
> 外圓線為圜之界。內形為圜[12]
>
> The external round line is the circle's boundary. The figure inside is the circle.
>
> 一說。圜是一形。乃一線屈轉一周、復于元處所作。如上圖甲丁線轉至乙丁。乙丁轉至丙丁。丙丁轉至甲丁。復元處。其中形即成圜[13]
>
> Explanation: The circle is a figure, which is constructed by a line rotating all the way around until it returns to its initial position. As in the figure above, line AD rotates to BD; line BD rotates to CD; line CD rotates to AD. When it returns to its initial position, the figure within is a circle.

Both the Latin and the Chinese text are accompanied by a figure in the Euclidian tradition to which the commentary refers (fig. 7.1). These figures are not

11 Here and in what follows, cyclical characters are rendered by capital letters.

12 Later editions do not all take up the distinction between 圜 and 圓; see, for example, the 1865 edition reproduced in ZKJDT, 5:1162, where this sentence reads: 外圜線為圜之界。內形為圜。

13 *Jihe yuanben* (1607), 1:5b–6a.

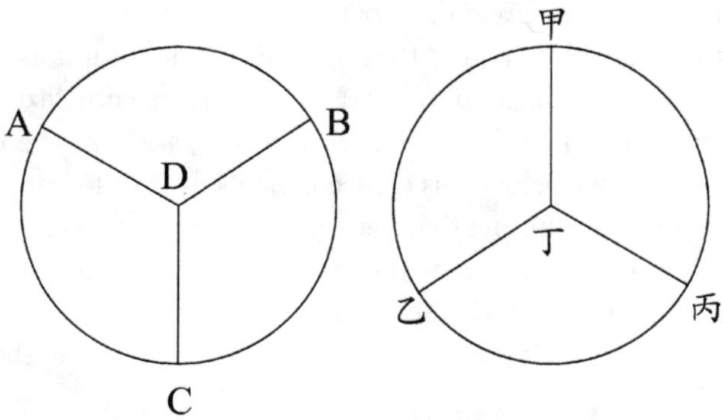

FIGURE 7.1 The figures illustrating the definition of the circle in Clavius's work and in the *Jihe yuanben*

identical. Points A, B, and C and *jia* 甲, *yi* 乙, and *bing* 丙 are not situated in the same positions on the circle. Moreover, they have been named clockwise in Latin and anti-clockwise in Chinese. One might nonetheless say that, as with the definitions themselves, the figures are similar enough. In the Chinese text, the figure is explained in the commentary using both the characters *yuan* 圜 (circle) and *yuan* 圓 (round) in a non-interchangeable way. This enables readers to ascertain that the *yuan* 圜 defined here is the same object as the one they may have been familiar with, called *yuantian* 圓田 (lit. round field) by Cheng Dawei. Still, even the final explanation (*shuo* 說) remains abstract: the rotation of a geometric object (namely a straight line) generates a circle, without any agency of a material object or human action.

Whereas, in the Latin work, Clavius's commentary is printed in italics to distinguish it from the Euclidian text, in the Chinese translation, the different authorship of the additions to this text is not made visible by the use of small characters or by the mention of the commentator's name, as was common in the Chinese textual tradition. Nevertheless, the translation distinguished the Euclidian text: any addition to it is made in separate paragraphs. This definition of the circle was written at a time when mathematics, rather than astronomy, lay at the core of the Jesuits' teaching of the profane sciences in China.

2 Mathematics for Astronomers

The 1607 translation of Euclid's *Elements*, however important it might seem in the context of the global history of mathematics, was hardly accessible without the help of a teacher. In the decades that followed its publication, another geometry textbook was composed. It focused on methods rather than definitions, theorems, and proofs. This second textbook would eventually be integrated into the mathematical knowledge to be used by astronomers.

Completed in 1623, it was entitled *Jihe yaofa* 幾何要法 (Essential methods of geometry). It was the result of collaboration between another Jesuit missionary, Giulio Aleni (1582–1649) and another convert, Qu Shigu 瞿式穀 (1593–?). At the time of the translation of Euclid's *Elements*, Xu Guangqi, who had just passed the metropolitan examination, was in the capital awaiting his first appointment. By contrast, Qu Shigu was a rather obscure literatus. The *Jihe yaofa* was composed in his native town, Changshu 常熟, in Zhejiang province, where Aleni spent some time. The work, first printed in 1631 in Fuzhou 福州, may well be more representative of the Jesuit missionaries' teaching of geometry to Chinese literati than the 1607 translation. It is constructed around extracts from the *Jihe yuanben*, focusing on geometric constructions.[14] The circle is defined at the beginning of chapter 2, which is devoted to it:

> 第一界
> 圓形於平地、居一界之間、為圜。
>
> Definition 1
> A round figure on even ground, occupying the space [within] a boundary, is a circle.
>
> 第二界
> 外圓線為圜之界。[15]
>
> Definition 2
> The external round line is the boundary of the circle.

Whereas one recognizes the kinship of these definitions with what we read in Book I of *Jihe yuanben*, the distinction between the Euclidian text and Clavius's commentary has been blurred: definition 1 is taken from the former, whereas definition 2 is taken from the latter. Regarding syntax, the convention followed in the definitions of the *Elements* and in the 1607 translation was to make the defined term the topic of the sentence, while the explanation of the term constituted the comment. This convention has been abandoned in the *Jihe yaofa*. One may also note that the distinction between the characters *yuan* 圜 (circle)

14 Jami, "Giulio Aleni's Contribution."
15 *Jihe yaofa* (1631), 2:1b.

TABLE 7.1 Topics and instruments in the *Jihe yaofa*

Chapter (卷)	Topic	Instruments
1	line (線)	ruler, iron/copper pen (尺，鐵／銅鑄筆)
2	circle (圜)	compass (規)
3	angle, triangle (角，三角形)	three-legged compass (三髀規)
4	quadrilateral (四邊形)	try square (矩)

and *yuan* 圓 (round) has now been incorporated into the definitions themselves, whereas the characterization of the circle by the equidistance of the points of its circumference to its center is simply omitted.

Moreover, the context of the definitions is quite different from that of the *Jihe yuanben*. The first four methods (*fa* 法) given in chapter 2 of *Jihe yaofa* are devoted not to geometric constructions but to the making of compasses (*zaogui* 造規); the four different patterns of compasses given are illustrated just like geometric constructions.[16] Only once equipped with the right instrument can the reader tackle the various geometric constructions in which compasses are used. A similar pattern is followed in other chapters of the *Jihe yaofa*, in which the instruments of geometry related to the main topic of each chapter are described in detail (see table 7.1). Thus, the *Jihe yaofa* includes not only geometrical knowledge and methods but also illustrated descriptions of the tools necessary for drawing geometric figures. Possessing the right tools is a prerequisite for the practice of geometry.

In 1629, when he proposed to the Chongzhen 崇禎 emperor (r. 1628–1644) that a calendar reform should be prepared using the Jesuits' methods, Xu Guangqi again emphasized the status of mathematics as a fundamental field. In agreement with him, the Jesuits who set out to prepare the reform relied on a number of European treatises, using the Tychonic astronomical model as a basis for their calculations.[17] This was a major departure from the Ptolemaic model found in their earlier works. It should be noted, however, that geometry is a foundation for both models, as both systems use circles in the modelling of orbits for the purposes of calculation.

Xu Guangqi also argued that the series of works written by the Jesuits to prepare this reform, which formed the *Chongzhen lishu* 崇禎曆書 (Astronomical

16 *Jihe yaofa* (1631), 2:4a–6b.

17 Hashimoto, *Hsü Kuang-ch'i*.

works of the Chongzhen reign, 1629–1634), did more than just set out a new system for calculating the calendar (that is, a set of algorithms and constants). Instead, they provided quantified circle-based geometrical models for solar, lunar, and planetary motion. This meant that if the calendar needed further reform in the future, one could return to this geometrical model and adjust constants according to observations. Thus, the circles of Euclidian geometry were explicitly placed at the foundation of calendrical astronomy. But at the same time, the methods introduced by the Jesuits had to be adapted to Chinese calendar-making: Xu Guangqi advocated "melting their material and substance to cast them into the *Datong* mould."[18] Firstly, this can be understood to refer to the need for a luni-solar calendar, much more complex to compute than the solar calendar used in Europe before and after Clavius's reform. Secondly, this also meant that knowledge imported from Europe was to contribute to reinforcing imperial power and dynastic stability by ensuring that the emperor could effectively perform his function as intermediary between the cosmos and the human realm by providing his people with an accurate calendar. In other words, whereas the Jesuits' original purpose in teaching mathematics and astronomy in China was to put mathematics in the service of evangelization, their work on astronomical reform effectively put their sciences in the service of a radically different cause: the upholding of the cosmology that underlay the imperial order.

The calendar reform proposed by Xu Guangqi was never implemented under the Chongzhen emperor. In 1644, when the Manchus took Beijing during their conquest of the Ming empire, they accepted the offer of one of these Jesuits, Johann Adam Schall von Bell (1592–1666), to calculate a new calendar for their new dynasty, and they put him in charge of the Astronomical Bureau. At the same time as his Calendar of Timely Modelling (*Shixian li* 時憲曆) was issued, a slightly modified version of the collection of works presented to the Chongzhen emperor was reprinted under the title *Xiyang xinfa lishu* 西洋新法曆書 (Works on astronomy according to the new Western methods).

Among the Jesuits' writings included in the *Chongzhen lishu*, there are five that present mathematical knowledge: *Chousuan* 籌算 (Calculating rods, 1628), which describes the making and use of Napier's bones;[19] *Bili guijie* 比例規解

18 熔彼方之材質，入大統之型模. *Datong* is the name of the astronomical system in use during the Ming dynasty (1368–1644); see Hashimoto and Jami, "From the Elements," 274–275.

19 On this manually operated calculating device and its introduction into China, see Cervera, *Las Varillas de Napier*.

(Explanation of the proportional compass, 1630), devoted to the instruments devised by Galileo; *Dace* 大測 (Great measurement, 1631); *Celiang quanyi* 測量全義 (Complete meaning of measurement, 1631), which deals with various geometric objects, including spherical triangles, and closes with a chapter on astronomical instruments; and *Geyuan baxian biao* 割圓八線表 (Eight trigonometric tables, 1635), which provides a major calculation aide for astronomy.[20] If one compares these works to the *Jihe yuanben* and other mathematical works by Ricci, they offer a very different picture of mathematics. It might be said that with the astronomical reform, mathematics had become instrumental in two senses. Firstly, it added descriptions of instruments into the written corpus of knowledge. Secondly, it was intended to provide calculation tools specifically for the astronomer.

This interpretation is strengthened by the fact that, when the Qing decided to implement the calendar reform initiated by Xu Guangqi, the geometrical treatise that was chosen for inclusion in the augmented version of the *Chongzhen lishu* (renamed *Xiyang xinfa lishu* 西洋新法曆書, Books on astronomy according to the new Western method), instead of the *Jihe yuanben*, was the *Jihe yaofa*, despite the fact that in the other works of the *Xiyang xinfa lishu* there are fourteen references to *Jihe yuanben* and none to *Jihe yaofa*. As we have seen, the latter work's approach to geometry can be characterized as instrumental in the double meaning evoked above. Thus, the astronomical reform resulted in a shift in the mathematics practiced and taught by the Jesuits, who had first presented geometry as the foundation of all their profane sciences. Whereas some of the building blocks of Clavius's version of Euclid's *Elements* can still be identified in the Jesuits' new texts, they are now used to construct a corpus in which the main purpose of mathematics was to use material objects (instruments) to perform calculations and measurements of other material objects (celestial bodies). This is in stark contrast with the mathematics introduced by Ricci, in which geometry and arithmetic are characterized as "discussing them (magnitude and number) in the abstract, casting off material objects" (*tuo yu wu qi er kong lun zhi* 脫於物體而空論之).[21] In keeping with this, beside the *Jihe yuanben*, Ricci also co-authored a work on calculation, the *Tongwen suanzhi* 同文算指 (Instructions for calculation in common script, 1614), which introduced written calculation for the first time. At a time when the abacus, on which the *Suanfa tongzong* rests, was universally used in China, this new technique enabled scholars to perform calculation using only

20 Jami, "Mathematical Knowledge."

21 Jami, *The Emperor's New Mathematics*, 26; see ZKJDT, 5:1151.

the "four treasures of the study" (*wenfang sibao* 文房四寶), namely brush, inkstone, ink stick, and paper—their distinctive tools, rather than the abacus, the tool of merchants.[22]

This change in the role and status of instruments in Western mathematics as practiced in China corresponds to another change that concerns the status of mathematics among the various fields to which it relates. When Xu Guangqi promoted the 1629 reform, he upheld Ricci's view on the matter: "magnitude and number" (*dushu* 度數, as he and Ricci had rendered the two instances of quantity that defined the two branches of mathematics) was a fundamental field of knowledge. It could be applied to ten other fields: astronomy as well as astrology and meteorology, surveying for the purpose of water conservancy works, music and harmonics, military engineering and equipment, finances and taxes, building and civil engineering, mechanical devices, cartography, medical hemerology, and time measurement, especially by means of clepsydras and dials.[23] This list reflects Xu Guangqi's concern with statecraft. On the other hand, when it came to structuring the body of knowledge contained in the *Chongzhen lishu*, Xu Guangqi proposed a fivefold classification: bases of the methods (*fayuan* 法原), numbers of the methods (*fashu* 法數), calculations of the methods (*fasuan* 法算), instruments of the methods (*faqi* 法器), and intercommunication (*huitong* 會通). This last category refers to the conversion of units. The treatise on the proportional compass pertains to the instruments of the methods, as does the last chapter of the *Celiang quanyi*, whereas the first nine chapters of this work belong to the bases of the methods, together with the *Dace*. Trigonometric tables as well as Napier's bones pertain to the numbers of the methods.[24] It is remarkable that this last device belongs with tables rather than with instruments. But the most important point here is that what we regard as a single field, namely mathematics (for which the field called *dushu* by Xu Guangqi can be regarded as one among many definitions found in historical sources), is no longer a relevant entity when it comes to organizing knowledge on the basis of its role in the astronomical compendium. Instead, as stated above, the function of mathematics is now merely to provide tools for constructing the astronomical knowledge that underlay the 1629 calendar reform.

22 On this point, see Jami, *The Emperor's New Mathematics*, 356–357; Jami, "La carrière de Mei Wending," 31.

23 Xu, *Xu Guangqi ji*, 2:337–338. See Engelfriet, *Euclid in China*, 349–350.

24 Xu, *Xu Guangqi ji*, 2:377–378; see Jami, "Mathematical Knowledge," 667–668.

3 Imperial Mathematics: Measuring the Heavens, but Also the Earth

When the Jesuits began to work on their reform project in 1629, their astronomy and the mathematics that underlay it were seen as highly technical fields—only a few Chinese scholars studied them. At the same time, these fields had little prestige compared to classical studies, which were the prerogative of the literati class. This began to change under the Kangxi emperor (r. 1662–1722), the second Manchu sovereign to rule China, as he took a personal interest in these sciences. Mathematical lecture notes were written for him by the Jesuits and eventually revised and incorporated into the *Shuli jingyun* 數理精蘊 (Essence of numbers and their principles, 1723).[25] The production of this imperial textbook is closely linked to efforts to summarize and, where relevant, to update the methods expounded in the *Xiyang xinfa lishu*, efforts which resulted in the *Lixiang kaocheng* 曆象考成 (Thorough investigation of astronomical phenomena, 1723). The period of compilation of these two works also overlapped with another major imperial project involving Western learning: the famous geodetic survey of the Qing empire that resulted in a complete map of the its territory, the *Huangyu quanlan tu* 皇輿全覽圖 (Overview maps of the imperial territories, completed in 1717). Since the *Shuli jingyun* was intended as a textbook for all scholars and officials, it is worthwhile revisiting it in the light of recent work on this map,[26] in order to understand how concern with measuring the Earth, as well as the Heavens, shaped Chinese mathematics in the early eighteenth century.

Whereas the imperial appropriation of the Jesuits' astronomy was motivated by the fact that it was crucial for the dynasty to be able to issue an accurate calendar, the emergence of mathematics as an imperially sponsored field of knowledge stemmed mainly from the personal taste of the Kangxi emperor. Towards the end of his sixty-year reign, during which he had studied mathematics as well as astronomy and music with the Jesuits, he commissioned a compendium, entitled *Yuzhi lüli yuanyuan* 御製律曆淵源 (Origins of the pitchpipes and the calendar, imperially composed, 1723), made up of three treatises representing the three fields. Beside the *Shuli jingyun* mentioned above, these treatises included the *Lixiang kaocheng* 曆象考成 (Thorough investigation of astronomical phenomena) and the *Lülü zhengyi* 律呂正義 (Correct interpretation of [standard] pitchpipes). Whereas the association between astronomy and harmonics is as old as dynastic histories—some

25 Jami, *The Emperor's New Mathematics*, 315–384.
26 Cams, *Companions in Geography*.

of which contain a monograph entitled *Lüli zhi* 律曆志 (Monograph on the pitchpipes and the calendar)—mathematics had not been the subject of an imperially-sponsored editorial project for more than a millennium, since the compilation of the *Suanjing shishu* 算經十書 (Ten mathematical classics) by Li Chunfeng 李淳風 (602–670) in 656. Kangxi's project further raised the status of mathematics by associating it with astronomy and harmonics, two fields that were deemed crucial in harmonizing the human realm with the cosmos.

Starting in 1708, that is, five years before the *Yuzhi lüli yuanyuan* was commissioned, another imperial project was launched: the cartographic survey of the Qing empire, an enterprise on a scale unprecedented worldwide, which resulted in the complete map of the empire entitled *Huangyu quanlan tu*. As with mathematics, astronomy, and harmonics, this project was the outcome of the emperor's reign-long interest in cartography. Since the 1680s, he had taken steps to ensure that he had maps of the Chinese provinces of his empire, as well as of the territories that lay beyond the Great Wall. For the former, he mainly relied on local officials. For the latter, various mapping expeditions were sent; moreover, he ordered several Jesuit missionaries to carry out measurements during their travels in his retinue or with various expeditions.[27] Comparing the maps collected in this way led to the realization that there was a discrepancy between the results yielded by traditional surveying methods and those derived by the Jesuits, who used French instruments and techniques to make observations in order to determine the latitude of places through which they travelled. Therefore, in 1702, the emperor sent a team led by one of his sons to measure one degree of a meridian, in order to fix a new standard for the *li* 里.[28] The first expedition, sent in 1708, carried out a survey along the Great Wall. Between 1709 and 1716, six expeditions then surveyed the Inner Territories, while other teams were sent to the Manchu homeland and to the Mongol frontier. Data on Korea were requested from its king, whom the Qing regarded as their tributary. A first version of the copperplate engraved *Huangyu quanlan tu* was completed in 1717. An expedition to Tibet, which set out in the same year, brought back data that was included in a revised version completed in 1719.[29] The surveying teams included at least two missionaries, one official from the Inner Court,

27 Cams, *Companions in Geography*, 44–48.

28 Cams, *Companions in Geography*, 76–81. This new standard was defined so that there were 200 *li* in a degree of meridian, which gives 1 *li* = 444 m (Jami, *The Emperor's New Mathematics*, 393).

29 Cams, *Companions in Geography*, 102–124, 201.

a military officer, a civil service official, and a specialist from the Astronomical Bureau.[30] Monitored directly by the emperor and managed from the Inner Court, these expeditions received support from local officials wherever they went. Given the scale of this project and the fact that the compilation of the *Yuzhi lüli yuanyuan* was also carried out within the Inner Court, one may ask whether this gigantic survey had any impact on mathematics, in a way similar to the impact that astronomical reform had from the 1630s on. No treatise devoted specifically to surveying and mapmaking was published in China at the time. Therefore, in what follows I would like to assess the ways in which imperial mathematics as found in the *Shuli jingyun* incorporated "mathematics for surveyors," as the *Chongzhen lishu* had incorporated "mathematics for astronomers."

As far as we know, the *Yuzhi lüli yuanyuan* was compiled by a large team of Chinese scholars, and it is difficult to ascribe the authorship of specific passages to one or more of them in particular.[31] The *Shuli jingyun* is largely based on the lecture notes prepared for the emperor by the Jesuits mentioned above. It is the result of considerable rewriting of these notes: here again no single authorship can be distinguished. The work presents a synthesis of mathematical knowledge available at the time. It is the largest mathematical treatise ever printed in imperial China.[32] Its fifty-three chapters are organized into three parts of very different lengths and layout. The first part, devoted to "Establishing the structure to clarify the substance" (*Ligang mingti* 立綱明體), contains five chapters; its text is divided into numbered items (*jie* 節). The second part, concerned with "Dividing into groups to put to use" (*Fentiao zhiyong* 分條致用), contains forty chapters, which consist of problems, each accompanied by a method (*fa* 法) for deriving the solution. The third part contains eight chapters of tables (*biao* 表).

The first part of the work includes three chapters entitled "Elements of geometry" (*Jihe yuanben*). This is not the 1607 translation. Instead, it is based on one of the many geometry textbooks written in seventeenth-century Europe that have the same title, Ignace Gaston Pardies's (1636–1674) *Elémens de géométrie* (1671).[33] It is followed by a chapter on the "Elements of calculation" (*Suanfa*

30 Cams, *Companions in Geography*, 139–140.

31 Jami, *The Emperor's New Mathematics*, 373–378.

32 Jami, *The Emperor's New Mathematics*, 315.

33 Pardies, *Elémens de géométrie*. This book's lengthy title, which can be rendered as *Elements of geometry, in which by a short and easy method one may learn what must be known of Euclid, Archimedes, Apollonius, and the most beautiful inventions of the ancient and modern geometers*, states the author's intention quite clearly.

yuanben 算法原本), mainly derived from Book VII and the beginning of Book VIII of Euclid's *Elements*. The second part of the *Shuli jingyun* is organized according to headings taken from both the Jesuits' teaching and the Chinese mathematical tradition, bringing out links between the two wherever relevant.[34] Thus, chapter eighteen, entitled "Measurement" (*Celiang* 測量), on which we will now focus, bears little resemblance to the *Celiang quanyi* mentioned above, as all the triangles discussed in chapter eighteen are plane. The chapter opens with a short introduction, which, among others, points to the broad applications of the techniques presented in it:

> 不特凡物之高深廣遠。可得而推。即七政之躔度。天地之形體。俱可得而測也。[35]
>
> Not only can the height, depth, width, and distance of things be derived, but also the trajectories of the Seven Governors (sun, moon, and planets), and the bodies of Heaven and Earth can all be measured.

The problems that follow, however, are solely concerned with measuring the height or distance of objects that are earthly, whether man-made or natural. They include flagpoles, pavilions, pagodas, walls, trees, mountains, rivers, and rocks. These are the features that surveyors measured and used as landmarks when surveying the empire. The problems share a feature that distinguishes them from the majority of problems found in the *Shuli jingyun*: the method for solving them starts with one or more measurements that yield numerical data not given in the problems at the onset; these data are then entered into the calculations that finally result in the magnitude sought. Altogether, there are eighteen problems in the chapter, nine on "measurement by base and altitude" (*gougu celiang* 勾股測量) and nine on "trigonometric measurement by magnitude and number" (*sanjiao dushu celiang* 三角度數測量). These require different instruments, as noted in small characters:[36]

> 勾股測量（凡用矩度。或立表杆。必用垂線。取其與地平成直角。以為準則。若地不平。須記取某處與人目所看相平為記。）[37]

34 Jami, *The Emperor's New Mathematics*, 317–339.

35 *Shuli jingyun xia juan* 18, 1b; ZKJDT, 3:607.

36 In the quotations below, the text in parentheses represents the text in small characters in the original.

37 *Shuli jingyun xia juan* 18,1b; ZKJDT, 3:607.

Measurement by base and altitude (In general, one uses a square, or one erects a pole and needs to use a plumb line to get it at right angles with the horizontal, so as to have a standard. If the ground is not even, one must take note of a place level with one's line of sight to use as a reference.)

三角度數測量（度數測量。必取資於儀器。全圜儀、半圜儀、象限儀、雖為體不同。其為用則一。以九十度為準。以定表遊表為二視線。其相距之度。即為所測之角。）[38]

Trigonometric measurement by magnitude and number (For measurement by magnitude and number, one must rely on instruments. Although the circle, the semi-circle, and the quadrant have different shapes, the method for using them is the same. One takes 90 degrees as the standard, and one takes the fixed pinnules and moving pinnules as the two lines of sight. The [number of] degrees that separate them is the angle one is measuring.)

The first problem of the chapter contains a note that further describes the quadrant one is to use and then refers the reader back to item XII.16 of *Jihe yuanben*, found in the first part of the *Shuli jingyun*, which was a reworking of the 1691 treatise written for the Kangxi emperor. This item gives a "method for making a surveying instrument graduated into divisions" (*zuo fenshu bili celiang yiqi fa* 作分數比例測量儀器法), in this case a semi-circle (fig. 7.2). The instructions are given in five successive steps:

1. Divide the semi-circle ACB into 180 degrees, and each of these degrees into 60 minutes.
2. Divide the half-square inscribed in this semi-circle into small squares by drawing parallel lines.
3. Mount two fixed pinnules at both ends of the diameter.
4. Mount a moving pinnule at the end of alidade with 100 divisions.
5. Mount a plumb line.[39]

This item is followed by two others that explain respectively how to make and use measurements to draw a map and how to reproduce a map on a smaller scale.[40] Here, one is undoubtedly in the realm of cartography. In other words, the *Shuli jingyun* version of the *Jihe yuanben*, unlike the 1607 translation, includes the technical description and use of surveying instruments. Whereas

38 *Shuli jingyun xia juan* 18, 25a; ZKJDT, 3:618.

39 *Shuli jingyun shang juan* 4, *Jihe yuanben* XII, 16; ZKJDT, 3:135–136.

40 *Shuli jingyun shang juan* 4, *Jihe yuanben* XII, 17–18; ZKJDT, 3:136–137.

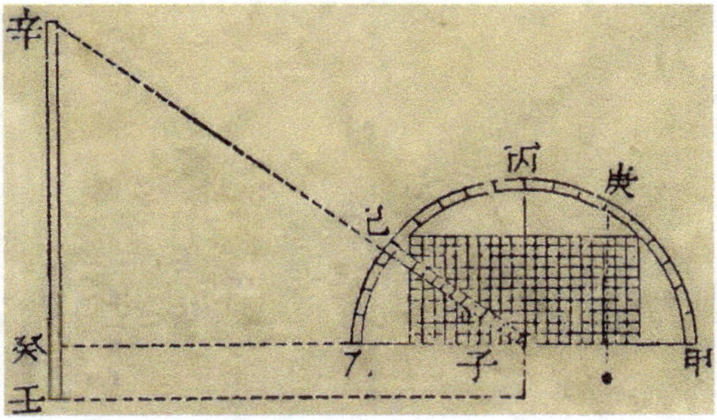

FIGURE 7.2
The surveying instrument described in *Jihe yuanben*. *Shuli jingyun shang juan* 4, 54a

one could characterize the mathematical knowledge found in the *Chongzhen lishu* as "mathematics for astronomers," the *Shuli jingyun* includes "mathematics for surveyors" as part of the knowledge it presents.

4 Defining the Circle in Imperial Mathematics

Did this integration of surveying methods, which Pardies included in his work as part of "Practical geometry" (Book IX of his *Elémens de géométrie*), have any impact on the way geometry itself was presented in the *Shuli jingyun*? In order to answer this question, let us examine the definition of the circle in this work. It is item I.5 of the *Jihe yuanben*, just after the definitions of point, line, surface, solid and angle.

> 第五
> 凡有一線。以此線之一端為樞。復以此線之一端為界。旋轉一周。即成一圜。如甲乙一線。以甲端為樞。乙端為界。旋轉復至乙處。即成乙丙丁戊之圜。此圜線謂之圜界。圜界內所積之面度。謂之圜面。[41]

> 5.
> Let there be a line. Take one of its ends as a pivot; again take [the other] end as boundary. Turning it through a complete rotation makes a circle. Thus, with line AB, take the A end as a pivot; take the B end as boundary. Turning it until it returns to the [original] place of B makes the BCDE circle. This circular line is called circumference of the circle. The area accumulated inside the circumference is called the area of the circle.

41 *Jihe yuanben* I.5, in *Shuli jingyun* II, 3b; ZKJDT, 3:22.

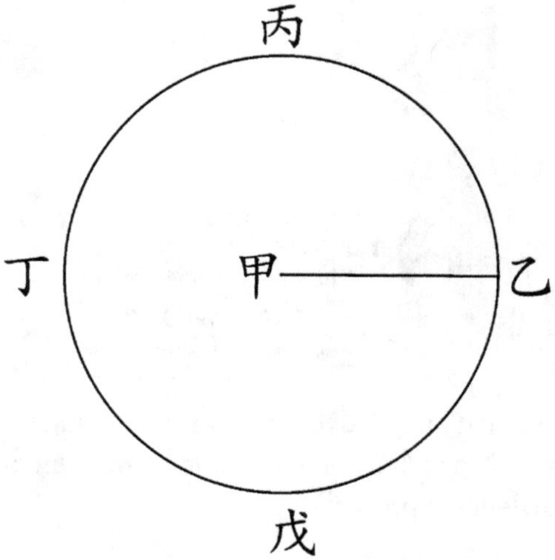

FIGURE 7.3
Illustration of the construction of a circle, after *Shuli jingyun shang juan* 2, 3b

There is no trace here of the Euclidian definition, and, although the circle is generated by the rotation of a line, in a way similar to that in Clavius's explanation translated in 1607 by Ricci and Xu, the style is somewhat different (fig. 7.3). Whereas that "Explanation" is a description of a line rotating around one of its ends, here it might be said that the repeated use of *yi ... wei* 以……為 (lit. take ... as) introduces a form of agency which involves the reader in a procedure. In order to explore this matter further, let us compare this definition to the one given in Pardies's textbook:

> 10. Si nous imaginons une ligne *ab* attachée par le bout *a* au milieu de la ligne *dc*, & que de plus nous fassions mouvoir cette ligne autour du point *a*; quand elle sera revenuë au lieu d'où elle avoit commencé à se mouvoir, l'extrémité *b* aura décrit une ligne courbe qui s'appelle Cercle, ou plûtost Circonférence du cercle: car à proprement parler, le Cercle est tout l'espace renfermé dans cette circonférence.[42]

> 10. If we imagine a line *ab* tied by its end *a* to the midpoint of the line *dc*, and that moreover we make this line move around point *a*; when it has returned to the point from which it had started to move, the end *b* will have described a curved line called a Circle, or rather Circumference of a circle: for strictly speaking the Circle is the whole space enclosed in this circumference.

42 Pardies, *Elémens de géométrie*, 3–4.

What is proposed here is not a definition of the circle that characterizes it by its properties, but a mental operation akin to a thought experiment that rests on imagining oneself exerting a concrete action on material objects, effectively causing the rotation of line *ab* that will generate a circle. This concrete aspect is strengthened by the figure that accompanies this definition: the rotating line is represented as a material object very much resembling an alidade such as the ones fixed on surveying instruments described further in the work (see fig. 7.4).[43] Compared to this visual representation, the *Shuli jingyun*'s figure is much more abstract: no material device enabling a rotation of the line is required.

Two intermediary texts are known between the French one and the Chinese one quoted above. Pardies's textbook was translated into Manchu in 1690; this formed part of an attempt by the Kangxi emperor to create a body of writings on the sciences in Manchu.[44] The following year, the Manchu text was in turn translated into Chinese, relying heavily on the terminology coined by Ricci and Xu in the 1607 translation. The editors of the *Shuli jingyun* worked from this last version. Let us now look at the Manchu translation of Pardies's definition:

> jakuci
> yaya emu jijun bifi, tere jijun i emu dube ba horgikū obume hadafi, jai emu ube be šurdehei sucungga aššame deribuhe bade isinaha manggi, uthai emu mu uheren jijun banjinambi. duibuleci, giya i seme emu jijun i giya seme babe orgikv obume hadafi, i seme dube be guribume, hadaha babe moselame ucungga aššame deribuhe i sere bade isinaha manggi, uthai emu i bing ding u seme muheliyen jijun banjimbi. ere i bing ding u i muheliyen jijun be muheren i jecen sembi. jecen de horiha babe muheren sembi.[45]
>
> 8.
> For any line, nail one end of this line to make a pivot. After you revolve the second end to where it first started, then a line in the shape of a ring is generated. For example, if we nail the endpoint called *jia* of a line *jia yi* to make a pivot moving the end called *yi*; after arriving back to the initial

43 Pardies, *Elemens de géométrie*, 105. In the *Shuli jingyun*, a similar surveying instrument is described; it is also used in some problems (ZKJDT, 3:135–136 and 618–638).

44 Jami, "Science in Manchu."

45 *Giho yuwanben bithe*, 1:3–4. I am grateful to Wang Qianjin 汪前進 for his help accessing this source, and to Mark Elliott for his correction of my translation.

> *yi* point, by grinding the nailed place, then a round line called *yi bing ding wu* is generated. This *yi bing ding wu* line is called circle border; the place enclosed by this border is called circle.

This Manchu text retains and amplifies the material character of the French original. The object and action are no longer a product of one's imagination. Moreover, the absence of mathematical terminology in Manchu, to which this translation was doubtless intended as a remedy,[46] prompted the translators to use verbs such as "nail" and "grinding",[47] which I have chosen to render literally. Let us now look at the Chinese translation of this Manchu definition:

> 第八
> 凡有一線。以線之一頭為樞。以線之一頭為界。旋轉復還原處。即成一圜。設如若甲乙線. 以甲為樞。以乙為界。旋轉復還乙處。即成乙丙丁戊之圜。此圜線謂之圜界。界線積處為圜面。[48]
>
> 8.
> Let there be a line. Take one of its ends as a pivot, and one of its ends as boundary. Turning it through a complete rotation makes a circle. Thus, with line AB, take the A end as a pivot; take the B end as boundary. Turning it until it returns to the [original] place of B makes the circle BCDE. This circular line is called circumference of the circle. The area accumulated inside the border line is called the area of the circle.

This only differs from the *Shuli jingyun*'s version by a few characters. The action of nailing has been replaced by the more abstract phrase *yi ... wei* 以……為 mentioned above; that of milling has been replaced by the rotation of the line. It seems that concrete images are no longer needed once one has safely landed into a language that already accommodates geometry in the Euclidian tradition, namely classical Chinese. As a result, the "material quality" of the original French definition, preserved and even emphasized in its Manchu translation, has been largely discarded. On the other hand, the figures illustrating the definitions are similar in the Manchu and Chinese versions (see fig. 7.5).

46 This is witnessed by the fact that Chinese characters are added next to geometrical terms and the names of points on the manuscript.
47 *Moselambi* (to mill, to grind) is used to indicate turning in the way a millstone does as it grinds.
48 *Jihe yuanben* 1691, 2b.

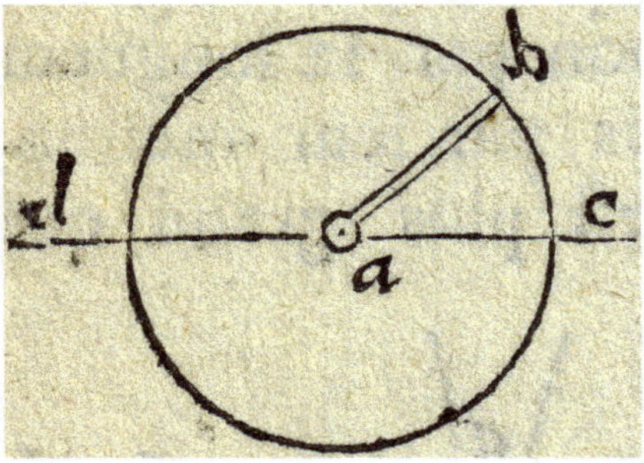

FIGURE 7.4
The construction of a circle in Pardies' *Elémens de géométrie*, 4

Even though the visual reference to the alidade, an essential element of surveying instruments, has been lost in the first translation, it remains that all four versions of this definition of the circle are constructions in a mechanical rather than geometric sense of the term, as they involve movement (fig. 7.5). This is an instance of the fact that Euclidian geometry as presented in the *Shuli jingyun* aims at being pedagogically effective. Like the *Jihe yaofa*, it integrates the making and use of instruments. However, the instruments integrated are no longer those of geometry itself (ruler, pen, compasses, and try square) but rather surveying instruments. As a whole, the *Shuli jingyun* includes mathematics for surveyors, and Pardies's geometry, with its emphasis on application to earthly measurements, met the needs of the Qing state for surveying techniques.

5 Conclusion

By focusing on the definitions of the circle in three mathematical treatises and on the way in which each of them relates to the body of knowledge of which they form part, we have provided an example of the major changes undergone by the content and style of geometry in the Euclidian tradition produced by Jesuit missionaries in China, including the changes in the institutional context in which they were produced and the main fields of expertise that these Jesuits put in the service of emperors.

I have argued that institutional changes were echoed in the way in which mathematics was written at two junctures. The first one was the calendar reform initiated in 1629, which reduced geometry in the Euclidian tradition to a prerequisite for astronomy, giving it an instrumental character. The second one came with the appropriation of another field of Western learning, cartography, by the Kangxi emperor. This new appropriation, occurring at a time when mathematics was gaining the status of an independent field of knowledge

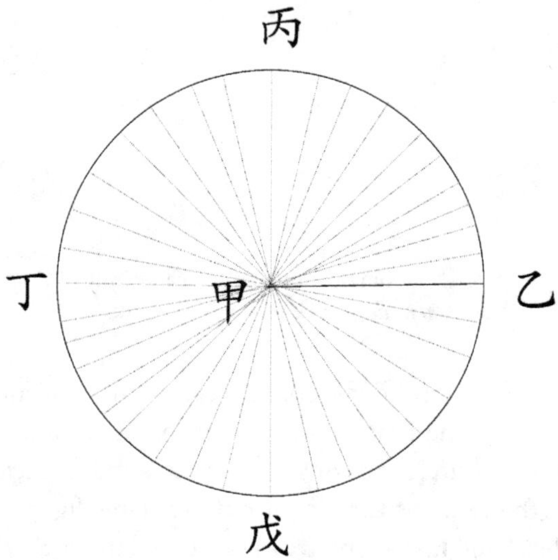

FIGURE 7.5
Illustration of the construction of a circle in the draft Chinese translation of the Manchu text *Giho yuwan ben bithe*

under imperial patronage, entailed the integration of instruments imported to China for the purpose of surveying the empire into the corpus of mathematical knowledge.

This has implications not only for the history of mathematics but also for the history of cartography: the techniques used by cartographers, although they were not presented as such in a separate treatise, were widely available in the *Shuli jingyun*. As with astronomy, the key to the imperial state's monopoly on cartography lay not in its retention of knowledge relevant to the field but in the exclusive possession of the instruments needed to practice cartography to the standards set by the imperial project. Measuring Heaven and Earth required not only the mastery of knowledge but also the ability to make, use, and maintain instruments.

The three definitions of the circle as a geometric object correspond to three different ways of ordering the world with which this field of mathematics was associated. In the first one, mathematics is part of Heavenly knowledge, taken to be the prerogative of the Christian worldview as construed by the Jesuits. In the second one, mathematics is one of the foundations of the knowledge that enables the emperor to act effectively as the intermediary between Heaven and Earth, by harmonizing the rhythm of human life with that of the cosmos. This might be called the temporal dimension of the political cosmology of imperial China. Finally, the third definition of the circle, in a context where mathematics fulfils the needs of imperial cartography, evokes what one might call the spatial dimension of this political cosmology. The survey of his dominions gave Kangxi an intimate knowledge of them, in a way evocative of the mythical sage king Yu

禹 the Great, founder of the Xia 夏 dynasty of Chinese antiquity, whose familiarity with his kingdom resulted from his lifelong work to control the floods.

This appropriation of Euclidian geometry and astronomical knowledge imported from Europe by China's emperors in the seventeenth and eighteenth centuries thus provides an example of the ways in which the scientific and technical knowledge underlying the cosmology prevailing in one region of the world can be put in the service of another cosmology in another region of the world. The circulating knowledge then constitutes the overlap between two cosmologies that remain radically different.

Acknowledgments

I am grateful to Christopher Cullen for reading through this article.

Bibliography

Cams, Mario. *Companions in Geography: East-West Collaboration in the Mapping of Qing China (c. 1685–1735)*. Leiden: Brill, 2017.

Cervera, José Antonio. *Las Varillas de Napier en China: Gioacommo Rho, S.J. (1592–1638) y su trabajo como matemático y astrónomo en Beijing*. Mexico City: El Colegio de México, 2012.

Cheng, Anne. *Histoire de la pensée chinoise*. Paris: Le Seuil, 1997.

Chu, Pingyi. "Trust, Instruments, and Cross-cultural Scientific Exchanges: Chinese Debate over the Shape of the Earth, 1600–1800." *Science in Context* 12, no. 3 (1999): 385–411.

Clavius, Christophorus. *Euclidis Elementorum libri X. Accessit XVI de solidorum regularium comparatione. Omnes perspicuis demonstrationibus, accuratisq[ue] scholiis illustrati*, Rome: Apud Vicentium Accoltum, 1574.

Cullen, Christopher and Catherine Jami. "Christmas 1668 and After: How Jesuit Astronomy Was Restored to Power in Beijing." *Journal for the History of Astronomy* 51, no. 1 (2020): 3–50.

Engelfriet, Peter M. *Euclid in China: The Genesis of the First Chinese Translation of Euclid's Elements, Books I–VI (Jihe yuanben, Beijing, 1607) and its Reception up to 1723*. Leiden: Brill, 1998.

Giho yuwanben bithe. Library of the Institute of History of Natural Sciences of the Chinese Academy of Sciences, Beijing, 子530.9917 337.

Graham, A.C. *Later Mohist Logic, Ethics and Science*. Hong Kong: The Chinese University Press, 1978.

Guo Shuchun 郭書春, ed., *Zhongguo kexue jishu dianji tonghui. Shuxue juan* 中國科學技術典籍通彙。數學卷. 5 vols. Zhengzhou: Henan jiaoyu chubanshe, 1993. (abbreviated as ZKJDT).

Hashimoto, Keizo. *Hsü Kuang-ch'i and Astronomical Reform: The Process of the Chinese Acceptance of Western Astronomy, 1629–1635*. Osaka: Kansai University Press, 1988.

Hashimoto, Keizo and Catherine Jami. "From the Elements to Calendar Reform: Xu Guangqi's Shaping of Mathematics and Astronomy." In *Statecraft and Intellectual Renewal: The Cross-Cultural Synthesis of Xu Guangqi (1562–1633)*, edited by Catherine Jami, Peter Engelfriet, and Gregory Blue, 263–278. Leiden: Brill, 2001.

Jami, Catherine. "La carrière de Mei Wending (1633–1721) et le statut des sciences mathématiques dans le savoir lettré." *Extrême-Orient, Extrême-Occident* 36 (2013): 19–47.

Jami, Catherine. *The Emperor's New Mathematics: Western Learning and Imperial Authority During the Kangxi Reign (1662–1722)*. Oxford: Oxford University Press, 2012.

Jami, Catherine. "Giulio Aleni's Contribution to Geometry in China: The *Jihe yaofa*." In *'Scholar from the West': Giulio Aleni S.J. and the Dialogue between Christianity and China*, edited by Tiziana Lippiello and Roman Malek, 553–572. Nettetal: Steyler Verlag, 1997.

Jami, Catherine. "Mathematical Knowledge in the Chongzhen lishu." In *Western Learning and Christianity in China: The Contribution of Johann Adam Schall von Bell, S.J. (1592–1666)*, edited by Roman Malek, 661–674. Sankt Augustin: China-Zentrum & Monumenta Serica Institute, 1998.

Jami, Catherine. "Science in Manchu in Early Qing China: Does it Matter?" In *Looking at it from Asia: The Processes that Shaped the Sources of History of Science*, edited by Florence Bretelle-Establet, 371–391. New York: Springer, 2010.

Jihe yaofa 幾何要法 (1631). Bibliothèque nationale de France, Chinois 4869; https://gallica.bnf.fr/ark:/12148/btv1b90062493.r=%E5%B9%BE%E4%BD%95%E8%A6%81%E6%B3%95?rk=21459;2

Jihe yuanben 幾何原本, 1607, National Library of China; https://www.wdl.org/en/item/17216/

Jihe yuanben 幾何原本, 1691, National Central Library, Taipei; https://www.wdl.org/fr/item/7103/

Pardies, Ignace Gaston, S.J. *Elémens de géométrie, ou par une methode courte & aisée l'on peut apprendre ce qu'il faut sçavoir d'Euclide, d'Archimede, d'Appollonius, & les plus belles inventions des anciens & des nouveaux Géomètres*. Paris: Sébastien Mabre-Cramoisy, 1671.

Shuli jingyun 數理精蘊, 1723. Facsimile reprint in ZKJDT, vol. 3.

Xu Guangqi 徐光啟, *Xu Guangqi ji* 徐光啟集. Edited by Wang Zhongmin 王重民. 2 vols. Shanghai: Guji chubanshe, 1984.

CHAPTER 8

The Buddhist Universe in Early Modern Japan: Cosmological Dispute and the Epistemology of Vision

D. Max Moerman

Japan, like many places in Buddhist Asia, inherited a classical Indian cosmology. According to this cosmology, the world is flat. Layers of wind, water, and golden earth support it from below, and a multitude of heavens rise above. At the center of this world is Mount Sumeru orbited by the sun and moon and surrounded by a radiating series of mountains and seas. In the outermost sea are four great continents, and the continent to the south, known as Jambudvīpa, is the realm of humans. This traditional Buddhist cosmology, described in scripture and systematized in commentaries, remained the standard Japanese view of the universe from the sixth to the seventeenth century (fig. 8.1). It was a cosmology that went unchallenged until the arrival of European Jesuits, who promoted a world defined by Aristotelian physics, Ptolemaic astronomy, and Roman Catholic theology. In the face of such global threats, Japanese Buddhist monks developed new modes of cosmological discourse throughout the eighteenth and nineteenth centuries. The development of this discourse was neither reactionary nor regressive but rather deeply transcultural: "a process of transformation that unfolds through extended contacts and relations between cultures," in "which forms emerge in local contexts within circuits of exchange."[1] This chapter presents an example of overlapping cosmologies as a transcultural process in which Buddhist and European views of the world and of the universe were promoted, refuted, contested, appropriated, and transformed.

Although occasioned by moments of cultural encounter, this transformation of Buddhist cosmology took place between multiple interlocutors and over the *longue durée*. While grounded in textual sources, it was conducted primarily through visual representation and was, in the end, a dispute over vision. It was not a struggle between East and West, or between the pre-modern and the modern, or between religion and science. Such simple dichotomies are more

1 Juneja and Kravagna, "Understanding Transculturalism," 24–25.

 | DOI:10.1163/9789004511675_010

FIGURE 8.1
Zontō 存統, illustration of Mount Sumeru universe from *Expanded Image of the World* (*Sekai daisōzu* 世界大相圖), 1821. Woodblock print with hand coloring, 195 × 64 cm
LIBRARY OF CONGRESS

than false. They obscure what was common to all participants: the practices of image making and the claims to represent the invisible. The emphasis on vision reveals the debate to have been as much about epistemology as it was about cosmology. In Buddhist terms, cognition and vision are co-articulated in a single term (Skt. *jñāna-darśana*, Jp. *chiken* 智見). Vision is epistemological; knowledge is optical: Buddhist wisdom is described in such terms as "limitless vision" (Skt. *samanta-darśin*), and "the universal eye" (Skt. *samanta-cakṣus*).[2] The mental attainments of Buddhist adepts are expressed in a similarly ocular vocabulary. In addition to the physical eye, advanced practitioners gain the heavenly eye (Skt. *divya-cakṣus*), which sees the causes and effects of all things; the wisdom eye (Skt. *prajñā-cakṣus*), which sees the emptiness of all things; the dharma eye (Skt. *dharma-cakṣus*), which sees the impermanence of all things; and the Buddha eye (Skt. *buddha-cakṣus*), which includes them all.[3] These five eyes (Skt. *pañca-cakṣūṃṣi*) constitute the five modes within the Mahayana classification of discriminative knowledge. This traditional language of universal vision, a vocabulary rooted in classical Buddhist theories of knowledge, was deployed by Buddhists to uphold their cosmology in the face of Christian and European critique throughout the eighteenth and nineteenth centuries. Buddhist monks advanced their claims by reformulating an ancient Indian cosmology into contemporary astronomical treatises and using explanatory diagrams to present the empirical reality of a Buddhist universe. In an age of overlapping cosmologies, Buddhist monks defended their worldview by adopting, adapting, appropriating, and at times inverting the methods of their opponents.

1 Ptolemaic Cosmology and the Jesuit Critique

The astral sciences not only provided the conditions of possibility for European travel to Japan, they were also integral to the Jesuit strategy of conversion. "It would be good," wrote Francis Xavier (1506–1552), if missionaries, "knew something about the celestial sphere, since the Japanese are delighted with learning about the movements of the heavens, the eclipses of the sun, the waxing and waning of the moon, and how rain, snow and hail, thunder, lightning, comets and other natural phenomena are produced. The explanation of such matters is a great help in gaining the good will of the people."[4] The Jesuit preoccupation

2 McMahan, *Empty Vision*, 2.

3 Wayman, "Buddhist Theory of Vision."

4 *To Father Ignatius Loyola, in Rome from Goa, April 9, 1552*. Costelloe, *Letters and Instructions*, 385. Schurhammer and Wicki, *Epistolae S. Francisici Xaverii*, 373.

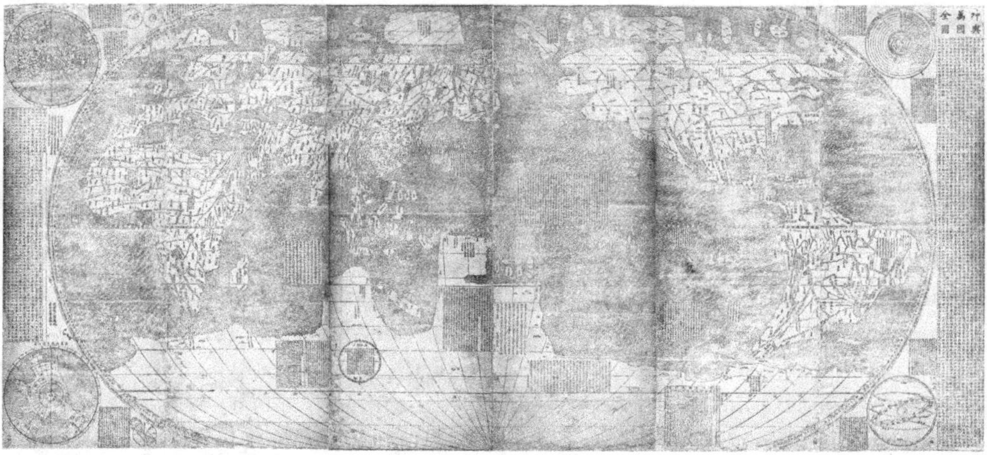

FIGURE 8.2 Matteo Ricci, *Complete Map of Myriad Countries*, 1602. Six-panel monochrome woodblock print, 171 × 361 cm
COURTESY OF THE JAMES FORD BELL LIBRARY, UNIVERSITY OF MINNESOTA

with cosmology is apparent in the principal textbook produced for Japanese seminarians, the *Compendium Catholicae Veritatis*, compiled by the Spanish Jesuit Pedro Gómez (1533–1600) in 1593. Its first section, entitled *De Sphaera*, presents the theory of a spherical earth and also explains, in words and images, "the nature of the heavenly bodies, the motion, number, and order of the heavens, solar and lunar eclipses, and the magnitude of the stars and sky."[5] Gómez notes that he compiled the text to "clearly demonstrate the invisible attributes of God in visible things: namely, the machine of the world and the perpetual and immutable order of the heavens."[6]

The Italian Jesuit Matteo Ricci (1552–1610) also sought to "demonstrate the invisible attributes of God in visible things." In 1602, he produced an enormous six-panel woodblock-printed world map, a Ptolemaic projection of a spherical earth with cosmographic diagrams in its four corners, entitled *Complete Map of Myriad Countries* (Ch. *Kunyu wanguo quantu*, Jp. *Konyo bankokuzenzu* 坤輿萬國全圖, fig. 8.2). In the preface to his map, Ricci explains, "through the comprehension of heaven and earth, one may testify to the ultimate kindness, greatness, and oneness of the supreme power of the Lord who rules over heaven

---

5 Kirishitan Bunko Library, *Compendium Catholicae Veritatis*, 74–79. On the *Compendium*, see Üçerler, "Jesuit Humanist Education."

6 *Quia, ut Apostolus ait, visibilia haec, mundi scilicet machine, caelorumque pertetuus et immutabilis ordo, invisibilia Dei attributa maxime demonstrant.* Hiraoka, "Jesuit Cosmological Textbook," 108.

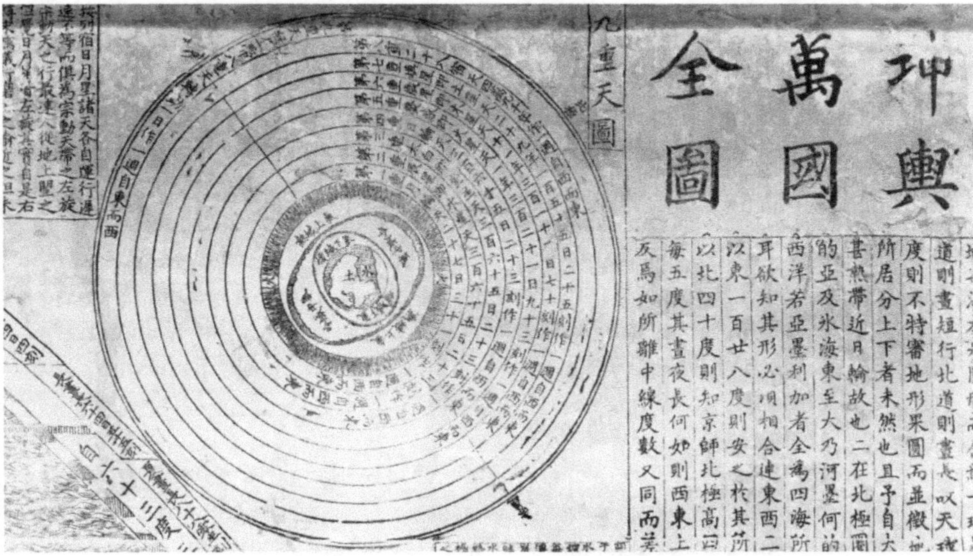

FIGURE 8.3 Matteo Ricci, *Complete Map of Myriad Countries*. Detail of nine-sphere universe showing planets revolving around Earth
COURTESY OF THE JAMES FORD BELL LIBRARY, UNIVERSITY OF MINNESOTA

and earth."[7] To aid in the comprehension of heaven and earth, Ricci's map includes diagrams of the northern and southern hemispheres, solar and lunar eclipses, and a spherical earth surrounded by two oval rings of cold and warm air and a flaming ring of fire. Nine heavens of orbiting celestial bodies surround the earth, each noted with the length and direction of their rotation: Moon, Mercury, Venus, the Sun, Mars, Jupiter, Saturn, and the twenty-eight constellations (fig. 8.3). Two additional heavens were added to the 1603 edition of the map, with the explanation: "The eleventh heaven is inhabited by the Supreme Ruler, Lord of Heaven, and the saints and angels of paradise, eternally silent and immovable."[8]

The refutation of Buddhist cosmology is central to Ricci's argument. In the preface to his map, Ricci charges: "the Buddhists' claim that the Middle Kingdom is located in the southern continent of Jambudvīpa is false, as is their claim about the height of Mount Sumeru."[9] The cosmic mountain at the center of the Buddhist universe and the form and location of the Buddhist world were inimical to Ricci's Christianity. Ricci's map arrived in Japan soon after being printed

7 Translation from Chen, "Human Body," 542.
8 D'Elia, "Recent Discoveries," 126.
9 Translation from Akin, "Printed Maps," 219–220.

in China and was used to teach geography in the Jesuit seminary in Kyoto. In 1605, Fukansai Habian (1565–1621), a Zen monk who converted to Christianity echoed Ricci's cosmological critique:

> What we call "paradise in Heaven" in Christian teaching has nothing to do with the Triple Realm of Śākyamuni's sutras. It lies in the eleventh layer of clear blue Heavens that we see above us, where the moon, sun, and stars are fixed. Śākyamuni, quite unaware that the heavenly bodies are in this sky, claimed that the moon, sun, and stars move around the center of Mount Sumeru, carried by the wind. ... It goes without saying that all of this is absolutely ludicrous. Do not identify any such silly ideas with Christian teachings.[10]

A simplified copy of Ricci's map was included in Wang Qi's 王圻 1609 *Collected Illustrations of the Three Realms* (*Sancai tuhui* 三才圖會), a Chinese encyclopedia widely influential in Japan. Wang quotes extensively, and without acknowledgement, from Ricci's preface, including Ricci's criticism of Jambudvīpa and Mount Sumeru. As the Jesuit author is nowhere acknowledged, his condemnation of Buddhist cosmology is removed from the realm of Christian polemic and assumes the voice of academic authority. A century later, the Japanese author Terajima Ryōan 寺島良安 would include the very same simplified print of Ricci's map in his 1712 *Japanese and Chinese Collected Illustrations of the Three Realms* (*Wakan sansai zue* 和漢三才圖會) and conclude that "the Buddhist claim that there are 30,000 worlds is the foolish talk of those who have not carefully investigated matters."[11]

Even after the Jesuits were expelled and their texts were proscribed early in the seventeenth century, European astronomical theory continued to circulate in Japan. Stripped of Christian ideas, material from Gómez's *De Sphaera* also reappeared in the mid-seventeenth-century *Outline of Terrestrial and Celestial Globes* (*Nigi ryakusetsu* 二儀略説) by Kobayashi Yoshinobu 小林義信 (a.k.a. Kentei 謙貞, 1601–1684).[12] Another work to propound Aristotelian and Ptolemaic cosmology was the four-volume *Heaven and Earth with Commentaries* (*Kenkon bensetsu* 乾坤辯説), which was based on a Latin treatise confiscated from the Jesuit mathematician and astronomer Francisco Cassola when he arrived in Japan in 1643. It was translated by the apostate Portuguese missionary

10 Baskin and Bowring, *The Myōtei Dialogues*, 182.

11 Terajima, *Wakan sansei zue*, 55:3b.

12 On the relationship between *De Sphaera, Nigi ryakusetsu*, and *Kenkon bensetsu*, See Obara, "Kirishitan jidai no kagaku shisō;" and Hiraoka, *Nambankei uchuron no genteki kenkyū*.

Christovão Ferreira (1580–c. 1650), who had taken the Japanese name Sawano Chūan, and edited with commentaries by the Confucian scholar Mukai Genshō 向井玄升 (1609–1677).[13]

The Ptolemaic cosmology of the Jesuits had appeared in Japanese astronomical treatises by the late seventeenth century. Iguchi Tsunenori's 井口常範 1689 *Astronomy Illustrated and Explained* (*Tenmon zukai* 天文圖解) explains the general principles of Chinese astronomy and calendrical science, includes the method of calculating the relative location of the sun, moon, and major planets, and illustrates these theories with diagrams of armillary spheres and star maps. Yet he also presents, and advocates, the European theory of the global earth. On facing pages, Iguchi pairs a traditional Buddhist image of Mount Sumeru with Matteo Ricci's diagram of a nine-sphere universe, noting their radii and periods of revolution, and asserts the supremacy of Western astronomy over Buddhist cosmology (fig. 8.4). Iguchi's critique extended to both the Buddhist image of the universe and the theory of vision on which it relied: "The Buddhists say that they can see the form of Mount Sumeru with the power of the heavenly eye. Astronomers, however, have only the power of the human eye and therefore must rely on theory. Yet the astronomers' explanations of the solar and lunar eclipses are correct and those of the Buddhists are not."[14] The "power of the heavenly eye" (Jp. *tengentsu* 天眼通; Skt. *divya-cakṣur-abhijñā*), by which, according to Iguchi, Buddhists claim to see Mount Sumeru, refers to a form of sight unobstructed by any obstacle, an ability to see far and near, past and future, inside and outside. It is one of the six supernormal powers (Jp. *jinzu* 神通; Skt. *abhijñā*) attained by those who have reached an advanced level of Buddhist meditation and one of the five modes of Buddhist vision and knowledge. Iguichi's critique of the Buddhist view of the universe thus extends to the totality of Buddhist metaphysics: cosmology, ontology, and epistemology.

2 Buddhist Cosmology and the Epistemology of Vision

Japanese Buddhists responded in earnest to such attacks on their theories of the cosmos, vision, and knowledge, by returning to the classical texts of Buddhist cosmology and reformulating them in detailed and illustrated contemporary treatises. In 1699, the monk Imai Shirō 今井氏老 published an *Illustrated*

13 For a translation and analysis of the *Kenkon bensetsu*, see Pinto dos Santos, "Study in Cross-Cultural Transmission."

14 Iguchi, *Tenmon zukai*, 4:59b.

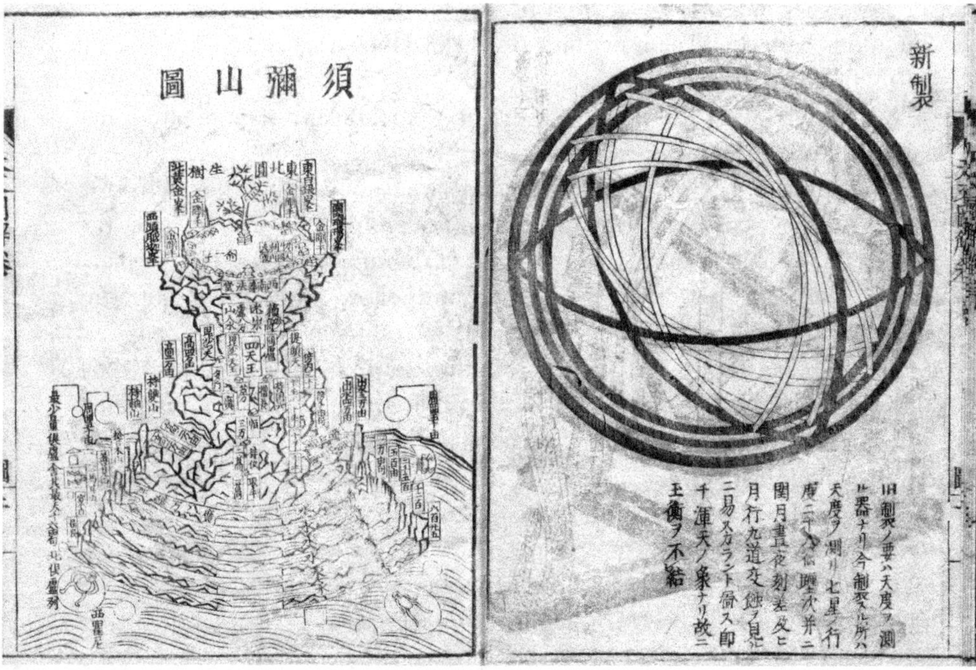

FIGURE 8.4 Iguchi Tsunenori, *Astronomy Illustrated and Explained*, 1689. Monochrome woodblock book illustration
NATIONAL INSTITUTE OF JAPANESE LITERATURE, NATIONAL INSTITUTES FOR THE HUMANITIES. CREATIVE COMMONS LICENSE CC BY-SA 4.0

Explanation in Japanese of the Solar and Lunar Orbits (*Nichigetsu sentenzu wagoshō* 日月旋転圖和語鈔), which represented the Buddhist theory of solar and lunar movement with extensive diagrams (fig. 8.5). One year later, in 1700, the monk Yūhan 宥範 published a similar text entitled *Illustrations of Solar and Lunar Movement as Explained in the Dhātu Chapter of the Abhidharmakośa* (*Kusha sekenbon nichigetsu kōdō zu shō* 俱舎世間品日月行動圖解, fig. 8.6). In 1707, in the Kegon monk Hōtan 鳳潭 (1659–1738) published a fourteen-fascicle edition of Yuanhui's 圓暉 eighth-century commentary on Vasubandhu's fifth-century *Abhidharmakośa* (*Kanchū kōen kusharon juso* 冠註講苑俱舎論頌疏) in which the orbits of the sun and moon around Mount Sumeru were explained and illustrated (fig. 8.7). The same year, Mori Shōken 森尚謙 (1653–1721) published a ten-volume *Treatise for the Defense of the Dharma* (*Gohō shijiron* 護法資治論), which sought to locate Mount Sumeru with a global earth (fig. 8.8), and in which he warned that "Western geography and astronomy are great disasters that must be defeated to ensure the future of Buddhism."[15]

15 Mori Shōken, *Gohō shijiron*, 2:16b.

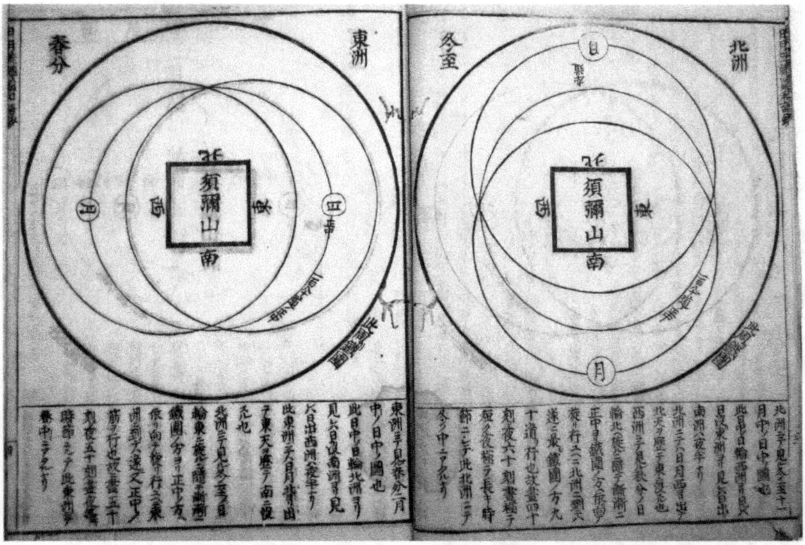

FIGURE 8.5 Imai Shirō, *Illustrated Explanation in Japanese of the Solar and Lunar Orbits*, 1699. Monochrome woodblock book illustration
YOKOHAMA CITY UNIVERSITY LIBRARY

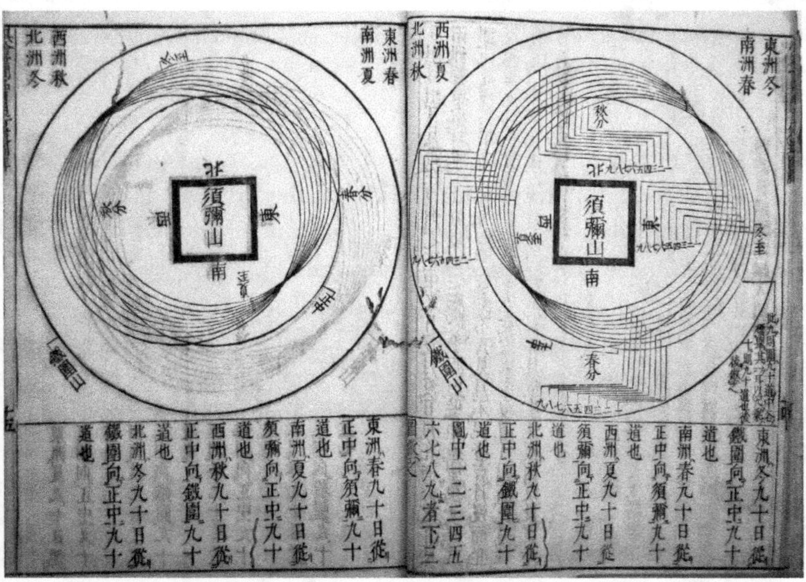

FIGURE 8.6 Yūhan, *Illustrations of Solar and Lunar Movement as Explained in the Dhātu Section of the Abhidharmakośa*, 1700. Monochrome woodblock book illustration
YOKOHAMA CITY UNIVERSITY LIBRARY

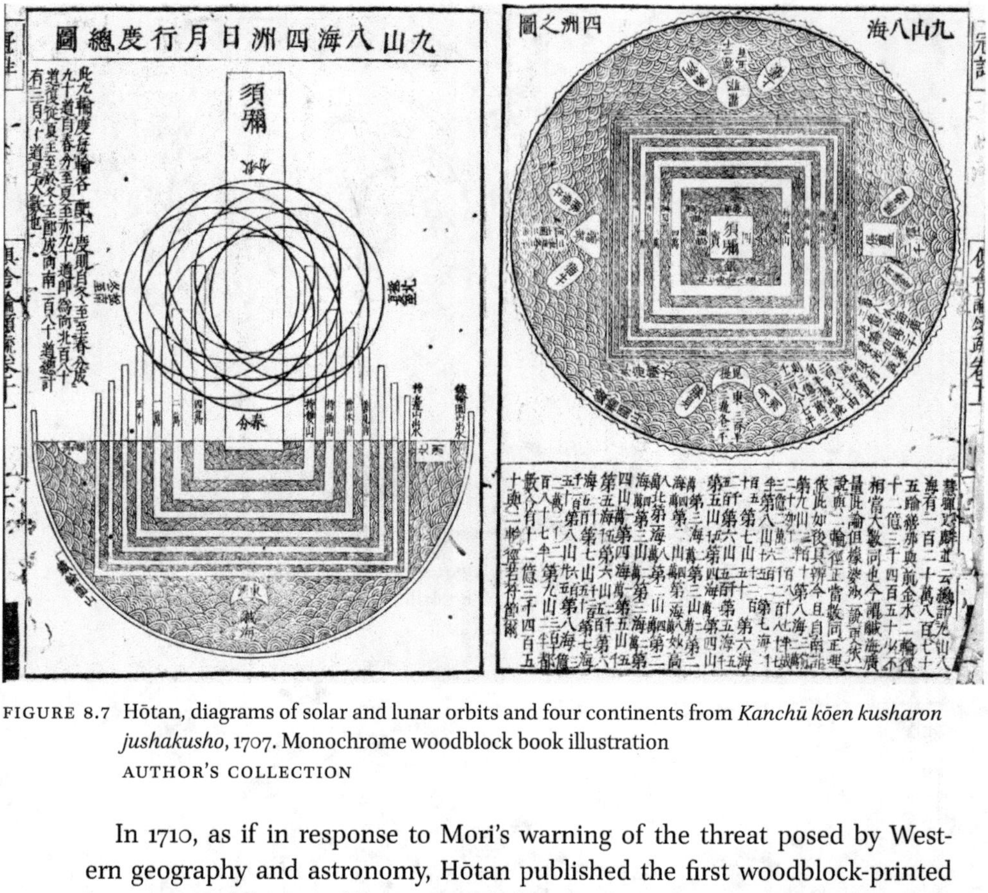

FIGURE 8.7 Hōtan, diagrams of solar and lunar orbits and four continents from *Kanchū kōen kusharon jushakusho*, 1707. Monochrome woodblock book illustration
AUTHOR'S COLLECTION

In 1710, as if in response to Mori's warning of the threat posed by Western geography and astronomy, Hōtan published the first woodblock-printed Japanese Buddhist world map, entitled a *Handy Map of the Myriad Countries of Jambudvīpa* (*Nansenbushū bankoku shōka no zu* 南瞻部洲萬國掌菓之圖) or more literally, a *Map of the Myriad Countries of Jambudvīpa* [*Held*] *like a Fruit in the Hand* (fig. 8.9). Hōtan's world map, like that of Ricci issued a century earlier, is an illustrated defense of religious cosmology. Hōtan's title denotes not only a Buddhist vision of the world but also the very quality of Buddhist vision itself. "Fruit in the hand," (*shōka* 掌菓) is a scriptural term for that which is easy to see and comprehend, and signals Hōtan's deployment of the traditional Buddhist equation of vision and knowledge. As the map's preface explains, "The wisdom eye of the sage (*hijiri no egan* 聖慧眼) is far more powerful that the human eye, and sees the boundless ten-thousand-fold world, just like a fruit held in the hand." This phrase has a long history in Buddhist literature. In the *Vimalakīrti-nirdeśa-sūtra*, Aniruddha, the disciple of the Buddha foremost in divine sight, is asked, "How far can this heavenly eye of yours see?" He replies, "I can see the boundless ten-thousand-fold world as though I were peering down at a fruit

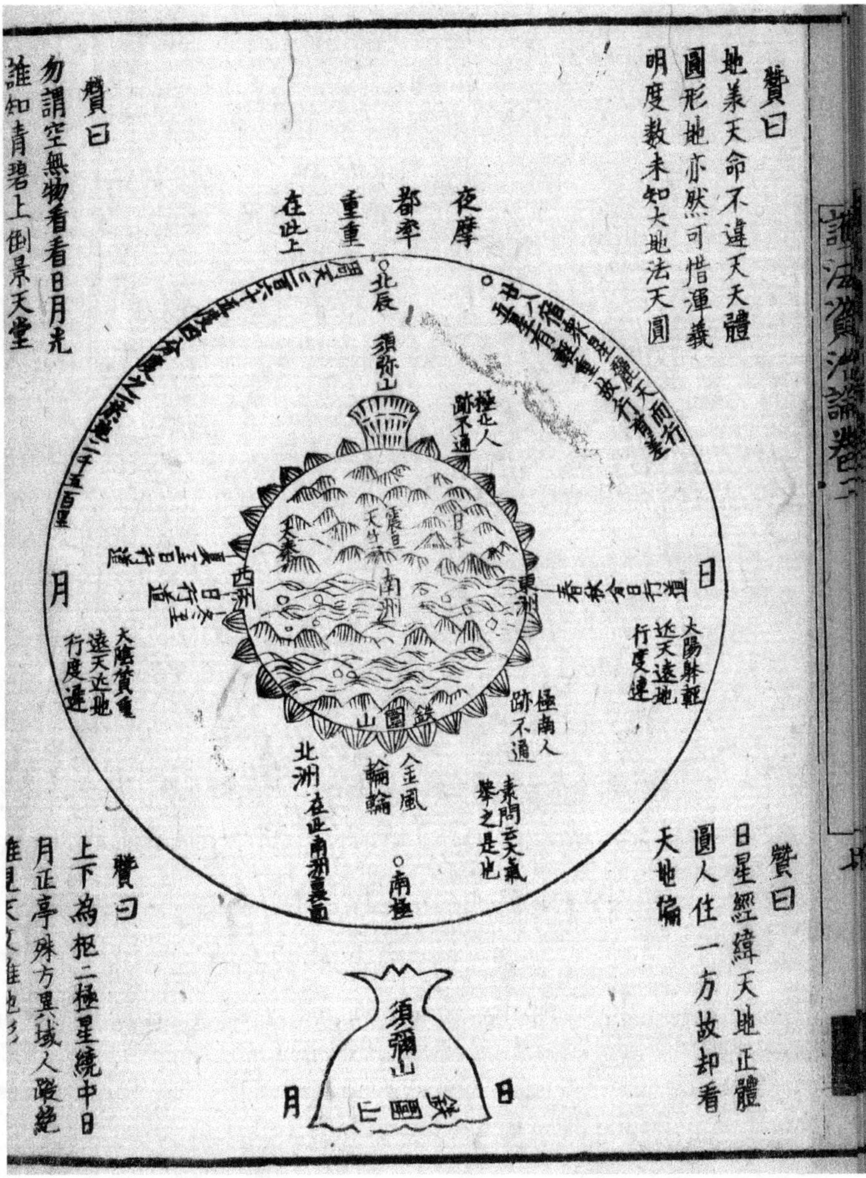

FIGURE 8.8 Mori Shōken, *Diagram of Mount Sumeru and Global Earth*, from *Defense of Buddhism*, 1707. Monochrome woodblock book illustration
AUTHOR'S COLLECTION

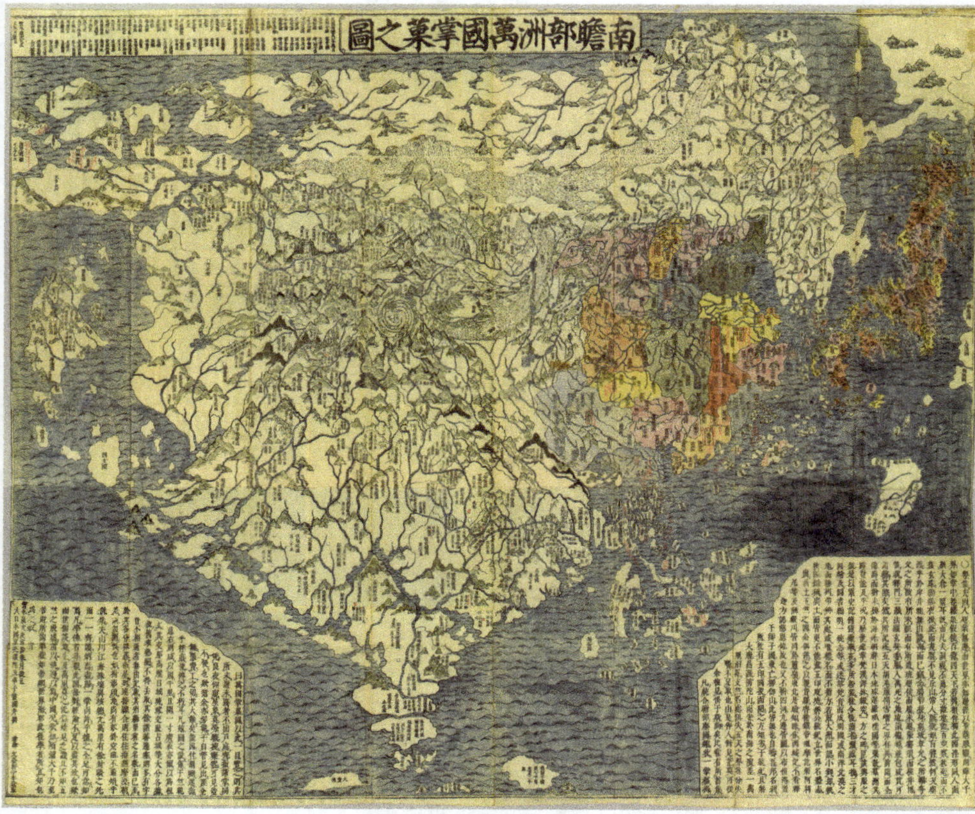

FIGURE 8.9 Hōtan, *Handy Map of the Myriad Countries of Jambudvīpa*, 1710. Woodblock print with hand coloring, 121 × 144 cm
COURTESY, DAVID RUMSEY MAP COLLECTION, DAVID RUMSEY MAP CENTER, STANFORD LIBRARIES

in the palm of my hand."[16] The *Commentary on the Sutra on Contemplation of Amitāyus* (Jp. *Kanmuryōjukyō gisho*, Ch. *Guan wuliangshou jing yishu* 觀無量壽經義疏), by Yuanzhao 元照 (1048–116), similarly states: the "heavenly eye sees the thousand-fold world like a fruit in the hand."[17]

16 T 14:522c29–523a1. 三千大千佛國如於掌中觀寶冠耳. The phrase "demonstrable in the palm of the hand" (Ch. *zhizhang* 指掌) also appears in the titles of Chinese maps from the twelfth century. See De Weerdt, "Maps and Memory," 164n1. A similar phrase, "all places in the palm of the hand" is used by Yuan Taiyuan 阮泰元 in his preface to the 1603 edition of Ricci's map. D'Elia, "Recent Discoveries," 144.

17 T 37.290b11: 天眼觀大千界如觀掌果. The expression is also found in Zhiyi's commentary on the *Lotus Sutra, Miaofa lianhuajing wenju* 妙法蓮華經文句 (T 34:15c16–17), the

Informed by such scriptural claims, Hōtan explains the qualitative difference between the cosmological vision of Buddhist and of ordinary humans: "There are innumerable realms, as countless as the leaves of mustard grass, beneath the four heavens. Our realm of Jambudvīpa is like a single grain within a great storehouse of millet. … The vision of an ordinary person is as far from the vision of the wisdom eye as that of a blind person is from the sighted. He can say nothing of the worlds as numerous as atoms." Hōtan's argument, that the view of the universe from the perspective of Buddhist wisdom is qualitatively different from that of human vision, relies on the classical Buddhist equation of vision and knowledge. Hōtan's "wisdom eye" (Skt. *prajñā-cakṣus*) refers to the power of vision to discern the true nature of all things. In both his title and preface, Hōtan is thus promoting a particular Buddhist view of the world as well as a particular Buddhist view of knowledge: a new way of seeing—total, unencumbered, and commanding. Hōtan explains: "With this map one can take in the entire world in a single glance; one can visit distant places without ever traveling beyond one's garden gate; one can point out the various countries of the world just as easily as pointing out the stars in the night sky."[18]

Ricci had described his world map as "enabling the spectator to travel about while reclining at ease in his study," and "to be able to scan all of the countries of the world in turn without going out of doors."[19] Such unobstructed vision is, for Hōtan as for Ricci, in the service of religious cosmology. Hōtan concludes: "We must seek as much understanding of distant lands as we do our own, and even more so of Mount Sumeru at the center of the universe and the vast trichiliocosm itself, just as Sudhana sought the Flower Realm and the vast world of Indra's Net." With a gesture toward the larger cosmic vision of the Buddhist tradition, Hōtan invokes the pilgrim Sudhana, whose journey toward enlightenment is the subject of the *Gandavyūha*, the final section of the *Avataṃsaka* or *Kegon sūtra*, the eponymous scripture of Hōtan's lineage. Sudhana achieved his quest for insight with a visionary experience of the entire structure of the Buddhist universe. "Stripped of a delusion," Sudhana "became clairvoyant without distortion" and, "with the unobstructed eye of liberation, saw all objects without hindrance."[20] "He saw a billion world universes, in which he saw a hundred

Shoulengyan yishu zhu jing 首楞嚴義疏注經 (T 39:848b9–848b17), the *Great Collection Sutra* 大方等大集經 (T 13:136a9), and elsewhere.

18 The expression "visiting distant places without ever traveling beyond one's garden gate" comes from the *Yi jing* commentary on hexagram 60. Lynn, *The Classic of Changes*, 519. See also Lynn, *Classic of the Way*, 141.

19 Giles, "Chinese World Map," 369–370.

20 Cleary, *Entry into the Realm of Reality*, 366.

million sets of the Four Continents, with a hundred million Jambudvīpas and a hundred million heavens of contentment."[21]

Hōtan's advocacy of the Buddhist cosmology was echoed by the Pure Land monk Monnō 文雄 (1700–1763) (a.k.a. Musō 無相) who published *A Vindication of the Theory of the Nine Mountains and the Eight Seas* (*Kusen hakkai gechōron* 九山八海解嘲論) and *A Refutation of Inquiries into the Classic of Heaven* (*Hi tenkyō wakumon* 非天経或問) in 1754.[22] Monnō advanced the now-familiar formulation of Buddhist vision: "The Mount Sumeru universe is what the Buddha Śākyamuni sees from the perspective of absolute truth, it cannot be seen by ordinary humans."[23] With copious diagrams and illustrations, he explained the movement of the sun and moon, the mechanisms of lunar and solar eclipses, and the process of seasonal change (fig. 8.10). Citing such classical texts of Buddhist cosmology as the *Sutra of the Great Conflagration*, the *Sutra on the Arising of Worlds*, the *Lokasthānābhidharma-śāstra*, and the *Abhidharmakośa*, Monnō argued: "The world is flat and round and contains within it vast seas like water in a basin." And in response to the Chinese text that popularized the principles of Western astronomy throughout East Asia, he wrote, "The spherical earth spoken of in the *Inquiries into the Classic of Heaven* is an absolute heresy!"[24]

In 1689, Iguchi had written, "the Buddhists say that they can see the form of Mount Sumeru with the power of the heavenly eye. Astronomers, however, have only the power of the ordinary human eye and therefore must rely on theory. Yet the astronomers' explanations of the solar and lunar eclipses are correct and those of the Buddhists are not." In 1754, Monnō had written, "the Mount Sumeru universe is what the Buddha Śākyamuni sees from the perspective of absolute truth, it cannot be seen by ordinary humans."[25] And in 1806, the monk Mitsuan Shakusōji 密庵釈僧慈 asserted, "the explanations of Buddhist texts rely on the heavenly eye. Astronomers cannot see by such means and have only the laws of computation."[26] For over a century, Buddhist cosmological debate centered on vision and knowledge. Whereas terms of the debate may have remained discrete, the modes of discourse, display, and demonstration were soon to become increasingly complex.

---

21 Cleary, *Entry into the Realm of Reality*, 369.

22 Monnō's *Refutation* is a critique of *Inquiries into the Classic of Heaven*, a popular Chinese introduction to European astronomical theory, which had been published in a Japanese edition in 1730.

23 Cited in Nishimura, "Shuminsen to chikyūsetsu," 128.

24 *Kusen hakkai kaichōron*, 11a.

25 Cited in Nishimura, "Shuminsen to chikyūsetsu," 128.

26 Unno, *Nihonjin no daichzō*, 205.

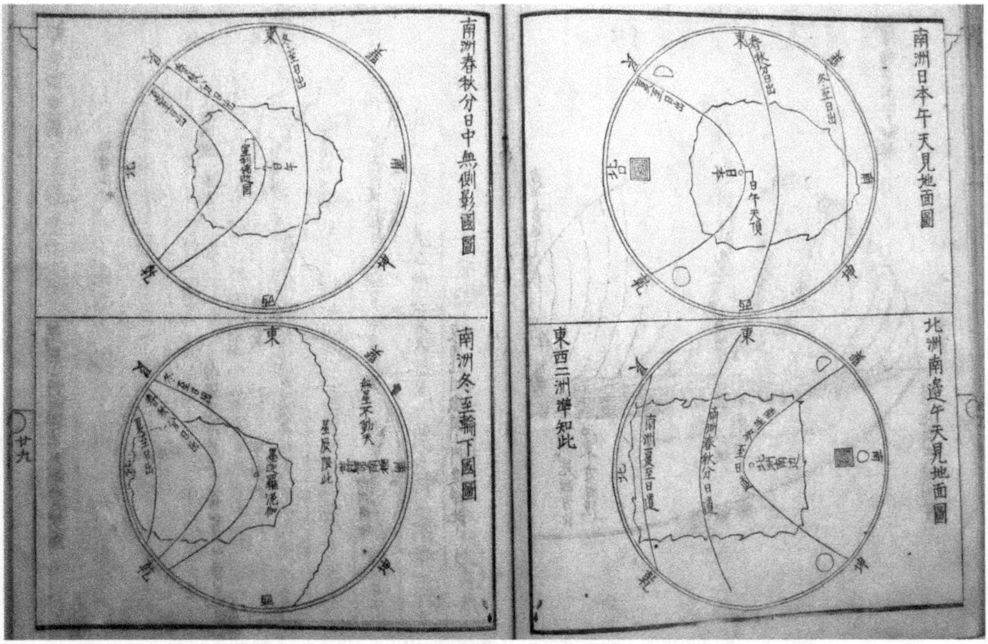

FIGURE 8.10 Monnō, *Discourse on the Theory of the Nine Mountains and the Eight Seas*, 1754. Monochrome woodblock book illustration
YOKOHAMA CITY UNIVERSITY LIBRARY

The central figure in Buddhist cosmological discourse in nineteenth-century Japan was the Tendai monk Fumon Entsū 普門円通 (1755–1834) who led a movement to create a distinctly "Buddhist astronomy," known as Bonreki 梵暦 or Butsureki 佛暦.[27] Entsū founded a society for Buddhist astronomy (Bonrekisha 梵暦社), and composed more than forty works, including the five-volume *Astronomy of Buddhist Countries* (*Bukkoku rekishōhen* 佛國暦象編) of 1810 (fig. 8.11). Entsū promoted a unique formulation of Buddhist cosmology based on an exhaustive reading of scriptural sources; on the comparative analysis of Chinese, Arabic, and European astronomical theories; and on his own detailed mathematical calculations. Entsū's sources were multiple and relied on earlier Chinese theorists even as he argued for the priority of Indian traditions. For example, he borrowed Mei Wending's claim for the Chinese origin of Western astronomy but substituted India for China. He also argued for the similarity between the Sumeru model and the Gaitian model based on Zhoubi. Yet, how-

27 My discussion of Entsū and Bonreki is deeply indebted to the work of Masahiko Okada. See Okada, *Wasurerareta Bukkyō tenmongaku*.

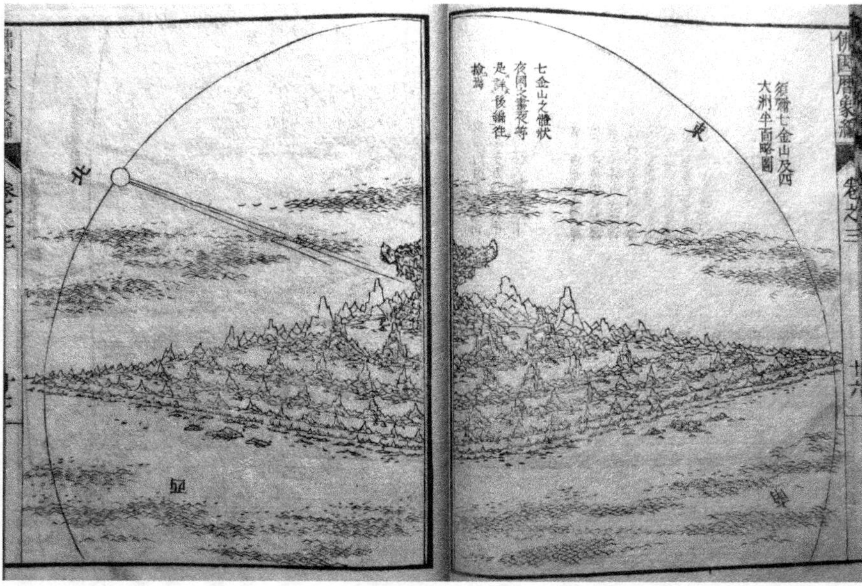

FIGURE 8.11 Fumon Entsū, illustration of the sun's rays on Mount Sumeru from *Astronomy of Buddhist Countries*, 1810. Monochrome woodblock book illustration
MASAHIKO OKADA COLLECTION

ever dependent on Chinese sources and innovative his practical method, Entsū still relied on the classical Buddhist epistemology of vision and enlightened sight:

> The unsurpassed enlightenment attained by the Buddha, the World Honored One, clearly sees (洞視 *dōshi*) the innumerable worlds. Calendrical science and the structure of phenomena were one of the five fields of learning in ancient India and are an essential part of the Great Way. The laws of karma explain the universe, even the creation and destruction of Mount Sumeru and the Triple World, and the movement of the sun, the moon, and the stars. These forms of Indian knowledge are far superior to the reasoning of contemporary scholars, which, compared to the wisdom of the Buddhas and bodhisattvas, is as insubstantial as the leaves of bamboo scattered by a whirlwind. The practitioners of Indian knowledge apprehend true reality and perceive everything in the world with the power of the heavenly eye.[28]

28 Entsū, *Bukkoku rekishōhen*, vol. 1, Preface (first), 1a–b.

Entsū concludes his *Astronomy of Buddhist Countries* by articulating the classical Buddhist synthesis of seeing and knowing, with the neologism "optical knowledge" (*genchi* 眼智).[29] In his final chapter, Entsū introduces what he calls "the law of optical knowledge (*genchi no hō* 眼智之法)." After presenting the theories of vision and knowledge from various scriptural sources, Entsū explains that only Buddhist astronomy represents the universe from the perspective of "the heavenly eye and the five modes of vision," whereas Western astronomy sees it merely from the limited perspective of the physical eye.[30]

These two kinds of vision, Entsū explains, produce two kinds of knowledge: the physical eye that perceives the world as represented in Western astronomy, on the one hand, and the heavenly eye—what Hōtan had described as "the vision of the wisdom eye" that apprehends "the worlds as numerous as atoms"—on the other. This theoretical bifocality, like that of Hōtan a century earlier, is grounded in the theory of two truths (Skt. *satya-dvayam*, Jp. *nitai* 二諦), a categorization of phenomena fundamental to all forms of Mahayana since at least Nagarjuna. Conventional truths (Skt. *saṃvṛti-satya*, Jp. *sezokutai* 世俗諦) "refer to objects of ordinary experience that involve misperceptions tainted by ignorance" whereas ultimate truths (Skt. *paramārtha-satya*, Jp. *shintai* 真諦) are "those realities that exist as they appear and whose direct perception can lead to liberation."[31] Entsū, however, took this method one step further. Like his predecessors, he applied this fundamental Buddhist epistemological distinction to the theories and findings of modern astronomy. But he also sought to reconstruct classical Buddhist cosmology within a contemporary context. He did so by appropriating the very terms of his intellectual opposition, commandeering their quantitative approach and rhetoric of mathematical verification to advance the Buddhist cause. Recognizing the advantage that calculation and demonstration provided the proponents of heliocentrism, Entsū insisted that Buddhists present their worldview in the same terms: "People believe in the theory of a spherical earth because it is scientifically demonstrated through astronomical calculation. To dispel people's doubts about the Mount Sumeru world, we must demonstrate it using astronomical calculation as well."[32]

29 Entsū, *Bukkoku rekishōhen*, 1:3a.
30 Entsū, *Bukkoku rekishōhen*, 5:52a.
31 Buswell and Lopez, *Princeton Dictionary of Buddhism*, 623, 788.
32 Entsū, *Jikken shumikai setsu*, 5:3a.

3 Buddhist Cosmology and the Machine of the World

In order to demonstrate how the horizontal orbit of the sun and moon around Mount Sumeru can explain seasonal change and solar and lunar eclipses, Entsū designed a three-dimensional model of the Buddhist universe. His first model no longer survives, but a print of it, dated 1813, remains extant (fig. 8.12).[33] Entitled *Model of Mount Sumeru Inscribed and Explained* (*Shumisengi mei narabini jo* 須彌山儀銘並序), it depicts a squat cylindrical device set on a decorative wooden base. The sides of the device are painted in horizontal bands and annotated with textual inscriptions, indicating the discs of wind, water, and golden earth that support the Mount Sumeru universe. On the surface are the eight mountain ranges and seas with the hourglass-shaped Mount Sumeru rising from the center. In the outermost sea, bounded by a circular range of iron mountains, lies the four great continents with smaller landforms scattered around them. Above the surface layer of the model and seas are eight small discs representing the positions of constellations, arranged in pairs aligned with each of the four continents, surrounded by a ring marked with astrological notations. The model itself moves. A system of fixed, elevated, circular tracks is colored red, white, and gold and marked with regular increments. Two wire rings, suspending the sun and moon, travel as they orbit the cosmic mountain. Positioned vertically against one side of the device is a black wooden pole with a gold-colored pulley, escapement, and counterweight. This simple clockwork mechanism regulates the measured movement of the sun and moon as they circle Sumeru.

To further explain the workings of his model, Entsū also published a two-volume text, entitled *Model of Mount Sumeru Explained in Japanese* (*Shumisen gimei narabini jo wage* 須彌山儀銘並序和解, 1813), which provided a line-by-line Japanese commentary of the print's classical Chinese preface and provided exhaustive numerical measurements of every element of the model and their corresponding scriptural sources. The details of Entsū's model and the preci-

33 The painting is reproduced in Ryūkoku daigaku Ōmiya toshokan, *Bukkyō no uchūkan*, 18. Entsū inscribed the text and Genshun 玄俊 painted the image. A undated hand-colored woodblock print measuring 109.8 × 49.4 cm entitled simply *Image of Mount Sumeru* (*Shumisenzu* 須彌山図) may represent an earlier depiction of such a model. Completed perhaps ca. 1810, it depicts Indra's palace and assembly hall at the top of Mount Sumeru, the sun and moon as palatial structures connected to Mount Sumeru by curved metal rods, and the twenty heavens of the Realm of Desire and the Realm of Form. The print is reproduced in black and white in Oda and Kunikida, *Kochizu ni miru sekai to Nihon*, 23 and in color in Miyoshi, "Japanische und europäische Kartographie," 38.

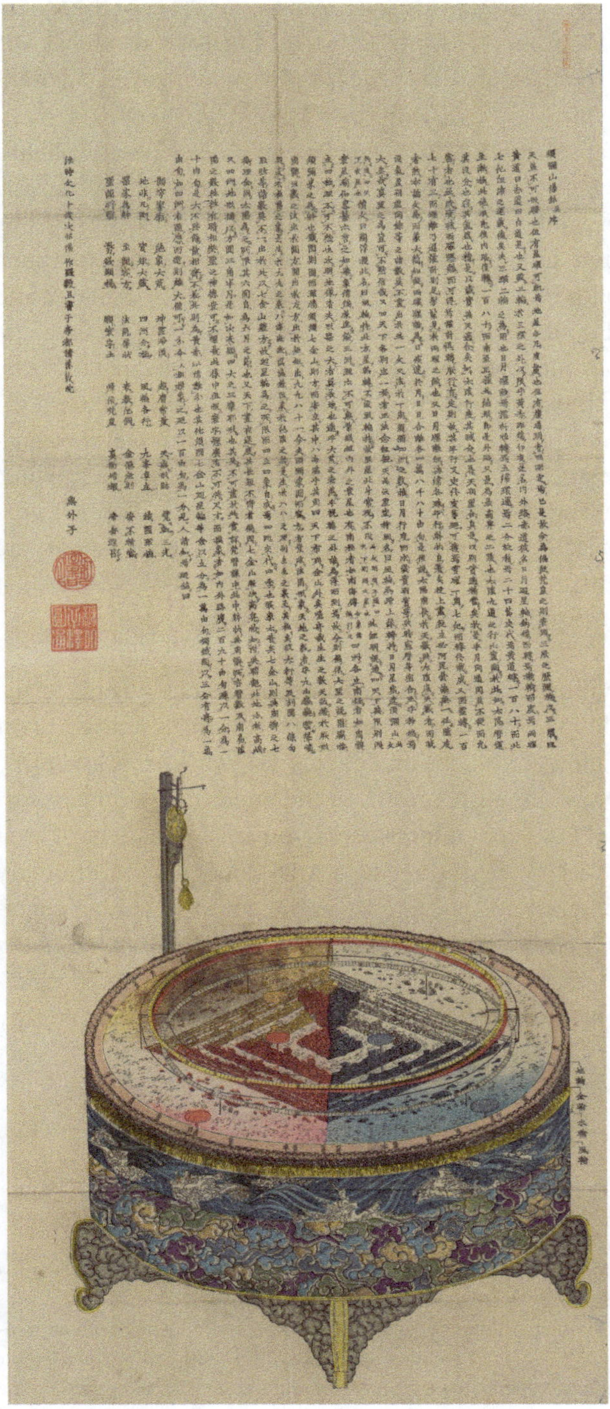

FIGURE 8.12
Entsū, *Model of Mount Sumeru Inscribed and Explained*, 1813. Color woodblock print, 137 × 62 cm
YOKOHAMA CITY UNIVERSITY LIBRARY

sion of his calculations were crucial to his defense of the accuracy of the Mount Sumeru world. "If the Mount Sumeru world is correct," Entsū wrote, "then it will be in accordance with calendrical calculations; if the calendrical calculations are correct, they will be in accordance with the solar and lunar eclipses."[34]

Entsū's second explanatory print, *Model of Relative Phenomena Explained* (*Shukushōgi setsu* 縮象儀説, 1814), demonstrates the movement of the sun and moon as they pass over Jambudvīpa (fig. 8.13). The *Model* represents just one quarter of the Mt. Sumeru universe, a ninety-degree section of the outermost sea surrounding the southern world continent. The layers of wind, water, and gold that support the Buddhist world are here abbreviated as decorative elements on the tripod that supports the device. At its center is a circular metal band, inscribed with astrological notations. It is bisected by three semicircular metal bands colored red, white, and gold, which mark the location of the equator and the path of the sun and moon; and by two thinner metal arcs, which mark the sun's orbit during the summer and winter solstice.[35] The viewer is nevertheless assured that "the astronomical and terrestrial measurements are based on a careful investigation of the sutras and commentaries."

In the accompanying inscription, Entsū explains that the device

> shows a single continent of the Mount Sumeru world as it is perceived by the power of human vision. … It represents the view of the Mount Sumeru world of nine mountains and eight seas, two wheels and three rings, as perceived by the sense organs, or the mind of an ordinary person. It is a view that Buddhists understand as limited. It is like the view of the world represented by celestial and terrestrial globes or by an armillary sphere, a physical object perceived through the power of human cognition. The *Model of Relative Phenomena* and the *Model of Mount Sumeru* are paired models of the Buddhist universe, similar to a pair of armillary spheres. They are constructed to correct the deluded view of ordinary people.[36]

Entsū contrasts this deluded view with the power and clarity of Buddhist vision, which is

---

34 Entsū, *Bonreki sakushin*, 23.

35 Yamada, "Ryūkoku daigaku Ōmiya toshokan shozō shukushōgi," 67.

36 The inscription is transcribed in Yamada, "Ryūkoku daigaku Ōmiya toshokan shozō shukushōgi," 64–65. The "two rings and three rings" refer to the sun's path during the summer and winter equinox and to the equator and the solar and lunar orbits identified by the metal bands and rings that arc over the surface of the device.

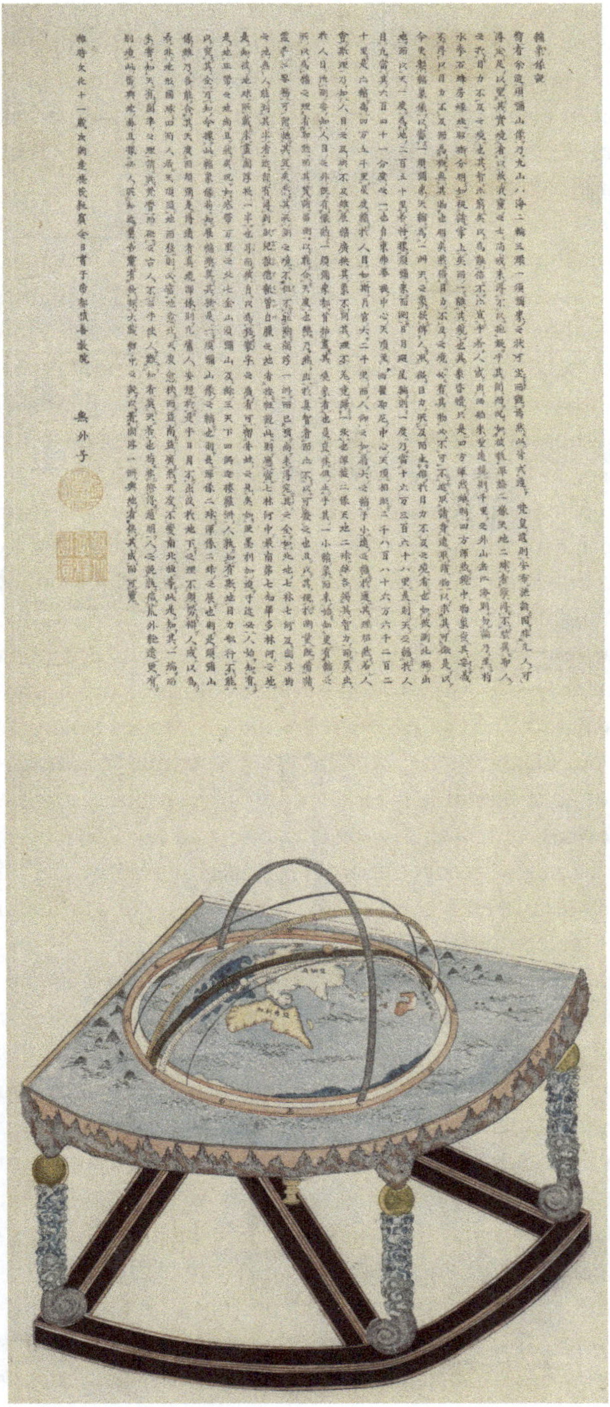

FIGURE 8.13
Entsū, *Model of Relative Phenomena Explained*, 1814. Color woodblock print, 129 × 53 cm
YOKOHAMA CITY UNIVERSITY LIBRARY

> like a person using a telescope imported from the West that allows them to see mountains and seas a great distance away, or to see the constellations as clearly as he might see an anthill, or as clearly as the palm of one's hand, entirely and in every direction. Even the entire world in the four directions appears as clear as an object in a mirror. This mode of vision cannot be compared to that of ordinary humans.

Deploying a Buddhist vocabulary of epistemological clarity—"as clearly as the palm of one's hand … as clearly as an object in a mirror"—Entsū echoes Hōtan's claims, made a century earlier. Both have produced visual prosthetics that allow one, in Hōtan's words, to "take in the entire world in a single glance; to visit distant places without ever traveling beyond one's garden gate; and to point out the various countries of the world just as easily as pointing out the stars in the night sky." Entsū, however, has radically expanded the scope of Hōtan's project from a geographic to a cosmographic scale. As Entsū explains, "The reduction of large to small and of far to near allows the world to be comprehended by human reason and human vision just as terrestrial and celestial globes allow one to comprehend phenomena that cannot otherwise be apprehended. For the majesty of a single Mount Sumeru world lies beyond the capacity of human vision and human understanding." Entsū, moreover, has formulated a Buddhist cosmology at once within and against the terms of the European critique: mathematical calculation, diagrammatic illustration, and empirical demonstration of such astronomical phenomena as solar and lunar eclipses. He has designed mechanical devices comparable to the terrestrial and celestial globes, armillary spheres, orreries, and telescopes in order to "clearly demonstrate," in the words of the Spanish Jesuit Pedro Gómez, "the machine of the world and the perpetual and immutable order of the heavens."

The world that appears on the surface of Entsū's *Model of Relative Phenomena*, however, is unlike any presented before. It is a hybrid world combining cartography of European origin with the cosmology of Buddhist texts. It depicts the three continents of Asia, Africa, and Europe, much as they appear on European-style maps, each named and printed in separate colors, compressed to approximate the traditional form of Jambudvīpa. "The seven forests and seven rivers in the northern area of Jambudvīpa," Entsū explains, "are what are now identified as Greenland and the polar regions." This unique reconfiguration of European cartographic conventions to approximate the shape of the Buddhist world-continent illustrates Entsū's claim that "the three continents of Asia, Europe, and Africa that appear on the maps that Westerners have brought to Japan are what is called Jambudvīpa."[37] If the *Model of Mount*

37 Entsū, *Bukkoku rekishō hen*, 3:6–7.

FIGURE 8.14
Entsū, *Model of Mount Sumeru*, 1824. Wood, metal, plaster, lacquer, color. Ryūshinji, Shizuoka
PHOTO COURTESY OF MASAHIKO OKADA

Sumeru demonstrates the structure and movement of the world as absolute phenomena, then the *Model of Relative Phenomena* demonstrates the workings of the world as relative phenomena. With these two mechanical instruments, Entsū seeks to reconcile the discrepancy between these two perspectives.

Dated to 1824, the earliest extant example of one of Entsū's astronomical models is housed at the Shizuoka temple of Ryūshinji 龍津寺 (fig. 8.14).[38] Commissioned by Entsū himself, or perhaps by a disciple, this articulated instrument is nearly identical to the model depicted in the painting and print of 1813. Powered by a single counterweight, an internal clockwork mechanism is calibrated to regulate the movement of solar and lunar orbs along metal rings, encircling a detailed model of Mount Sumeru, at the rate of one rotation a day. The clockworks further regulate the rotation of eight elevated discs, bearing representations of the constellations, so that the machine demonstrates both the passage of day and night and the change of the seasons. Such mechanical models of the Buddhist world would be refined and reproduced by Entsū's disciples for the next half century.

The mechanical model of Mount Sumeru designed by one of Entsū's students, the Rinzai monk Kanchū 環中, is represented in a color woodblock print published in 1848 (fig. 8.15). In style and format, it resembles Entsū's hanging scroll of 1813, but it represents changes that Kanchū had by then made to his master's design. The most prominent of these is the replacement of the pole-mounted counterweight and escapement mechanism with fully automated clockworks. Kanchū's designs were fabricated by Tanaka Hisashige—a pioneering engineer of automata, clockworks, steam engines, and telegraphy. These

38 For a comprehensive survey of this and related devices, see Okada, "Kindai Bukkyō to Shumisengi."

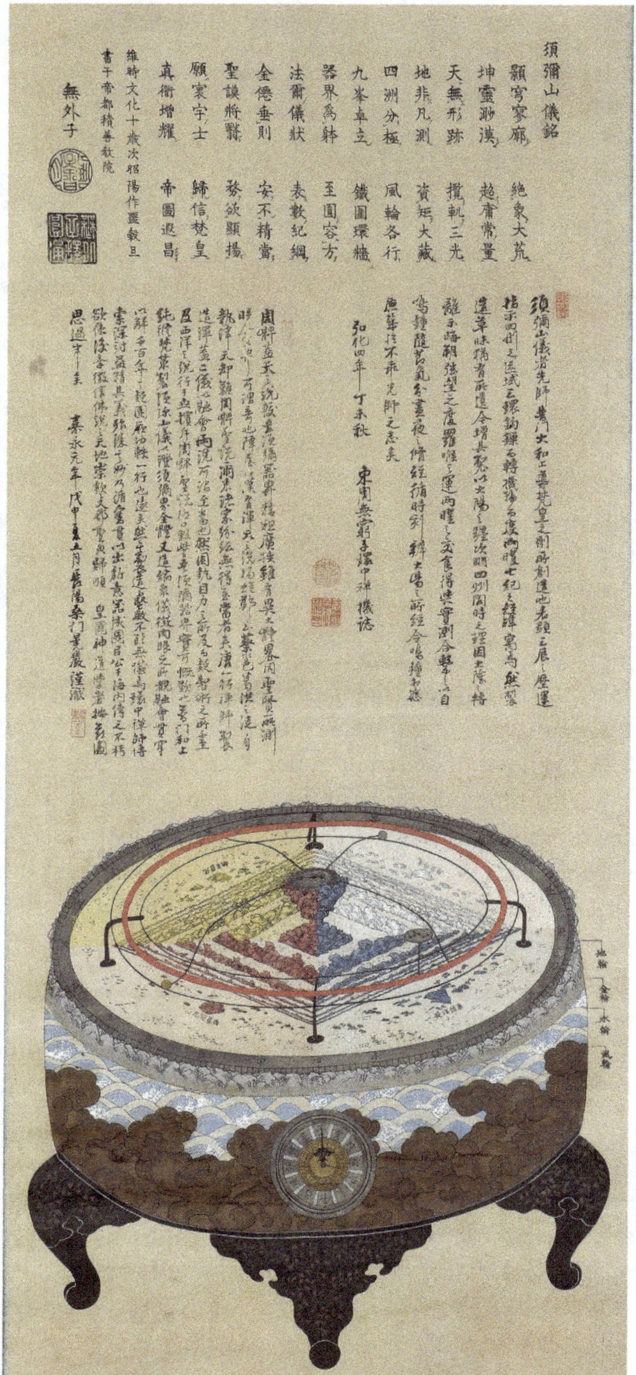

FIGURE 8.15
Kanchū, *Model of Mount Sumeru Inscribed*, 1848. Woodblock print with hand coloring, 118 × 54 cm
RYŪKOKU UNIVERSITY LIBRARY

FIGURE 8.16
Kanchū and Tanaka Hisashige, *Model of Mount Sumeru*, ca. 1847–1850. Wood, brass, lacquer, colors, 33×34 cm (including stand)
IMAGE FROM MODY, *A COLLECTION OF JAPANESE CLOCKS*

new, intricate machines shared the complex technologies of Tanaka's famous chronometers.[39] Made of wood, metal, crystal, lacquer, mother of pearl, and clockworks, Tanaka's Buddhist astronomical clocks were precision instruments (and luxury goods) that marked the movement of the sun and moon, the date and time, the twenty-four divisions of the solar year, the passage of the four seasons, and the location of the twenty-eight constellations.

Between 1847 and 1850, Tanaka produced a number of clockwork models of the Mount Sumeru universe.[40] The earliest example, no longer extant, shows the phases of the moon, the rotation of the sun, the movement of the stars, the ebb and flow of the tides, and the twenty-four seasonal periods (fig. 8.16).[41] On one side of the base is a metal clock face with hands and dial, and the entire instrument is operated by a large clockwork mechanism housed within the base. In a second version from 1850, now in the Seiko Museum, a large brass key is provided to wind the clockworks (fig. 8.17). A third version, in the collection of the Toshiba Science Museum, has three additional black metal rings, which indicate the orbits of the sun during the solstices and equinoxes (fig. 8.18). Even more elaborate and twice the size, a final version, signed by Tanaka and dated to 1850, is in the collection of the Ryūkoku University Library (fig. 8.19).

39 On the clockworks of Tanaka Hisashige, see Hashimoto, "Mechanization of Time."
40 Tanaka Ōmiō Kenshokai, *Tanaka Ōmi daien*, 55–56.
41 The instrument, formerly in the collection of Higashi Honganji, is reproduced in Mody, *Collection of Japanese Clocks*, plate 111, figs. 1–3, 40–42.

FIGURE 8.17 Kanchū and Tanaka Hisashige, *Model of Mount Sumeru*, ca. 1847–1850. Wood, brass, lacquer, glass, colors. Seiko Museum
PHOTOGRAPH BY AUTHOR.

FIGURE 8.18
Attributed to Kanchū and Tanaka Hisashige, *Model of Mount Sumeru*, 1850. Wood, brass, glass, lacquer, gold, and colors. Toshiba Science Museum
PHOTO PROVIDED BY THE TOSHIBA SCIENCE MUSEUM

FIGURE 8.19
Kanchū and Tanaka Hisashige, *Model of Mount Sumeru*, 1850. Wood, brass, lacquer, glass, mother-of-pearl, colors, 55 × 66.5 cm
RYŪKOKU UNIVERSITY LIBRARY

In 1850 as well, Kanchū produced a second matching scroll, entitled *Model of Relative Phenomena Illustrated* (*Shukushōgizu* 縮象儀図, fig. 8.20). Bearing an even longer inscription from Entsū, it represents Kanchū's modifications of Entsū's 1814 *Model of Relative Phenomena*. Compared with Entsū's image of a simple tabletop model produced thirty-five years earlier, Kanchū's image shows a fully mechanized instrument, which was manufactured by Tanaka as well, set within the same clockwork base of Tanaka's final *Shumisengi* (fig. 8.21). In Tanaka's updated device, the fixed surface of Entsū's map has been replaced by a swiveling disc, across which the passage of the sun and moon are calibrated to the twenty-four seasonal divisions of the year. The geography of Jambudvīpa has changed as well. Kanchū has retained Asia, Africa, and Europe, as depicted on Entsū's *Model of Relative Phenomena*, but here they have been fused into a single great continent and moved to the northwest. Regnant at the very center of the map, enlarged to the size of Africa, is the Japanese archipelago. To the south is a group of islands and a large continent colored a pale pink, to the west is a long narrow continent, perhaps the Americas, colored green. Twelve lines, drawn and annotated, radiate out to the eastern and western edges of the disc. The lines converge at the exact center of the world map, the orientation point of astronomical observation, the imperial capital of Kyoto.[42]

42 The *Model of Relative Phenomena* that Tanaka produced for Kanchū is also in the Ryūkoku collection. The map of the world on the surface of the disc is no longer extant, but the device is otherwise identical to Kanchū's print.

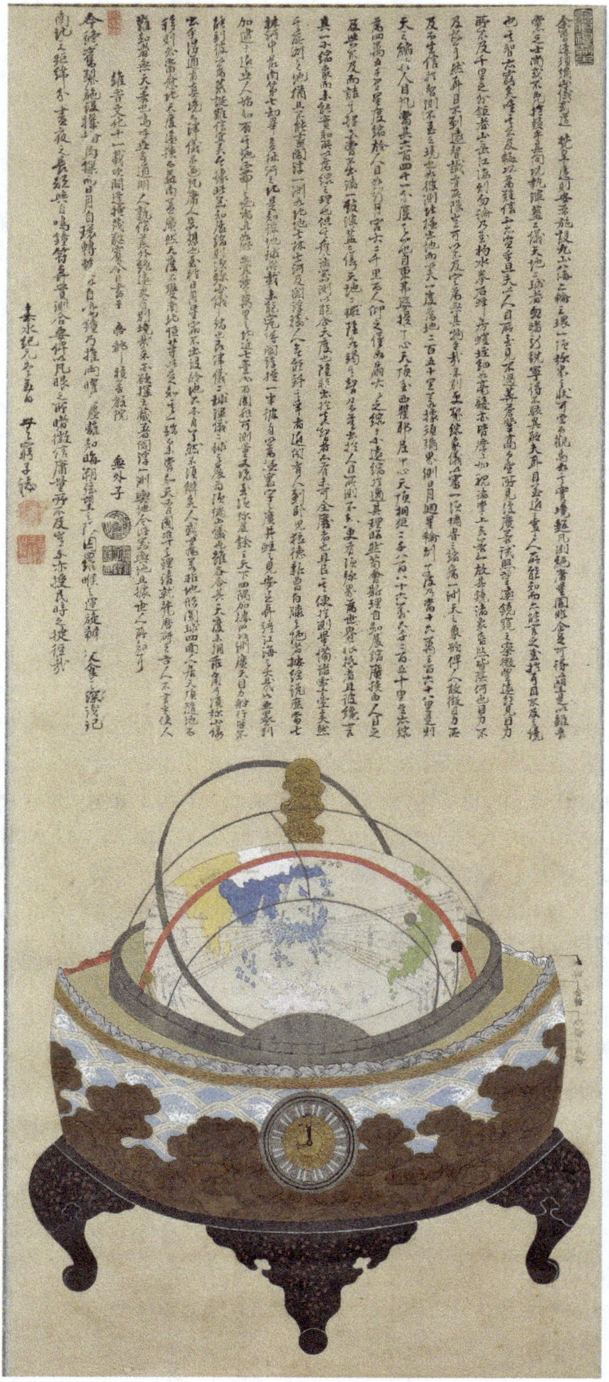

FIGURE 8.20
Kanchū, *Model of Relative Phenomena Illustrated*, 1848. Color woodblock print, 118 × 54 cm
RYŪKOKU UNIVERSITY LIBRARY.

FIGURE 8.21
Kanchū and Tanaka Hisashige, *Model of Relative Phenomena* 1847–1850. Wood, brass, lacquer, glass, mother-of-pearl, colors, 64.5×64.5 cm
RYŪKOKU UNIVERSITY LIBRARY

Sada Kaiseki 佐田介石 (1818–1882), a True Pure Land monk and student of Kanchū, designed his own mechanical model of the Mount Sumeru universe and had it produced by in 1855 by Tanaka Hisashige, the engineer who had previously produced Kanchū's devices. Sada's astronomical clock is more streamlined, and less topographically literal, than the mechanical Mount Sumerus of Entsū and Kanchū. The axial mountain has now been reduced to a tall central shaft around which a ring of heavenly bodies orbit. The four great continents surrounding Sumeru are represented by hemispherical cages, indicating the domes of air which, according to Sada, obscure the true position and movement of the heavenly bodies. His first model was destroyed in a fire, but he published diagrams for a second device in 1877 and had it produced by Tanaka the same year (fig. 8.22).[43] Sada's device was proudly exhibited at Japan's First National Industrial Exhibition, held in Tokyo's Ueno Park in the summer of 1877.

In designing their instruments, Entsū and Kanchū had recognized the inherent difficulty of depicting and explaining the discrepancy between the apparent and the true forms of the world. A certain aporia characterizes their models of the Mount Sumeru universe in its relative and absolute forms. It is as if their models, like the modes of visuality they represent, must remain apart, as a

43 It is illustrated in Sada, *Shijitsu tōshōgi ki*, 8a, 17a, 18b. A later edition was issued in 1880, entitled *A Detailed Explanation of a Device to Represent the Equivalency of the Apparent and the Real* (*Shijitsu tōshōgi shōsetsu* 視実等象儀詳説).

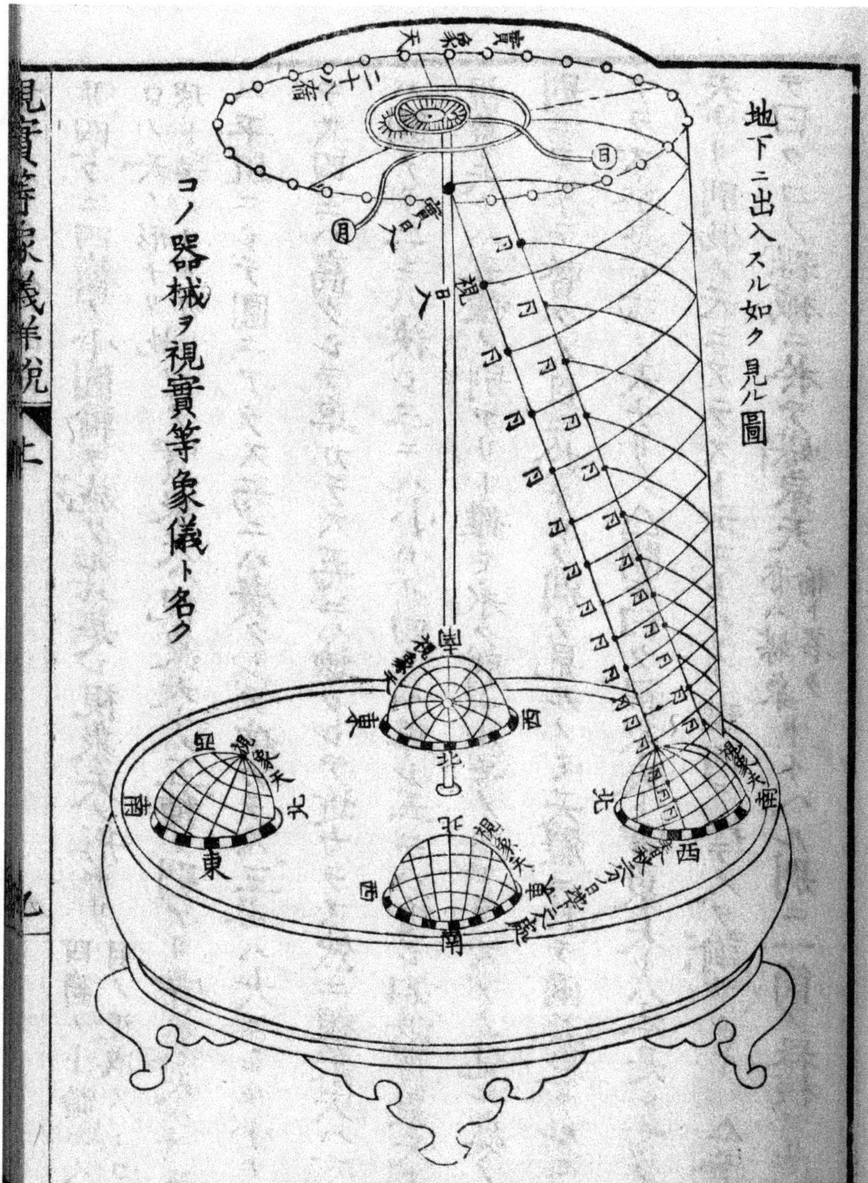

FIGURE 8.22 Sada Kaiseki, *A Detailed Explanation of a Device to Represent the Equivalency of the Apparent and the Real*, 1880. Monochrome woodblock book illustration
AUTHOR'S COLLECTION

kind of double vision that cannot easily be brought into focus. Sada followed a similar understanding of Buddhist cosmology and visuality. He explained that "astronomy and geography could be perceived in two ways: as apparent phenomena (*shishō* 視象), and as true phenomena (*jishō* 実象)." Apparent phenomena were, he claimed, "things as they appear" (*minasu koto ni*), and true phenomena, "the form of things as they are" (*ari no mama no katachi*).[44] Like Entsū, Sada distinguished apparent phenomena, "what appears to the human eye and is explained according to celestial and terrestrial globes," from true phenomena, "what the sages explain as the flat earth and the horizontal orbit of the sun and moon."[45] Although Sada's mechanical orrery was created to render visible the workings of the Buddhist universe, the very name of the invention underscores the epistemological gap implicit in the Buddhist theory of vision. Entitled a *Device to Represent the Equivalency of the Apparent and the Real* (*Shijitsu tōshōgi*), it was designed to bridge the chasm between the appearance of the cosmos and its ultimate reality. Sada's device thus sought at once to represent and to reconcile the field of difference between the paired instruments of Entsū and Kanchū.

Like Hōtan and Entsū before him, Sada understood the differences between Buddhist and Western views of the world in the terms of classical Buddhist theories of vision. "Mount Sumeru cannot be apprehended from the perspective of the physical eye of ordinary humans," explained Sada. "The true phenomena of the Mount Sumeru universe may be perceived not with the physical eye, but only with the heavenly eye. This refers not to the eyeball, but rather to spiritual perception."[46] According to Sada, "the vision of the heavenly eye is unobstructed by any barriers. Even microscopes and telescopes cannot compare to its power of vision. The Buddha Śākyamuni acquired it only through ascetic practice. It is the ability to see all—from the heavens of Mount Sumeru above to the hells below—as clearly as an object held in the hand."[47] Sada's advocacy of Buddhist vision over that of European optical devices shares not only the arguments but also the vocabulary of his predecessors. His description of the Buddha's visual command of the entire universe, "seen as clearly as an object held in the hand," had appeared in the title and preface of Hōtan's world map. Sada's distinction between the normal perception of ordinary humans and the supernormal perception enjoyed by Buddhas and advanced medita-

44 Sada, *Tsuchi chikyū setsuryaku*, 2:47b.
45 Sada, *Tsuchi chikyū setsuryaku*, 2:51b.
46 Sada, *Shijitsu tōshōgi shōsetsu*, 1:2b–3a.
47 Sada, *Shijitsu tōshōgi shōsetsu*, 2:11a–b.

tors adopts the same classical taxonomy of vision outlined by Entsū in the final chapter of his *Astronomy of Buddhist Countries*.

By the late-nineteenth century, classical Buddhist theories of vision had joined the modern discourse on optics and the mechanics and technologies of human perception. Sada's references to microscopes and telescopes were more than rhetorical flourishes. For the previous two hundred years, Japanese Buddhist representations of the earth and the universe had to contend with foreign technologies and modes of representation: longitudes and latitudes, Ptolemaic and Mercator projections, terrestrial and celestial globes, orreries and planispheres. The mechanical models of Buddhist cosmology, which Tanaka Hisashige manufactured for Kanchū and Sada, are marvels of modern technology. Founder of what has become the multinational electronics conglomerate, Toshiba Corporation, Tanaka stood at the forefront of contemporary Japanese science and engineering. He first gained recognition as the foremost inventor of clockwork-driven automata—mechanized figures that serve tea, write calligraphy, and perform archery—and later developed the first domestically produced steam locomotives, warships, munitions, lighting, and telegraphic equipment for the Meiji state. The fact that the greatest engineer of the age, responsible for the most technologically sophisticated, westernized, and industrialized modes of transportation, communication, and warfare, also produced mechanized devices to prove the accuracy of ancient Indian cosmology reveals that Buddhist cosmology in nineteenth-century Japan was anything but atavistic. Rather, it was a self-consciously contemporary discourse, cognizant of the range of astral science of the ancient and modern world and articulated through the latest forms of technical knowledge.

From the sixteenth-century assertion of the Ptolemaic and Aristotelian world by Jesuit missionaries to the nineteenth-century re-assertion of the Mount Sumeru world by Buddhist monastics, Japanese monks used maps, models, and machines to defend their worldview against the onslaught of Western cosmology. This should be seen not a rejection of scientific discourse but rather as its appropriation. As Donald Lopez has argued, "that these events occurred in the course of Christian missions to Buddhist Asia suggests that Buddhist claims about science originated in polemic, with Buddhists arguing that their religion is not superstition but science."[48] The Japanese Buddhist defense of their world, from Hōtan's map of Jambudvīpa to Sada's mechanical Mount Meru, was far from unique. Rather, it belonged to a wider phenomenon of overlapping cosmologies across Asia in which local Buddhist leaders con-

48 Lopez, *Buddhism and Science*, xi.

fronted the forces of Christian colonialism and scientific modernism. Like Migettuwatte Guṇānanda (1823–1890) and Anagarika Dharmapala (1864–1933) in Ceylon, Tendzin Trinlé (1789–1838) and Gendün Chöpel (1903–1951) in Tibet, and Taixu (1890–1947) in China, Japanese monks redefined cosmology through the construction of a scientific Buddhism and the formulation of indigenous and alternative modernities.

Bibliography

Akin, Alexander Van Zandt. "Printed Maps in Late Ming Publishing Culture: A Trans-Regional Perspective." PhD diss., Harvard University, 2009.

Baskin, James and Richard Bowring, eds. *The Myōtei Dialogues: A Japanese Christian Critique of Native Traditions*. Leiden: Brill, 2016.

Buswell, Robert E. Jr. and Donald S. Lopez Jr, eds., *Princeton Dictionary of Buddhism*. Princeton: Princeton University Press, 2014.

Chen, Hui-hung. "The Human Body as a Universe: Understanding Heaven by Visualization and Sensibility in Jesuit Cartography in China." *The Catholic Historical Review* 93, no. 3 (July 2007): 517–552.

Cleary, Thomas, trans. *Entry into the Realm of Reality: A Translation of the* Gandavyuha, *the Final Book of the* Avatamsaka Sutra. Boston: Shambala, 1989.

Costelloe, M. Joseph S.J., trans. *The Letters and Instructions of Francis Xavier*. St. Louis: The Institute of Jesuit Sources, 1992.

D'Elia, Pasquale M. "Recent Discoveries and New Studies (1938–1960) on the World Map in Chinese of Father Matteo Ricci SJ." *Monumenta Serica*, vol. 20 (1961): 82–164.

De Weerdt, Hilde. "Maps and Memory: Readings of Cartography in Twelfth- and Thirteenth-Century Song China." *Imago Mundi* 62, part 2 (2009): 145–167.

Entsū 円通. *Bonreki sakushin* 梵暦策進. N.p., 1816.

Entsū 円通. *Bukkoku rekishōhen* 佛國暦象編. 5 vols. Osaka: Yamato Yashirobei 大和屋四郎兵衛, 1815.

Entsū 円通. *Jikken shumikai setsu* 實驗須彌界説, 3 vols. Kyoto: Nakashima Rizaemon 中島利左衛門, 1821.

Giles, Lionel. "Translations from the Chinese World Map of Father Ricci." *The Geographic Journal* 52, no. 6 (1918): 367–385.

Hiraoka Ryuji 平岡隆二. "Jesuit Cosmological Textbook in 'Christian Century' Japan: *De sphaera* of Pedro Gómez (Part I)." *SCIAMVS* 6 (2005): 99–175.

Hiraoka Ryuji 平岡隆二. *Nambankei uchuron no genteki kenkyū* 南蛮系宇宙論の原典的研究. Fukuoka: Hana shoin, 2013.

Iguchi Tsunenori 井口常範. *Tenmon zukai* 天文圖解, 5 vols. Ōsaka: Kashiwaharaya Yozaemon, Itamiya Mohē 柏原屋與左衛門, 伊丹屋茂兵衛, 1689.

Juneja, Monica and Christian Kravagna. "Understanding Transculturalism." In *Transcultural Modernisms*, edited by Fahmin Amir et al., 22–33. Berlin: Sternberg Press, 2013.

Kirishitan Bunko Library, ed. *Compendia Catholicae Veritatis*. 3 vols. Tokyo: Ozorasha, 1997.

Lopez, Donald S. Jr. *Buddhism and Science: A Guide for the Perplexed*. Chicago: University of Chicago Press, 2008.

Lynn, Richard John, trans. *The Classic of Changes*. New York: Columbia University Press, 1994.

Lynn, Richard John, trans. *The Classic of the Way and Virtue*. New York: Columbia University Press, 1999.

McMahan, David L. *Empty Vision: Metaphor and Visionary Imagery in Mahayana Buddhism*. London: Routledge Curzon, 2002.

Mody, N.H.N. *A Collection of Japanese Clocks*. London: Kegan Paul, 1932.

Monnō 文雄. *Kusen hakkai gechōron* 九山八海解嘲論. N.p., 1754.

Mori Shōken 森尚謙. *Gohō shijiron* 護法資治論, 10 vols. Kyōto: Sawada Kichizaemon 澤田吉左衛門, 1766.

Miyoshi Tadayoshi. "Japanische und europäische Kartographie vom 16. bis zum 19. Jahrhundert." In *Japan und Europa, 1543–1929*, edited by Doris Croissant and Lothar Ledderose, 37–45. Berlin: Argon, 1993.

Nishimura Ryō 西村玲. "Nihon ni okeru shumisen ronso no tenkai" 日本における須弥山論争の展開. *Indogaku Bukkyōgaku kenkyū* 61, no. 2 (2013): 679–680.

Nishimura Ryō 西村玲. "Shuminsen to chikyūsetsu" 須弥山と地球説. In *Shizen to jin'i: "shizen" kan no hen'yō* 自然と人為:「自然」観の変容, edited by Karube Tadashi 苅部直 et al., 117–141. Tokyo: Iwanami Shoten, 2013.

Obara Satoru 尾原悟, "Kirishitan jidai no kagaku shisō: Pedoro Gomesu-cho Tenkyūron no kenkyū" キリシタン時代の科学思想：ペドロ・ゴメス著「天球論」の研究. *Kirishitan Kenkyū* キリシタン研究, vol. 10 (1965): 101–178.

Oda Takeo 小田嶽夫, and Kunikida Akiko 国木田明子, eds. *Kochizu ni miru sekai to Nihon* 古地図にみる世界と日本. Kobe: Kobe Shiritsu Hakubutsukan, 1983.

Okada Masahiko 岡田正彦. "Kindai Bukkyō to shumisengi: kindaiteki shizenkan to Bukkyō" 近代仏教と須弥山儀: 近代的自然観と仏教. *Nihon Bukkyō Gakkai Gakujitsu Taikai Happyō* 日本仏教学会学術大会発表 70 (09/03/2011): 1–12.

Okada Masahiko 岡田正彦. *Wasurerareta Bukkyō tenmongaku: jūkyū seiki no Nihon ni okeru Bukkyō sekai zō* 忘れられた仏教天文学: 十九世紀の日本における仏教世界像. Nagoya: Buitsu Soryushon, 2010.

Pinto dos Santos, José Miguel. "A Study in Cross-Cultural Transmission of Natural Philosophy: The Kenkon bensetsu." PhD diss., Universidade Nova de Lisboa, 2011.

Ryūkoku Daigaku Ōmiya Toshokan 龍谷大学大宮図書館, ed. *Bukkyō no uchūkan* 仏教の宇宙観. Kyoto: Ryūkoku Daigaku Ōmiya Toshokan, 2009.

Sada Kaiseki 佐田介石. *Shijitsu tōshōgi ki: ichimei tenchi kyōwagi ki* 視実等象儀記: 一名天地共和儀記. Tokyo: Fujita Furūme 藤田古梅, 1877.

Sada Kaiseki 佐田介石. *Tsuchi chikyū setsuryaku* 鎚地球説略. 2 vols. N.p., 1862.

Schurhammer, G., and I. Wicki, ed. *Epistolae S. Francisici Xaverii aliaque eiusscripta*. 2 vols. Rome: Monumenta Historica Societatis Iesu, 1996.

Takehiko Hashimoto. "The Mechanization of Time and the Calendar: Tanaka Hisashige and His Myriad Year Clock and Cosmological Models." *University of Tokyo Center for Philosophy Bulletin*, vol. 6 (2006): 47–55.

Tanaka Ōmiō Kenshokai 田中近江翁顕彰会編, ed. *Tanaka Ōmi daien* 田中近江大掾. Kurume: Tanaka Ōmiō kenshokai, 1931.

Terajima Ryōan 寺島良安. *Wakan sansai zue* 倭漢三才圖會. 107 vols. Okada Saburoemon, 1713–1715.

Üçerler, M. Antoni. "Jesuit Humanist Education in Sixteenth-Century Japan: The Latin and Japanese MSS of Pedro Gómez's Compendia on Astronomy, Philosophy, and Theology (1593–1595)." In *Compendia Catholicae Veritatis*, 11–60.

Unno Kazutaka 海野一隆. *Nihonjin no daichzō: seiyō chikyū no joyō wo megutte* 日本人の大地像—西洋地球説の受容をめぐって. Tokyo: Taishukan, 2006.

Wayman, Alex. "The Buddhist Theory of Vision." In *Buddhist Insight: Essays by Alex Wayman*, edited by George R. Elder, 153–161. Delhi: Motilal Banarsidass, 1984.

Yamada Keiji 山田慶兒. "Ryūkoku Daigaku Ōmiya toshokan shozō shukushōgi—ei yaku oyobi mokei—ni tsuite" 龍谷大学大宮図書館所蔵縮象儀—図・説および模型—について. *Nihon kenkyū* 16 (September 30, 1997): 59–71.

CHAPTER 9

Buddhist Cosmology in Bhutanese Murals: A Visual Negotiation Between *Abhidharma* and Kālacakra Systems

Eric Huntington

1 Introduction

Cosmology includes a wide range of subjects, from the processes of creation and destruction of the universe to the physical principles that allow it to operate and the roles of various places, objects, and beings within the cosmic system. In religious contexts, these topics take on special meaning, since one of the central functions of religion is to make sense of the world by defining its workings and especially the role of humans within it. Nowhere is this more true than in South Asian and Himalayan Buddhism, where even the highest forms of enlightenment can be described as insights into the true nature of reality and the hidden mechanisms of the universe.[1] At the same time, cosmological knowledge is also crucial for the most ignorant, since it can help motivate lay people and non-Buddhists to follow the Buddhist path. If one truly understands one's place in the world as Buddhists describe it, so the logic goes, one naturally adopts the Buddhist path as the most beneficial reaction to this situation.

Given such importance of cosmology in Buddhism, cosmological concerns are not just a matter for astronomers, philosophers, and intellectuals. Rather, they become one of the main methods of interface between the world outside Buddhist practice and the world within it. This relationship has been described for the well-known cosmological image of the wheel of existence (*bhavacakra, saṃsāracakra*), which depicts the workings of the cycle of endless rebirth in *saṃsāra*, its causes and effects, and especially the promise of escape from inevitable suffering provided by the Buddhist path.[2] Paintings of this subject are frequently depicted at the entrances to monasteries with the purpose of educating visitors about their existence within the cosmos as a means of introducing them, both ideologically and physically, to the space of Buddhist practice within the monastery.

---

1 Strong, *The Legend of King Aśoka*, 147.

2 Teiser, *Reinventing the Wheel*.

 | DOI:10.1163/9789004511675_011

This is only part of the story, however, since equally common at monastic entrances are murals that depict a model of the Buddhist physical cosmos. While the wheel of existence emphasizes existential concerns of *karma* and rebirth, these other murals depict the universe as a geographic space, with oceans, continents, mountains, and heavens all mapped into an explicit order. These murals vary immensely, conveying different understandings of the nature of the world, its creation, functioning, and significance as a public expression of Buddhist cosmology. Broadly speaking, such variations often arise due to differences in artistic lineage, ritual function, or didactic message.[3] Further complicating the issue, however, is the fact that Buddhism contains multiple, different authoritative models of the cosmos.

The two most prominent Himalayan Buddhist cosmologies are based respectively on the early Buddhist *abhidharma*, a kind of pre-tantric natural philosophy of the world, and the later Buddhist Kālacakra tantric system, a highly complex description that underlies esoteric rituals and meditations for enlightenment. Developed with radically different agendas, these two cosmologies do not simply offer alternative accounts of shared geography or astronomical observations but rather define the very constituents of reality and the natural laws that make sense of specific systems of philosophy and ritual. So obviously different are they in even their most basic facts that early commentators took pains to explain how they could both be considered authentic Buddhist teachings, and many great authors since have characterized, justified, and challenged their numerous contradictions.[4]

Such commentaries are, for both practitioners and scholars, the obvious place to turn in attempts to understand relationships between these two competing cosmological traditions, but the prominence of imagery of the cosmos on monastic walls allows for an alternative approach. Isolating pictorial examples where both cosmologies occur together—apparently painted by the same artists at the same times on neighboring monastic walls—one can identify the particular visual languages of these cosmologies within local traditions and establish these extraordinarily public and visible murals as sites of negotiation between conflicting authoritative models of the world. Indeed, the remarkable compromises and conflations in these murals exemplify a uniquely visual way of mediating contradictory cosmologies that does not occur in texts or other media. By mixing the visual characteristics of competing models of the cosmos, these paintings obviate differences in the particulars to reveal deeper relationships between contradictory cosmologies.

3 Huntington, *Creating the Universe*, 190–225.

4 Newman, "Outer Wheel of Time," 474–475; Tayé, *Myriad Worlds*.

Several important modern murals in Bhutan perfectly exemplify the clear mixing of visual languages from these two distinct cosmological traditions. In fact, Bhutanese architecture is perhaps unique in the Buddhist world for so completely adopting both the *abhidharma* and Kālacakra models for their murals. The intimate connection between these two cosmologies in Bhutan presents a rare opportunity to see how the competing logics of the two systems are accommodated in public and visual ways. While most analyses of these systems have relied upon their primary texts and commentaries, these visual records of cosmological thinking provide qualitatively different insights into the negotiations and assimilations that can occur between these two competing cosmologies.

The following analysis proceeds in three steps. First, the *abhidharma* and Kālacakra cosmic models are introduced by summarizing their descriptions in canonical texts, focusing on the features that appear in visual depictions. Then a brief overview illustrates some of the kinds of variations that are common in visual depictions of the cosmos throughout the Himalayas, giving a sense of the range in which the Bhutanese murals lie. Finally, in the main argument, a detailed formal and iconographic analysis of several specific murals parses their visual variations as ways of mediating the two compelling cosmologies being depicted side-by-side.

2 Two Textual Traditions

Before delving into the visual imagery, it is important to understand the details and differences of these two cosmological systems as recorded in authoritative texts. While it is quite clear that these texts were not always consulted in the production of cosmological imagery, and further that they were themselves based on preexisting traditions of cosmology, philosophy, and ritual, they do reveal the underlying logic behind each cosmological model. This logic—the structuring principles of each universe—is precisely what is negotiated in the murals.

The *abhidharma* cosmic model is most closely associated Vasubandhu's *Abhidharmakośa*, a fourth- to fifth-century text that reflects attitudes towards Buddhist science from a time when many Indic traditions were concretizing their cosmologies based on earlier precedents.[5] It still remains one of the

5 For more on the developments of Indic cosmology during this period and relationships between different textual traditions, see Sircar, *Cosmography and Geography*.

primary canonical sources for Buddhist cosmology,[6] and its study is a major stage in monastic education, representing the culmination of exoteric analysis of the world before proceeding to tantric systems.[7] Conveying a deeply analytic overview of the natural world (including the functioning of the elements, *karma*, mental activity, and the stages of meditation), it plays a major role in Mahayana discourse as well as being a basis for entry into Vajrayāna practice.

The *Abhidharmakośa*'s physical description of the cosmos begins with a minimalist account of its creation, which also explains the basic structural features that appear in visual depictions (fig. 9.1).[8] Within the void of space, the intentional actions (*karmas*) of sentient beings cause a great circle of wind (the most subtle of elements) to arise. In the upper reaches of this wind, clouds condense and precipitate into a cylinder of water millions of miles across. This water is churned by the motions of the winds until a dense layer of gold floats on top, like butter churned from milk. It is this disc of gold that forms the substrate for our inhabited world, which sits on the flat, circular, upper surface (fig. 9.2). At the center of the surface of this disc is the enormous mountain Meru, which serves as the home of various gods. This is surrounded by successively smaller square perimeters of mountains and oceans, and then finally by a vast ocean and the four major, but relatively small, continents in the four cardinal directions. Our human civilization resides on the southern continent, Jambudvīpa. Below Jambudvīpa are the hells where evil beings are reborn, and above Meru are the heavens where one finds ever greater and more ethereal beings with increasing altitude. The sun and moon are said to orbit Meru at half its height, with night created when the sun is hidden behind the enormous mountain from our local perspective. This entire world-system is delimited by a circle of iron mountains, called the Cakravāla, at the very edge of the surface of the disc. This mountain range is considered the barrier between the world and the non-world, and this edge or perimeter figures crucially in the analysis of the Bhutanese murals.

The Kālacakra cosmos shares features with the *abhidharma* version in consisting of a disc-shaped world surrounded by mountains with the enormous Meru in the center, but it also represents a radical redesign of Buddhist cosmology for Vajrayāna ritual purposes, along with significant borrowings from

6 Extended scholarly treatments of the cosmology of the *Abhidharmakośa* can be found in Kloetzli, *Buddhist Cosmology*, 24–50; and Sadakata, *Buddhist Cosmology*.

7 Dreyfus, *The Sound of Two Hands Clapping*, 114–118.

8 The following description relies on La Vallée Poussin, *L'Abhidharmakośa de Vasubandhu*, vol. 2; La Vallée Poussin, *Abhidharmakośabhāṣyam*, vol. 2; and Vasubandhu, *Abhidharmakośam*.

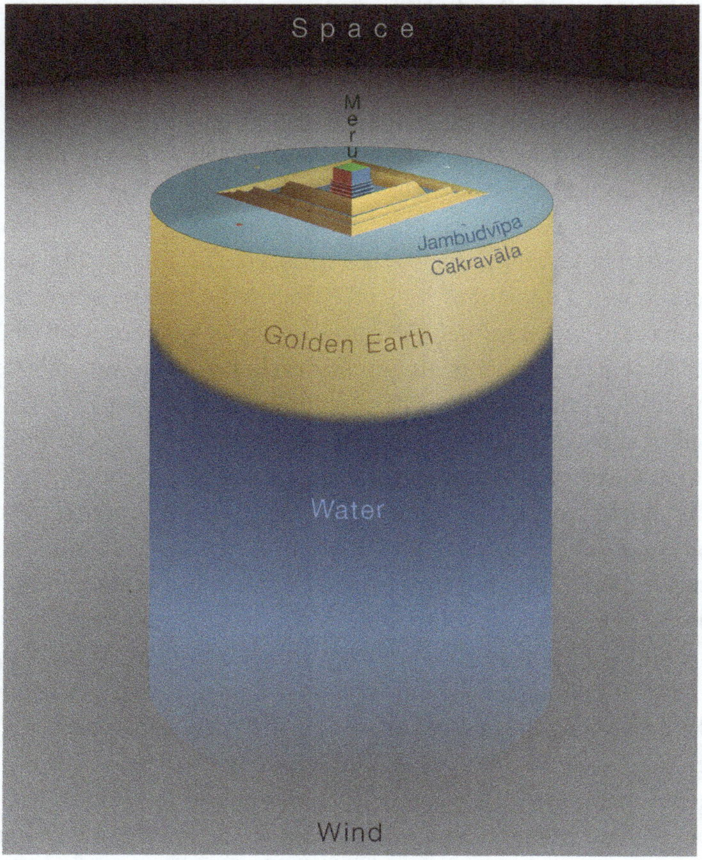

FIGURE 9.1 Perspective view of the *abhidharma* cosmos

other Indic traditions, especially Jainism.[9] This model is based on the literature of the *Kālacakra Tantra*, which originated in India in the early eleventh century and was quickly transmitted to Tibet and other regions.[10] One central premise of the Kālacakra system is that it is meant to articulate a fully consistent and holistic explanation of all of reality, from the structure of the cosmos to the workings of the body and the functioning of meditation.[11] Unlike the *Abhidharmakośa*, however, which does not address Vajrayāna ideology, the Kālacakra must also incorporate the full spectrum of advanced tantric practice. In order to accomplish this complex task, its analysis operates on many separate

9 Newman, "Outer Wheel of Time," 130; Wallace, *The Inner Kālacakratantra*, 70.
10 Newman, "Outer Wheel of Time," 75.
11 Berzin, *Taking the Kalachakra Initiation*, 41–42.

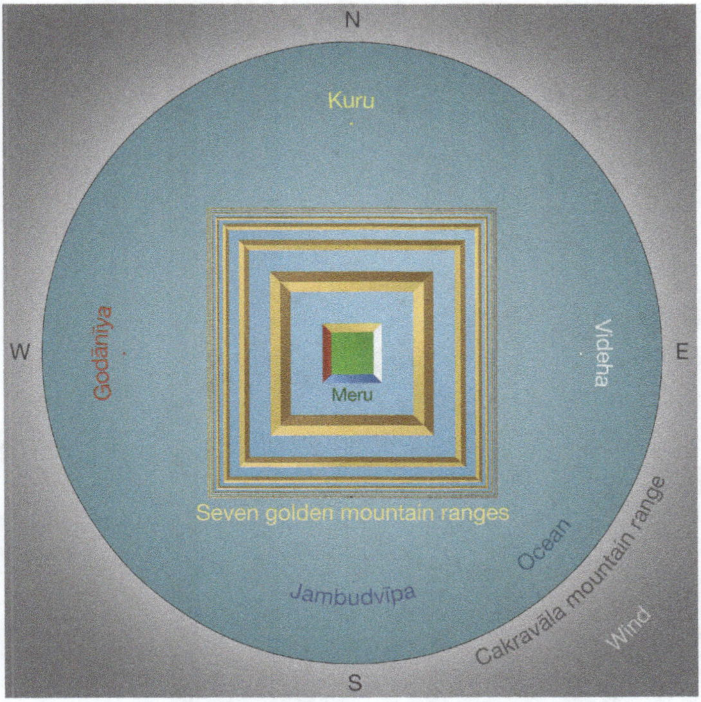

FIGURE 9.2 Plan view of *abhidharma* cosmos

levels of symbolism, so that each subject can potentially be understood in terms of the others. All aspects of reality are structured according to comparable systems of logic, with parallel explanations being divided into the basic categories of Outer (relating to the physical world), Inner (concerning the human body), and Other (describing the purification of the former through meditation).[12]

In the Outer Kālacakra, the features of the physical universe that commonly appear in visual imagery are described as follows (fig. 9.3):[13] Within vast space lies a circle of wind. Above this lie concentric and successively smaller elemental circles of fire, water, and earth. Each of these cylinders is divided into abodes where various sorts of beings dwell, and these layers taken as a whole are considered to correspond to the human figure below the waist. From the center of the surface of the earth rises enormous Meru, which is compared to the human spine. Above this are various additional realms and heavens compared to the

12 Jackson, "Kalachakra in Context," 31.

13 The following description is based largely on Gyatso, *Ornament of Stainless Light*, and Newman, "Outer Wheel of Time."

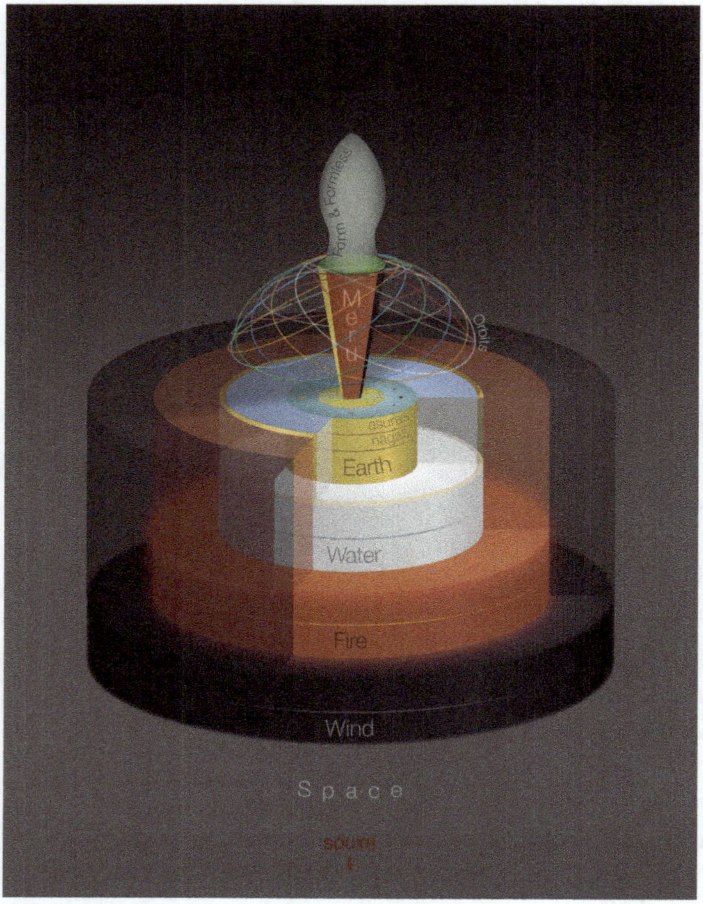

FIGURE 9.3 Perspective view of the Kālacakra cosmos

neck and head.[14] Such correspondences between world and body suggest an identity of macrocosm and microcosm that has powerful implications for meditative practice.[15] Surrounding Meru on the disc of land are concentric rings of mountains and oceans surrounded by a great circular continent, known as Greater Jambudvīpa, which is divided radially into twelve sections. Each of these sections is filled with oceans and has within it one lesser continent, upon which various beings live. Our human home is the peripheral continent known in this system as Lesser Jambudvīpa. In order to explain how the great

14 An illustration of these correspondences can be found at Brauen, *Mandala*, 156–157.

15 For more on the connections between Outer and Inner Kālacakra models of the cosmos and human body, see Wallace, *The Inner Kālacakratantra*, 64–108.

salt ocean may surround the inhabited continents, some commentators suggest that the unobstructed portions of the water, fire, and wind circles rise up to meet the level of the surface of the earth (Greater Jambudvīpa), forming an essentially planar world—even though these elemental discs are canonically understood as stacked.[16] The elemental disc of water, like the vast ocean of the *abhidharma* system, is surrounded by a ring of *vajra* mountains representing a barrier at the edge of the world.

One other noteworthy feature of the Kālacakra system that relates directly to Himalayan visual depictions is its characterization of motions of bodies in the sky, especially the sun's changing path through the seasons and the twelve signs of the zodiac.[17] Interpretation of these details is complicated by the abstruseness of the canonical texts, peculiarities of the geometries, and the possibility that traditional astronomers may have relied on calculation techniques that do not strictly correspond to the Kālacakra's physical model.[18] For understanding the Bhutanese murals, however, the primary concern is the depiction of an easy-to-recognize series of overlapping, non-concentric circles surrounding Meru like the petals of a flower (fig. 9.4). Each ring relates to the sun's path through the houses of the zodiac at different points in the year.[19] Such models of astral motion, being adopted from Indian astronomy, are not necessarily incompatible with the *abhidharma* system, but in the pre-modern Tibetan and Bhutanese contexts that draw so much astronomy from the Kālacakra transmission, visual depictions of these rings are far more commonly associated with the Kālacakra model.[20]

As is evident from the preceding descriptions and accompanying figures, the world models of the *Abhidharmakośa* and Kālacakra systems, while more closely related to each other than to the modern scientific model, differ quite obviously. From the very elements that make up the cosmos to the dimensions of the world, the shapes and colors of the continents, and the motions of astral bodies, any student of these systems would immediately recognize their contradictions. One important commentary on the Kālacakra system notes preemptively that any such model of the cosmos is nothing more than provisional, being suited to the needs of particular audiences at particular times rather

16 Tayé, *Myriad Worlds*, 148, 152.

17 Cornu, *Tibetan Astrology*, 129; Henning, "Orbital Confusion."

18 Henning, "The Kālacakra World Model Is Not Mechanical." For more on mathematical calculations, see Henning, *Kālacakra and the Tibetan Calendar*.

19 Henning, "Orbital Confusion;" Gyatso, *Ornament of Stainless Light*, 114–121, 616.

20 See examples at Bryant, *The Wheel of Time*, 239; Henss, *The Cultural Monuments of Tibet*, 1:186; and Thurman and Weldon, *Sacred Symbols*, 117.

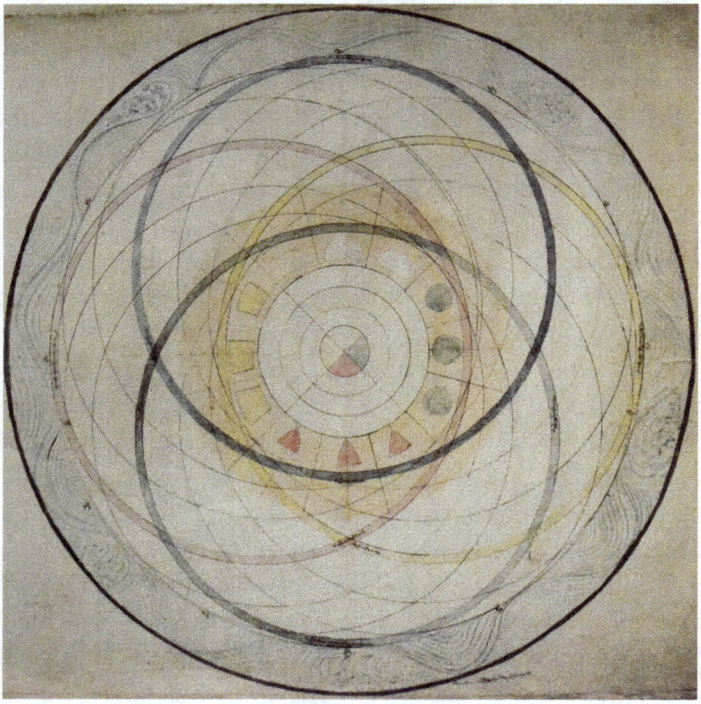

FIGURE 9.4 Detail from a set of paintings of Kālacakra cosmology. Tibet, sixteenth century, pigments on cloth. Rubin Museum of Art C2009.9 (HAR 61200)

than reflecting objective reality.[21] This is, in fact, one of the reasons why it is possible for both the *abhidharma* and Kālacakra systems to be portrayed side-by-side in Bhutanese murals.

3 Artistic Representations

Turning now to visual material, it is first important to understand that there are many variations in how each cosmos can be depicted in artwork. These originate for diverse reasons, including ritual symbolism, artistic lineage, and educational practices. A twentieth-century mural of the *abhidharma* cosmos from Lekir Monastery in Ladakh (fig. 9.5), for example, builds on the imagery of an offering ritual to place enormous emphasis on central Meru, depicting it as a

21 Newman, "Outer Wheel of Time," 472.

FIGURE 9.5 Cosmological mural, Lekir monastery, Ladakh, India

large white rectangle towering over smaller mountains and continents.[22] In an earlier painting from Mindroling in Tibet (fig. 9.6), by contrast, Meru is nearly unrecognizable amidst the extraordinarily detailed depictions of the palaces and heavens that accentuate a geographic specificity. While it is possible that the individual artists who made these paintings might not have even known of the possibility of such different forms of depiction, trained as they were in their own artistic lineages,[23] a comparative examination allows for the identification and analysis of particularly salient characteristics across broader traditions.

22 Huntington, *Creating the Universe*, 155–160.

23 Tsering Wangdus (Tib. Tshe ring dbang 'dus), personal communication, 2011.

FIGURE 9.6 Cosmological mural, Mindroling monastery, Tibetan Autonomous Region

Indeed, common to both of these examples, as with many others from the Tibetan tradition, is the termination of the cosmos at the Cakravāla, the ring of mountains that separates the world from the non-world of empty space in the *abhidharma* description. In the Lekir painting, the Cakravāla is painted as a thick, black, oval line around the continents and golden mountains. At Mindroling, the line of the Cakravāla is less visible, but its overall circular perimeter is highly apparent as the boundary of everything in the image (except the heavens that extend upwards toward the ceiling in elevation view). This circular perimeter suggests a plan view of the cosmos, but such images almost never depict the cosmos to scale in accordance with textual description. In the painting from Mindroling, for example, the continents and Meru are vastly oversized, presumably to allow the depiction of more details inside their borders (compare the red circles of the western continents against the to-scale illustration in fig. 9.2).

In addition to examples such as these, visual depictions of the *abhidharma* cosmos take many other forms in diverse circumstances in Himalayan Buddhist culture. Meru and the basic features of the cosmos also frequently appear within the wheel of existence as part of the realm of the *devas*, who live atop the mountain. Here, the Cakravāla perimeter is completely omitted, since the point is to depict the abode of the *devas* atop the central mountain. Meru and its immediately surrounding foothills also figure in narrative scenes of the Buddha's life story, especially his descent from the Trāyastriṃśa heaven atop Meru, where he taught his departed mother. In these cases, the height of Meru is often disproportionally exaggerated to emphasize the miraculous nature of the Buddha's visitation to Trāyastriṃśa, as well as the long staircase upon which he descended. Such visual changes, like the omission of the Cakravāla or the enlargement of Meru, help adapt each depiction to the ritual or narrative purposes at hand.

The Kālacakra cosmos is far less common to see. One type of image, depicting a plan view of the continents and overlapping circles of astronomical motion, has already been illustrated in figure 9.4, and as with the *abhidharma*, many other variations appear in other contexts. The painting in figure 9.7, for example, portrays a vertical division of the cosmos into realms of different beings. The bottom half of the image is taken up with hells filling semi-circular arcs that represent layers of the elemental discs below the surface of the earth (a combined elevation and plan view). The horizontal line in the center of the image, topped with small buildings, represents the surface of the earth where humans dwell. Towering above is the inverted cone of Meru, which itself is topped by layers of heavens rising to the upper edge of the painting. Such images are far from the only ways to depict the Kālacakra cosmos, but they do reveal some of the unique and distinguishing features of Kālacakra cosmos images, which include a significant emphasis on the elemental discs below the surface of the earth, the conical shape of Meru, and the astral circles surrounding the central mountain. As with the *abhidharma* cosmos, the Kālacakra cosmos also appears in numerous other contexts, such as diagrams explaining the correspondence between the Kālacakra cosmos and the human body.[24]

24 See Thurman and Weldon, *Sacred Symbols*, 115.

FIGURE 9.7 Painting of the Kālacakra cosmos, © Ethnographic Museum at the University of Zurich. Inv.-No. 21299
PHOTOGRAPH BY KATHRIN LEUENBERGER

4 Recent Bhutanese Murals

With a basic understanding of the two cosmological systems and some of the popular ways that these have been understood in visual imagery, some specific murals from Bhutan can now be investigated. All of these examples come from the entrance verandas of *dzongs*, which are a combination between a fortress and a monastery that also serves as a center of local government. Because of the active patronage of Buddhism in Bhutan, these murals have been repainted in recent history, but photographs from previous decades allow for some tentative

FIGURE 9.8 Overview of cosmological murals outside the assembly hall of Punakha Dzong, Bhutan

comparisons with earlier versions from the twentieth century at least. Significant changes seem possible in each new generation of painting, making longer term historicization difficult, but several kinds of visual negotiation between the *abhidharma* and Kālacakra cosmologies are clearly evident.

4.1 *Punakha Dzong*

The first set of paintings resides at Punakha Dzong, the former administrative center of Bhutan, which lies at the confluence of two important rivers in the region. The entrance to the main assembly hall of Punakha Dzong is decorated with four murals (each with smaller subjects depicted in the peripheries). In order from left to right, these are: a Kālacakra cosmos, an *abhidharma* cosmos, a wheel of existence, and a set of longevity symbols (fig. 9.8).

Attending only to the first two, one finds some familiar characteristics. The Kālacakra cosmos (fig. 9.9) depicts the astronomical circles around Meru, which is shown as a small green circle in the very center. The twelve lesser continents are depicted on a plain, blue circle that represents the ocean that fills Greater Jambudvīpa, while the elemental water layer is depicted as a wave-filled ocean surrounding this space. Outside the water circle and beyond the ring of vajra mountains (which is not depicted), we see iconic representations of the rings of fire (in red) and wind (in green) that form the foundation of

FIGURE 9.9 Kālacakra cosmos mural, Punakha Dzong, Bhutan

the cosmos. Rather than being depicted to scale, these elemental rings of fire and wind recall the elemental ring surrounding the exterior vajra fence of deity-maṇḍalas,[25] and indeed considering the function of the ring of vajra mountains surrounding the Kālacakra cosmos, the visual analogy is quite logical.

The center of the image presents the most obvious adaptation from the *abhidharma* system. The lesser continents, rather than being divided equally into twelve radial sections (see again fig. 9.4), are grouped in threes in the cardinal directions, as they are in the *abhidharma* (see again figs. 9.5 and 9.6). Furthermore, the shapes and colors of these continents, which are given differently in both sources, clearly match the *abhidharma* description, not the Kālacakra. However, unlike the *abhidharma* cosmos just next to it, which orients the continents so that south is at the bottom of the image, east lies at the bottom of this painting. This feature suggests that the artists were not simply copying elements from one depiction to the other. Rather, they did understand

25 See, for example, Brauen, *Mandala*, 22.

FIGURE 9.10 *Abhidharma* cosmos mural, Punakha Dzong, Bhutan

certain important structural differences between the two cosmologies. Indeed, the Kālacakra cosmology is usually depicted with east at the bottom in accordance with deity-maṇḍala imagery, and the *abhidharma* cosmos frequently appears with south at the bottom to emphasize the viewer's place in southern Jambudvīpa.

Turning to Punakha Dzong's *abhidharma* cosmos (fig. 9.10), one sees many of the expected features, including the squarish Meru, the sun and moon orbiting at half its height, the square surrounding mountain ranges, the various inhabited continents in the four directions (with south at the bottom), and layers of heavens above. Surprisingly, however, there is a relatively enormous depiction of the Cakravāla mountain range (grayish perimeter) and the circle of wind surrounding the cosmos (blue on green background). As mentioned previously, it is not particularly common to depict the Cakravāla in significant detail, and it is even less common to depict the wind disc in such expansive measure. One of the main reasons for the inclusion of such a wind disc here seems to be the importance of similar elemental discs in the Kālacakra model, which form not only a pictorial border but also a substantive part of its elemental system. Painted alongside each other, the two murals begin to share a similar logic of depiction. They represent a moment of assimilation and mutual influence, even as they depict mutually contradictory cosmologies.

While these preceding images were photographed in 2009 and relatively recently after the murals had been completed, photographs by Jürgen Schick from 1984–1986 record slightly earlier paintings that exemplify other modes of mutual connection and hint at a development of pictorial details over time. These murals are also from Punakha Dzong, although they were not painted on precisely the same walls as the more recent ones. Among these, the plan view of the Kālacakra (fig. 9.11) shows a clear understanding of the correct radial alignment of the continents. Although the shapes are somewhat difficult to discern, the colors of the continents also match the Kālacakra (although interestingly with south at the bottom). This painting also includes the ring of vajra mountains (in blue), surrounded by the circles of fire (red) and wind (yellow). Unlike the more recent example that simplifies Meru into a green circle, here Meru is divided into five colored sections, four for the colors of the cardinal directions and a central green circle, in correspondence with the canonical 5-part division of Kālacakra cosmology.[26] While it is difficult to quantify precisely how faithful each depiction is to its textual precedents or how much each borrows from other traditions, this earlier mural seems to be far less influenced by *abhidharma* visuality than the later version.

A second representation of the Kālacakra system in split plan/elevation view, also recorded by Schick, shows the heavens above and the circles of elements below (fig. 9.12). This version was not included in the more recently painted cosmological set at all. As with the other paintings at this site, the circles of fire and wind are reduced in scale compared to the disc of water, suggesting that they are conceived more as the borders of a maṇḍala than to-scale representations of cosmic geometry, allowing the expansion of the interior to highlight continents and central geography. The southern continents here are correctly red for the Kālacakra system, but their shape matches the *abhidharma* rather than Kālacakra, in which they are normally triangles. The continents to the west are yellow and square, as is canonical in the Kālacakra system, although they are incorrectly mirrored in the East, where the continents are normally black semi-circles. This change is not reflective of *abhidharma* cosmology, however, and probably results from a simplification by the artist based on the overall symmetry of the design.

---

26 See Wallace, *The Inner Kālacakratantra*, 72.

FIGURE 9.11 Plan Kālacakra cosmos mural, Punakha Dzong, Bhutan
COURTESY OF JÜRGEN SCHICK

Finally, the earlier version of the *abhidharma* cosmos is actually quite similar to the modern version, with the exception of missing the significant presence of the wind-disc surrounding the Cakravāla (fig. 9.13). This cosmos ends at the rather enlarged depiction of the Cakravāla as a blue ring of mountain peaks at the perimeter. In this way, the painting is more similar to the ones from Lekir and Mindroling than the later examples from Bhutan, which include the wind disc through apparent influence from the Kālacakra diagrams. Inside the boundary of the Cakravāla, the continents are depicted with their cor-

FIGURE 9.12 Kālacakra cosmos mural, Punakha Dzong, Bhutan
COURTESY OF JÜRGEN SCHICK

rect *abhidharma* shapes. Interestingly, the height (rather than base) of Meru is depicted in the center of the circle in such a way that the four oceans in the four quadrants are not the same size, with the southern ocean greatly compressed at the bottom of the image.

Clearly, the modern murals portray the cosmos quite differently than these decades-old paintings, in some places reflecting a mutual influence between the *abhidharma* and Kālacakra cosmologies and in others maintaining a continuity with previous traditions. The reasons for such changes are open to debate, and it is unclear how these changes might relate to actual cosmological knowledge on the part of the artists and practitioners or longer term trends. In the two plan views of the Kālacakra, at least, there seems to be a growing conflation with *abhidharma* elements over time. Broadly speaking, however, other recent

FIGURE 9.13 *Abhidharma* cosmos mural, Punakha Dzong, Bhutan
COURTESY OF JÜRGEN SCHICK

changes seem to be accompanied by a generally *greater* interest in canonical accuracy, especially for the *abhidharma*. Take again, for example, the enormous disc of wind that encircles the modern version of the *abhidharma* cosmos (see again fig. 9.10). While the presence of this disc of wind seems to be informed by the elemental rings around the Kālacakra cosmos, its enormous size far exceeds the elemental discs of the Kālacakra depictions. Instead, this seems to be an acknowledgement of the vastness of the wind disc precisely as described in the *Abhidharmakośa*, the authoritative textual source for this cosmology. This mural also exhibits especially detailed inscriptions labeling each part of the cosmos accurately, which none of the earlier murals photographed by Schick possess. There is also a recent and rare example at Simtokha Dzong that clearly shows an unparalleled concern for textual accuracy in its measured propor-

FIGURE 9.14 Cosmological mural, Simtokha Dzong, Bhutan

tions (fig. 9.14, compare against plan view of cosmos in fig. 9.2).[27] This mural also includes a depiction of astral rings that seems to accord with *abhidharma* rather than Kālacakra geometries, apparently another example of *abhidharma* cosmology being depicted in a visual language normally associated with the Kālacakra. While it is clear that visual representations of the two cosmoses are changing in conversation with each other, it cannot be assumed that this is because of a lack of knowledge or authenticity.

27 See also Imaeda, "Peintures cosmiques du Bhoutan," 628–631.

4.2 *Trongsa Dzong*

In pursuing these themes of negotiation, assimilation, and proportional or textual accuracy, two more murals from another site in Bhutan, Trongsa Dzong, are informative. One, a decades-old mural of the *abhidharma* cosmos (fig. 9.15), generally presents the expected characteristics, including a detailed depiction of the Cakravāla mountains. Beyond this edge, however, lies a rare interpretation of the disc of wind below the cosmos. In this version, rather than being portrayed as a disc at all, the layer of wind is depicted as essentially limitless, extending to the very edges of the wall on which the mural is painted. In this regard, this is the most textually accurate version yet, for while the *Abhidharmakośa* specifies a finite depth for the disc of wind, it defines its diameter as immeasurable.[28] This mural of the *abhidharma* system accordingly depicts the world as being surrounded in all directions by an infinite mass of winds, another move in the direction of canonical accuracy in contrast to other traditions of cosmological depiction across the Himalayas. Since this elemental circle of wind is the only major feature of this mural that departs from such traditional images, however, and since this mural appears right alongside the Kālacakra cosmos, it seems likely that this particular change may again be related to the Kālacakra's emphasis on elemental substrata.

One final example of the assimilation between *abhidharma* and Kālacakra systems may be the most subtle and significant. Where the paintings examined so far either use the visual language of one cosmology to depict aspects of another (such as the portrayal of the *abhidharma* circle of wind in the style of Kālacakra imagery) or conflate elements that overlap between the two systems (such as the continents in a Kālacakra mural appearing according to their *abhidharma* description), this example sees the inclusion into the *abhidharma* cosmos of an elemental foundation of fire that exists only in the Kālacakra and other tantric systems, not in the traditional *abhidharma* model.

In the previous cases, one important factor in the influence of the two cosmologies was the fact that paintings of the two were directly juxtaposed on the exterior walls of assembly halls within the *dzong*. In such close proximity, a spillover of visual vocabulary is easy to imagine. It is also very common, however, to have a single mural of the geographic cosmos depicted at the most exterior entrance to the entire monastery. For ritual reasons associated with offerings and patronage, this is nearly always an *abhidharma* version of the cosmos,[29] and at Trongsa Dzong, this mural does essentially portray the

28 Vasubandhu, *Abhidharmakośam*, 2:399.

29 Huntington, *Creating the Universe*, 139, 201.

FIGURE 9.15 *Abhidharma* cosmos mural, Trongsa Dzong, Bhutan

abhidharma model (fig. 9.16). At the edge, however, there is a radical change (fig. 9.17). While a circle of wind usually surrounds the Cakravāla of the *abhidharma* cosmos, here a circle of fire is inserted between the two, precisely matching similar depictions of the Kālacakra system (see again figs. 9. 9 and 9.11).

Like the Punakha *abhidharma* mural, this painting is also inscribed in detail with textual labels of the features of the cosmos, so it seems unlikely that the inclusion of the fire disc would be a simple mistake. Rather, perhaps because there is only a single version of the cosmos here, some ritual, philosophical, or political logic has forced the inclusion of tantric cosmology into the *abhi-*

FIGURE 9.16 Cosmological mural, entranceway, Trongsa Dzong, Bhutan

dharma image. The elemental substrata that include fire are common to many tantric systems beyond the Kālacakra,[30] so it is impossible to say whether this image is intended to evoke the Kālacakra model in particular or merely a tantricization of the *abhidharma*. There seems to be at least a visual or stylistic con-

30 See, for example, Skorupski, *The Sarvadurgatipariśodhana Tantra*, 26–27.

FIGURE 9.17 Detail of cosmological mural, entranceway, Trongsa Dzong, Bhutan

nection, however, since such elemental circles are commonly depicted around Kālacakra cosmoses but not around other tantric maṇḍalas. As there are no other cosmological murals in the surrounding architectural space, it also makes sense that this single mural might be adapted to express both *abhidharma* and tantric ideologies, possibly providing the foundation for both the *abhidharma* cosmology most significant in the entranceway and the more esoteric tantric rituals that take place inside the *dzong*. While an admittedly small detail like an extra circle around the cosmos might go unnoticed by many visitors entering the site, this combination of two cosmoses appears necessary for the symbolism and efficacy of practice there.

5 Conclusion

Such visual relationships between the two cosmological systems of Kālacakra and *abhidharma*, while subtle, are not minor. Although images of the cosmos are among the most varied of any subject in Buddhist art, the principles of Buddhist iconography generally remain rather stable. Changes in the shapes and colors of continents or the elemental substrata of the universe represent specific moments of conflation, influence, and accommodation between relatively separate traditions. Furthermore, the relationships revealed by these visual emblems of each cosmology are very different than the types that might be understood through comparisons of textual sources or other means of investigating overlapping cosmologies. The expansion of the wind disc in the *abhidharma* depictions, for example, does not actually represent any revision of the cosmological model at all, but rather the application of the visual logic of Kālacakra imagery to a preexisting feature. More radical assimilations, like the inclusion of the fire disc in the *abhidharma* cosmos at Trongsa Dzong, demonstrate a subtle way of adopting two contradictory cosmologies simultaneously, without actually concretizing these changes in a cosmological treatise that would challenge the accepted canon. The prominent place of cosmological murals in architecture also reveals a very different context for expression and adaptation than the written commentaries of scholars, with some changes in painted imagery arising from artistic principles of symmetry and style rather than logical or mathematical concerns. Such assimilations and negotiations are reconsidered each time one of these murals is repainted, providing a regular and very public venue for the mediation of overlapping cosmologies in the Buddhism of Bhutan.

Bibliography

Berzin, Alexander. *Taking the Kalachakra Initiation*. Ithaca, New York: Snow Lion Publications, 1997.

Brauen, Martin. *Mandala: Sacred Circle in Tibetan Buddhism*. Stuttgart: Arnoldsche Art Publishers, 2009.

Bryant, Barry. *The Wheel of Time Sand Mandala: Visual Scripture of Tibetan Buddhism*. San Francisco: HarperSanFrancisco, 1992.

Cornu, Philippe. *Tibetan Astrology*. Translated by Hamish Gregor. Boston: Shambhala, 2002.

Dreyfus, Georges B.J. *The Sound of Two Hands Clapping*. Berkeley: University of California, 2003.

Gyatso, Khedrup Norsang. *Ornament of Stainless Light: An Exposition of the Kālacakra Tantra*. Translated by Gavin Kilty. The Library of Tibetan Classics. Boston: Wisdom Publications, 2004.

Henning, Edward. *Kālacakra and the Tibetan Calendar*. New York: American Institute of Buddhist Studies, 2007.

Henning, Edward. "The Kālacakra World Model Is Not Mechanical." Sarnath, India, 2009. http://www.kalacakra.org/calendar/kc_model.htm.

Henning, Edward. "Orbital Confusion." Accessed August 17, 2016. http://www.kalacakra.org/phyinang/golconf.htm.

Henss, Michael. *The Cultural Monuments of Tibet: The Central Regions.* 2 vols. Munich: Prestel, 2014.

"Himalayan Art Resources," http://www.himalayanart.org/.

Huntington, Eric. *Creating the Universe: Depictions of the Cosmos in Himalayan Buddhism.* Seattle: University of Washington Press, 2018.

Imaeda, Yoshiro. "Peintures cosmiques du Bhoutan." In *Orientalia Iosephi Tucci memoriae dicata* 2: 627–634. Serie orientale Roma 56. Rome: Istituto italiano per il Medio ed Estremo Oriente, 1985.

Jackson, Roger R. "Kalachakra in Context." In *The Wheel of Time: The Kalachakra in Context*, edited by Beth Simon, 1–50. Ithaca, New York: Snow Lion Publications, 1985.

Kloetzli, Randy. *Buddhist Cosmology: From Single World System to Pure Land.* Delhi: Motilal Banarsidass, 1983.

La Vallée Poussin, Louis de. *Abhidharmakośabhāṣyam.* Translated by Leo M. Pruden. Berkeley: Asian Humanities Press, 1990.

La Vallée Poussin, Louis de. *L'Abhidharmakośa de Vasubandhu.* Paris: Paul Geuthner, 1923.

The Living Religious and Cultural Traditions of Bhutan: Catalogue. New Delhi: National Museum, 2001.

Newman, John Ronald. "The Outer Wheel of Time: Vajrayāna Buddhist Cosmology in the Kālacakra Tantra." PhD diss., University of Wisconsin, 1987.

Sadakata, Akira. *Buddhist Cosmology: Philosophy and Origins.* Translated by Gaynor Sekimori. Tokyo: Kosei Publishing Company, 2004.

Sircar, D.C. *Cosmography and Geography in Early Indian Literature.* Calcutta: Indian Studies: Past & Present, 1967.

Skorupski, Tadeusz. *The Sarvadurgatipariśodhana Tantra: Elimination of All Evil Destinies.* Delhi: Motilal Banarsidass, 1983.

Strong, John. *The Legend of King Aśoka: A Study and Translation of the Aśokāvadāna.* Princeton Library of Asian Translations. Princeton: Princeton University Press, 1983.

Tayé, Jamgön Kongtrul Lodrö. *Myriad Worlds: Buddhist Cosmology in Abhidharma, Kālacakra, and Dzog-Chen.* Ithaca, New York: Snow Lion Publications, 1995.

Teiser, Stephen F. *Reinventing the Wheel: Paintings of Rebirth in Medieval Buddhist Temples.* Seattle: University of Washington, 2006.

Thurman, Robert A.F., and David Weldon. *Sacred Symbols: The Ritual Art of Tibet.* New York: Sotheby's, 1999.

Vasubandhu. *Abhidharmakośam (svopajñabhāṣyasahitaṃ sphuṭārthavyākhyopetaṃ ca).* Edited by Swāmī Dwārikādās Śāstri. 2 vols. Bauddha Bharati 5–8. Varanasi: Bauddha Bharati, 1998.

Wallace, Vesna A. *The Inner Kālacakratantra: A Buddhist Tantric View of the Individual.* Oxford: Oxford University Press, 2001.

CHAPTER 10

Overlapping Heavens in the Wall Paintings of Mogao Cave 285 at Dunhuang: An Art-Historical Study of the Syncretistic Images on Its West Wall and Ceiling

Satomi Hiyama

1 Introduction

Mogao Cave 285, one of the most celebrated Buddhist cave temples in Dunhuang due to the extraordinary artistic quality of its wall paintings and its possessing the earliest datable inscriptions at the site, presents a unique fusion of celestial images associated with both Indian and Chinese origins (fig. 10.1). The elucidation of the iconographic contents of these images is not a simple task, since some are dated earlier than their relevant textual sources, others are unparalleled elsewhere, and many appear to carry double meanings, allowing for syncretistic readings.

Through the comparison of these images with textual references and with fifth- and sixth-century Buddhist and non-Buddhist visual arts in Gansu Province 甘肅省, Pingcheng 平城 (present-day Datong 大同), and Luoyang 洛陽, this cave can be understood as part of an ideological trend of Northern Wei in which novel religious concepts imported from the West were adopted by referencing traditional Chinese concepts. This resulted in the creation of new images based on ideas originating in different religious contexts that were melded together.

The syncretistic heavenly images in this cave contain multiple depictions of the same heavenly bodies, indicating that the decorators of this cave did not intend to systematically depict a single coherent cosmology. The involvement of multiethnic donor groups in the commission of this large-scale cave provides additional clues to a possible political goal of uniting local people with different cultural and religious backgrounds through a cultural project of creating a combined religious space. This cave temple thus serves as an interesting case study to show how fragmented elements of contemporary cosmologies, especially their visual imagery, could have been exchanged between different cultural realms, perhaps even separately from the relevant bodies

 | DOI:10.1163/9789004511675_012

FIGURE 10.1 Mogao Cave 285 (including the western wall, northern wall, and ceiling)
PHOTOGRAPH BY WU JIAN 吴健 © DUNHUANG ACADEMY 敦煌研究院

of scientific knowledge, but remained a driving force behind particular political intentions and religious needs.

2 Structure and Layout of Mogao Cave 285

Mogao Cave 285 is a rectangular cave with an eastward entrance, furnished with eight cells along its side walls and covered by a truncated pyramidal ceiling (fig. 10.2). The size of this cave (width 6.4 m, depth 5.9 m, height 4.3 m) allows it to hold a large number of devotees at once. Even though this type of cave structure is generally classified as a *vihāra*, the extensive decoration of the main chamber and the presence of a square platform (2 m on each side) in the middle of the floor has led scholars to believe this cave served a ritual function.[1]

1 The present platform, dated to around the eleventh or twelfth century, may have replaced one from an earlier period. Some scholars have interpreted it as an ordination platform. See Wang, *Dunhuang Fojiao*, 117. For a recent study on the function of this cave, see Yamabe, "'Zenjō-kutsu' saiko."

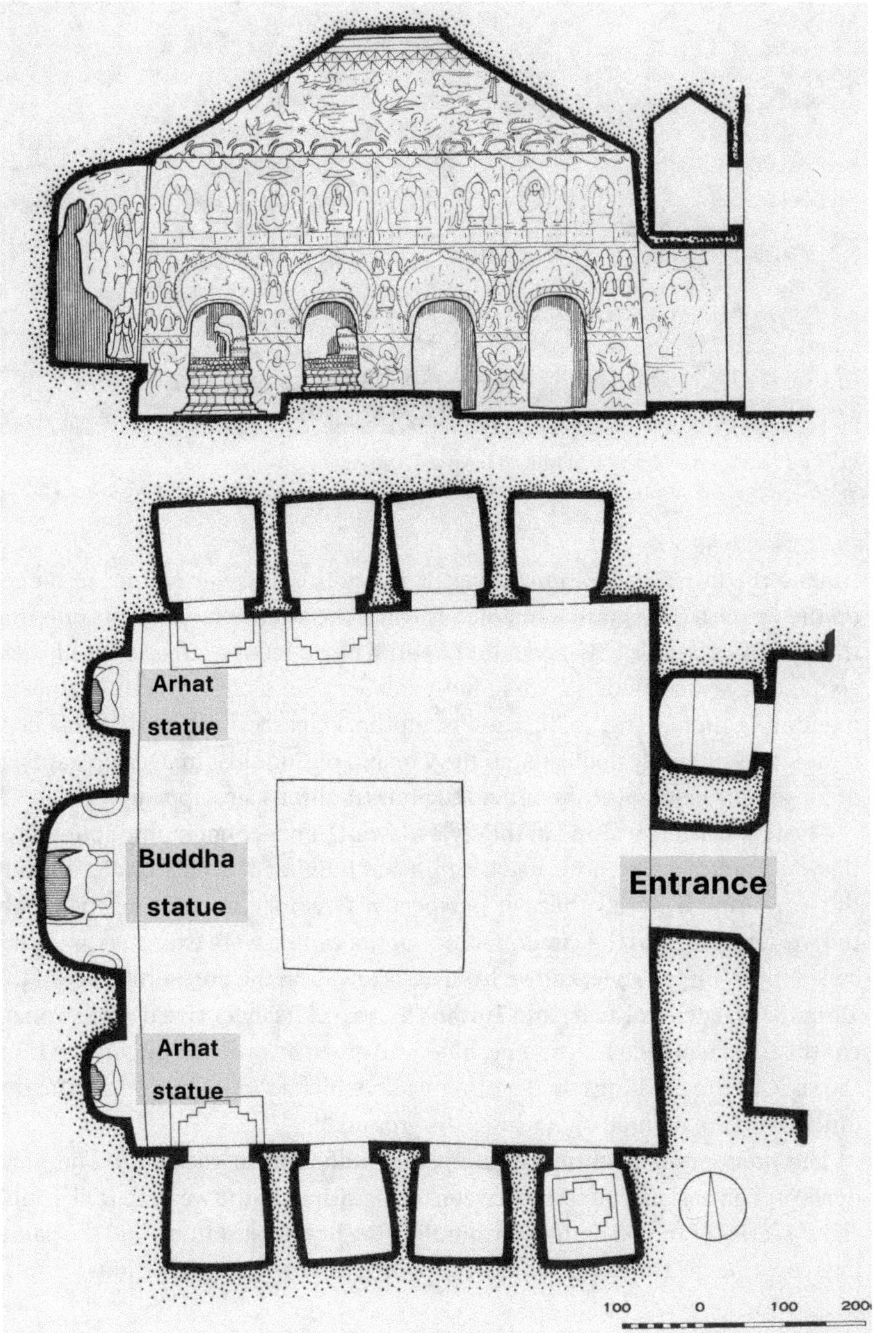

FIGURE 10.2 Plan of Mogao Cave 285, Dunhuang
MODIFIED BY THE AUTHOR AFTER *DUNHUANG GROTTOES* (I), 225

FIGURE 10.3 Western wall of Mogao Cave 285, Dunhuang

Among the lavish decorations covering its walls, three clay statues in niches on the western (rear) wall immediately catch the attention of visitors entering this cave. The large, richly-decorated, central niche holds a larger-than-life-size statue of a seated Buddha, while both smaller side niches contain statues of meditating monks (fig. 10.3). These sculptures clearly indicate that this cave temple was centrally dedicated to the worship of Buddhist imagery, regardless of how many elements from other religious traditions also appear.

The inconsistency found in the styles, layouts, and iconographic contents of the wall paintings has motivated a significant number of art historians to study this cave from a wide variety of perspectives. As for the painting style, only the western wall and the circumference of the ceiling were executed in the so-called Indo-Iranian style, known from oasis towns on the northern fringe of the Tarim Basin such as Kucha and Turfan (fig. 10.3).[2] Its extensive use of strongly contrasting colors, such as orange, blue, and green, is easily distinguished from the style of the paintings on the other walls, which follow Chinese artistic traditions and are painted on a white background.[3]

The program of paintings is completely different on each wall. The most dynamic contrast can be seen between the paintings of the western wall, which show a series of Indian deities surrounding the Buddhist statues, and the paintings covering the truncated pyramidal ceiling, which represent various Chinese

2 See Waldschmidt, "Über den Stil."
3 For more on the Chinese style paintings, see Zhang Jianyu, "Dunhuang Xiwei"; Tabayashi, "Tonkō Bakkōkutsu dai 285 kutsu kaisaku," 515.

FIGURE 10.4 Fuxi, Nüwa, and a *maṇi* jewel. Wall painting on the east side ceiling of Mogao Cave 285
PHOTOGRAPH BY SUN ZHIJUN 孫志軍 © DUNHUANG ACADEMY 敦煌研究院

deities, including the pair of creation gods Fuxi 伏羲 and Nüwa 女媧 (fig. 10.4). The paintings on the northern (right) and eastern (front) walls mainly feature devotional images of Buddhas surrounded by Bodhisattvas, with portraits of donors and inscriptions below. The southern (left) wall, in contrast, contains depictions of several *avadāna* stories. Because of the drastic differences in both iconography and style among the walls, it has been assumed that more than two workshops were involved in painting of this cave.[4]

Seven portrayals of multiethnic donor groups with votive inscriptions on the northern wall provide crucial hints for reconstructing the historical context of this extraordinary cave (fig. 10.5). The images show both ordained and lay people with family names indicating their Han (Yin 陰), Hephthalite (Hua 滑), Sogdian (Shi 史), and possibly Gaoche (高車丁零) origins.[5] Two of the donative inscriptions have the oldest identifiable dates in the Mogao caves, mentioning

4 While early studies regarded the paintings of the western wall as preceding the others, recent studies suggest that all the paintings may be contemporary. See Yagi, *Chūgoku bukkyō bijutsu*, 252–254; Tabayashi, "Tonkō Bakkōkutsu dai 285 kutsu no Bukkyō sekai," 242.

5 Ishimatsu, "Tonkō Bakkōkutsu," 53–55.

FIGURE 10.5 Amitābha Buddha with portraits of a Hephthalite donor group, with a votive inscription dated to 539 CE. Northern wall of Mogao Cave 285, Dunhuang © HERITAGE DIGITIZATION INSTITUTE, THE DUNHUANG ACADEMY 敦煌研究院文物数字化研究所

the years Dadai Dawei Datong 4 and 5 (大代大魏大統四・五年, 538–539 CE, during the Western Wei). As Ishimatsu demonstrated, the special format of the year used in this inscription was associated with the sutras copied by Yuan Rong 東陽王元榮 (Prince of Eastern Sun) and his daughter,[6] providing strong sup-

6 Ishimatsu, "Tonkō Bakkōkutsu," 66–68. Ishimatsu considers the donor group depicted in the

port for the widely accepted claim that Yuan Rong was the key figure involved in the commission of this monumental cave.[7]

3 Historical Context of Dunhuang in the Early Sixth Century

A brief overview of the biography of Yuan Rong reveals his special position to support the overlapping cosmologies found in the cave. As the fourth-generation grandson of Emperor Ming Yuan 明元帝 of the Northern Wei imperial family, Yuan Rong was most likely born before 488 CE in Pingcheng, the former capital of Northern Wei, and then moved to Luoyang, which was designated as the new capital by Emperor Xiaowen 孝文帝 in 494 CE.[8] He was then moved farther away to Dunhuang—which was a penal settlement for criminals at that time[9]—when he was appointed as a local governor of the area by the Northern Wei in around 525 CE.[10] Even after Dunhuang fell under the control of the Western Wei, he kept his position until the end of his life at the end of the 530s or the beginning of the 540s.

Dunhuang in the 530s CE seems to have enjoyed a relatively peaceful period. This is at least in part due to Yuan Rong's political skill in uniting the rich and powerful clans of the region to maintain stability, with Buddhism as the ideological backbone for his internal policy. In addition, Dunhuang in the late 530s was situated in a rare geopolitical condition in that the neighboring powers of Rouran 柔然 and Tuyuhun 吐谷渾 temporarily suspended their invasion

---

seventh (innermost) section closely related to Yuan Rong, while Wang (*Dunhuang Fojiao*, 260) suggests that they are possibly Yuan Rong and his wife.

7 Su, "Canguan Dunhuang Mogaoku," 211; Taguchi, "Hekiga ni tsuite," 53–54; Su, "Dongyangwang"; Rong, *Eighteen lectures*, 29–30; Tabayashi, "Tonkō Bakkōkutsu dai 285 kutsu kaisaku," 501–518; Wang, *Dunhuang Fojiao*, 241–248, 260.

8 The epitaph of Yuan Rong's younger sister Yuan Huaguang 元華光 indicates that she was born in 488 CE, suggesting that Yuan Rong must have been born prior to that year, which predates the relocation of the capital. See Zhao, *Han Wei Nanbeichao*, 165–166; Zhang Fan, "Chinese-Buddhist Encounter," 106.

9 The largest section of the upper half of the southern wall of Cave 285 is occupied by a depiction of the story of the Five Hundred Robbers, who became blind as a punishment for their sin, were converted by the Buddha, and became ordained monks. The selection of this narrative subject was likely related to the historical context of Dunhuang as a former penal settlement. See Sudō, "Zenjō Biku," 20–21; Tabayashi, "Tonkō Bakkōkutsu dai 285 kutsu kaisaku," 509.

10 For the study on the biography of Yuan Rong, see Zhao, "Wei zongshi," 7–12; Xiang, "Mogao, Yulin erku," 76–80; Wen, "Zailun Dongyangwang," 102; Tabayashi, "Tonkō Bakkōkutsu dai 285 kutsu kaisaku," 503–509; Zhang Fan, "Chinese-Buddhist Encounter," 106.

of Western Wei.[11] One factor that might have influenced the geopolitical landscape of the Eastern Silk Road during this period is the Hephthalites' occupation of Central Asia, called *Pax Hephthalica* by Yoshida.[12] In the period from around 509 to 557 CE, the Hephthalites, groups of nomadic people originating from Bactria, reached their maximum expansion, stretching from northern India to the eastern periphery of the Sasanian Empire and almost all of Central Asia, including the Tarim Basin.[13]

The portraits of the two groups of Hephthalite donor families with the surname "Hua" 滑 on the northern wall of Cave 285 are crucial (fig. 10.5). These not only indicate the presence of Hephthalite immigrants in Dunhuang at that time but also reveal their historical background, since the designation of the Hephthalites as Hua was specifically used in the Southern Dynasties.[14] Tabayashi pointed out that the use of the name "Hua" indicates the historical connection between these Hephthalite people and the kingdom of Tuyuhun, which controlled the area between Gansu and the Liang 梁 Dynasty at that time and mediated between the Hephthalites and the Liang.[15]

To establish good relations with the Hua Hephthalite people residing in this region was surely a strategic choice for the local governors of Dunhuang, since it not only helped to secure a diplomatic relationship with all the western regions controlled by the Hephthalites but also could have been an aid for maintaining peaceful relationships with other great powers in the east, such as Tuyuhun and Liang, both of which had ties with the Hephthalites. Furthermore, the Hephthalite connection seems to have been crucial for economic reasons, too. Yuan Rong's colophon in the *Humane King Perfection of Wisdom Sutra* (*Renwang bore jing* 仁王般若經, S 4528, copied in 531) referred to a donation of 4000 silver coins, which were the currency of Sasanian Iran and eventually of the Hep-

11 Tabayashi, "Tonkō Bakkōkutsu dai 285 kutsu kaisaku."

12 This term refers to the situation that the Hephthalites' occupation of Central Asia in the late fifth to the first half of the sixth century accelerated the traffic of people, goods, and ideas along the improved network connecting central and eastern Eurasia. Yoshida, "Some Reflections," and "Sogudojin," 25.

13 For studies on the Hephthalites' history, see Kuwayama, *Kāpishī*, 131–140; Grenet, "Regional Interaction," 209–218; De la Vaissière, *Sogdian Traders*, 101–111; Neelis, *Early Buddhist Transmission*, 159–170; Yoshida, "Sogudojin," 22–25; Vondrovec, *Coinage*, especially 1:399–406 for a definition of the Hephthalites based on numismatic evidence; Alram, *Das Antlitz des Fremden*, 63–122.

14 Enoki, "Katsukoku." In the Northern Dynasties, in contrast, the Hephthalites were called Yeda 嚈噠. Remarkably, these are the only inscriptions found in the Dunhuang area referring to the Hephthalite family name.

15 Tabayashi, "Tonkō Bakkōkutsu dai 285 kutsu kaisaku."

hthalites.[16] This indicates that commercial ties with western countries under Hephthalite control were indispensable for Yuan Rong to finance his Buddhist cultural projects.

These political considerations reveal an internal logic hidden behind the wall paintings showing overlapping cosmologies in Cave 285. The prestige position of the Hephthalite donors in the commission of this cave is evinced not only in the portraits officially honoring their virtuous deeds but also in the layout of the paintings. Among the various iconographic motifs that are painted in this cave, it is the Indian deities that have the most prestigious position, just next to the main devotional images on the western wall. The Hua Hephthalite donors are the most plausible candidates to be associated with the worship of these Indian images in this particular cave.

4 Buddhist/Chinese Heaven on the Ceiling

At the same time, Yuan Rong adopted another tactful visual strategy that was able to satisfy all the people involved in the commission of this cave, who had various ethnicities and religious affiliations, without disrespecting any of them: double or syncretistic imagery. Let us begin by observing the syncretistic nature of the iconography of heaven on the ceiling (fig. 10.4). The most widely accepted interpretation of the paintings on the ceiling associates them with *The Sutra of Four Regions of Sumeru* (*Xumi siyu jing* 須彌四域經), as first proposed by He Shizhe in 1980s.[17] This Buddhist apocryphal text is not extant, yet quotations in several Chinese Buddhist texts from the sixth century onward enable us to reconstruct its content.[18] It counts Fuxi and Nüwa, a pair of Chinese creator gods, as two Bodhisattvas who were sent to this world by Amitābha

16 Tabayashi, "Tonkō Bakkōkutsu dai 285 kutsu kaisaku," 509–510. Yuan Rong sponsored a large-scale sutra-copying project in the early 530s, traces of which can be found in several hundred Dunhuang manuscripts. For a list of the sutras copied at Yuan Rong's commission in chronological order, see Wang, *Dunhuang Fojiao*, 243–247.

17 He, "Guanyu 285 ku," 37–40; "Mogaoku di 285 ku kuding," 2–3.

18 See He, "Guanyu 285 ku," 38–39. Here I cite *The Collection of Passages Concerning Birth in the Pure Land* (*Anle-Ji* 安樂集) by Daochuo (道綽, 562–645 AD): "故須彌四域經云。天地初開之時。未有日月星辰。縱有天人來下。但用項光照用。爾時人民多生苦惱。於是阿彌陀佛遣二菩薩。一名寶應聲。二名寶吉祥。即伏犧女媧是。此二菩薩共相籌議。向第七梵天上取其七寶。來至此界造日月星辰二十八宿。以照天下。定其四時春秋冬夏。時二菩薩共相謂言。所以日月星辰二十八宿西行者。一切諸天人民盡共稽首阿彌陀佛。是以日月星辰皆悉傾心向彼。故西流也" (T195.18b).

Buddha. At the beginning of the world, there was no sun, moon, or stars. Fuxi and Nüwa then went to the seventh heaven to bring the seven treasures to this world to create the sun, the moon, and the twenty-eight *nakṣatras*, through which light and the four seasons were brought to the earth. Thus, it is said that the heavenly bodies move towards the west because they all pay homage to Amitābha Buddha—the Lord of the Western Paradise.

This description fits well with the overall layout of the painting on the ceiling. Fuxi and Nüwa are depicted at the center of the eastern (front) slope of the ceiling (fig. 10.4), with human upper bodies and reptilian lower bodies, respectively carrying in front of their chests a sun disk with a three-legged bird and a moon disk with a frog. The *maṇi* jewels (*maṇiratna*) depicted at the center of eastern and southern slopes of the ceiling can be understood as the seven treasures from which Fuxi and Nüwa created the heavenly bodies, while they hold the sun and moon. These figures, as well as most of the mythological creatures, lotus flowers, and clouds floating between them, show dynamic postures of moving toward the center of the composition, which is the central Buddha niche on the western wall. Thus they move toward the west.

At the same time, the identities of the various creatures filling most sections of the ceiling remain unexplained by this text. Instead, they share features with Chinese funeral art. These include winged immortals (*xianren* 仙人), divine animals such as phoenixes, and various mythical or monstrous creatures.[19] These figures were regarded as auspicious omens in China from an early period, especially in a Confucian context, and they were preferred motifs for funeral arts in the Luoyang area around the 520s.[20] In fact, in terms of both the architectural shape and iconographic contents, the ceiling of Cave 285 is closely related to the stone coffins in Luoyang as well as the tomb paintings in Gansu. The connection with funeral art is also significant for the representation of Fuxi and Nüwa; this pair of creator gods is frequently found in funeral contexts, not only in central China but also in Dunhuang and Turfan.[21] Based on

---

19 For iconographical studies of these representations, see Su, "Canguan Dunhuang Mogaoku," 209; He, "Guanyu erbawu ku," 37–40; He, "Mogaoku di 285 ku kuding," 1–13; Tabayashi, "Tonkō Bakkōkutsu dai 285 kutsu no bukkyō sekai," 233–241; Jiang, "'Tian' de tuxiang," 55–76.

20 See Lin, "Hokuchō jidai"; Tabayashi, "Tonkō Bakkōkutsu dai 285 kutsu no bukkyō sekai".

21 See Zhang Fan, "Chinese-Buddhist Encounter." He ("Guanyu erbawu ku," 38; "Mogaoku di 285 ku kuding," 2) pointed out that the iconographical features of the Fuxi and Nüwa in Mogao Cave 285 find their closest parallel in the painting on the coffin cover excavated at Xincheng 新城 Tomb no. 16 in Jiayuguan 嘉峪關, Gansu. For the representation of Fuxi and Nüwa in a tomb in Dunhuang dated to the third to fourth century, see Yin, "Tonkō Seishinbo." In the Astana tombs in Turfan, many silk paintings of Fuxi and Nüwa have

her extensive study of the iconography of Fuxi-Nüwa in the Northern Dynasties, Zhang argued that the combination of Fuxi-Nüwa and the *maṇi* jewel in Mogao Cave 285 followed an iconographic lineage developed in the Pingcheng area in the fifth century.[22] The hybrid design of the *maṇi* jewel, a hexagonal stone embedded in a lotus-like vegetal calyx with a long stem, is especially remarkable, as it allows for multiple simultaneous interpretations, including as the Buddhist magical jewel, the magic fungus *lingzhi* 靈芝 associated with immortality in Chinese mythology, and the lotus as the Buddhist symbol of rebirth.

The religious hybridity found in the Fuxi-Nüwa-*maṇiratna* image in Cave 285, bearing multiple meanings with both Chinese and Buddhist connotations, is further supported by reference to the wall paintings of the ceiling of Mogao Cave 249, another cave dated to the Western Wei dynasty that was likely patronized by Yuan Rong as well.[23] With the same architectural and painting styles as Cave 285, the ceiling is decorated with numerous images of various Chinese mythical creatures. Instead of Fuxi-Nüwa, though, an image of a mountain occupies the western (rear) slope of the ceiling and serves as the focal point of the whole composition (fig. 10.6). Previous studies have proposed two alternative interpretations of this image, either as the Buddhist *axis mundi* Sumeru, with the city of Sudarśana in the Trāyastriṃśa heaven on top and the Buddhist demi-god Asura standing in front, or as the mythical Kunlun Mountain, with the heavenly gate (*tianmen* 天門) connecting the earth with heaven on top and Fangxiang shi 方相氏, the exorcist-officer referenced in the *Rites of Zhou* (*Zhouli* 周禮), standing in front.[24] Furthermore, a pair of figures riding in mythical chariots and flying toward Sumeru/Kunlun, separately depicted on the northern and southern (right and left) slopes of the ceiling, have been interpreted either as Indra and Indrāṇī (Indra's consort) or as the King Father of the East (Dongwangfu 東王父) and the Queen Mother of the West (Xiwangmu 西王母).[25]

---

been excavated as well, roughly dated to the Tang period.

22 Zhang Fan, "Chinese-Buddhist Encounter".

23 For the study on the wall paintings of the ceiling of Mogao Cave 249 see He, "Dunhuang Mogaoku di 249 ku"; Saito, "Tonkō dai 249 kutsu"; Inamoto "Tonkō dai 249, 285 kutsu", 365–368; Tabayashi, "Tonkō Bakkōkutsu dai 285 kutsu no bukkyo sekai," 240–241; Wang, *Dunhuang Fojiao*, 249–254; Hiyama, "Transmission," 419–424.

24 While the majority of previous studies propose a Buddhist reading of these images (see for examples Saito, "Tonkō dai 249 kutsu," 51–54; Hiyama, "Transmission," 421–422), for the Taoist reading, see Ning, "Shangshi dengxian tu."

25 For a review of previous iconographic studies of these images, see Saito, "Tonkō dai 249 kutsu," 45–47.

FIGURE 10.6 Syncretistic image of a mountain. Wall painting of the west side ceiling of Mogao Cave 249
DRAWING BY THE AUTHOR

Previous studies focused on whether these representations had been intended as Buddhist or Taoist, with good iconographic reasons for both interpretations. These studies also included arguments that the images might have reflected a transitional phase of the localization of Buddhism that borrowed traditional visual motifs familiar to the Han-Chinese people. These conclusions need careful re-examination. None of the Mogao caves dated earlier than these two Western Wei examples include such a rich repository of Chinese mythical creatures in their decoration. The dynamic citation of Chinese iconographic conventions in the Buddhist art of these two caves was novel in the history of this site, and it was certainly related to contemporary cultural trends at Luoyang brought to Dunhuang through the movement of Yuan Rong and his retinues, which likely included skilled artisans.

The period when Yuan Rong lived in Pingcheng and Luoyang exactly overlaps with the rule of Emperor Xiaowen of the Northern Wei, who drastically advanced the cultural policy of Sinicization. At that time, various texts and visual arts were produced in which Buddhism was explained and visualized by borrowing conventional terms of Chinese classical literature and iconogra-

FIGURE 10.7 Sūrya with astral figures (likely Kṛttikā or the Southern Dipper). Wall painting of the upper left section of the western wall of Mogao Cave 285, Dunhuang © HERITAGE DIGITIZATION INSTITUTE, THE DUNHUANG ACADEMY 敦煌研究院文物数字化研究所

phy. A colophon of the *Humane King Perfection of Wisdom Sutra* copied by Yuan Rong in 530 also bears witness to the unique syncretism of Buddhist and traditional Chinese beliefs, as it expresses his wish to quickly become a Buddha and obtain longevity, like a Bodhisattva or Pengzu 彭祖; the latter is a Taoist deification of the South Pole symbolizing longevity.[26]

A consideration of such historical and ideological background behind the creation of Mogao Caves 249 and 285 helps develop a new evaluation of the cosmological representations of the ceilings of these two caves, as innovative double-images embracing overlapping conceptualizations, based on both Buddhist and Chinese cosmological pantheons.

5 Sino-Indian Astral Deities on the Western Wall

The strategy of a syncretistic double image was possibly applied to the representations of the astral figures on the western wall of Cave 285 (figs. 10.3, 10.7, 10.8).

These figures are found in a blue-colored strip at the uppermost section of the western wall, which is positioned between the Chinese heavenly realm on the ceiling and the Indian divine pantheons of the earthy realm in the main

26 "速早成佛、救護弟子延命壽命、上等菩薩、下齊彭祖." This manuscript is kept in Kyou Shoya 杏雨書屋 (no. 656) in Osaka. See Inamoto "Tonkō dai 249, 285 kutsu", 366; Wang, *Dunhuang Fojiao*, 243–247.

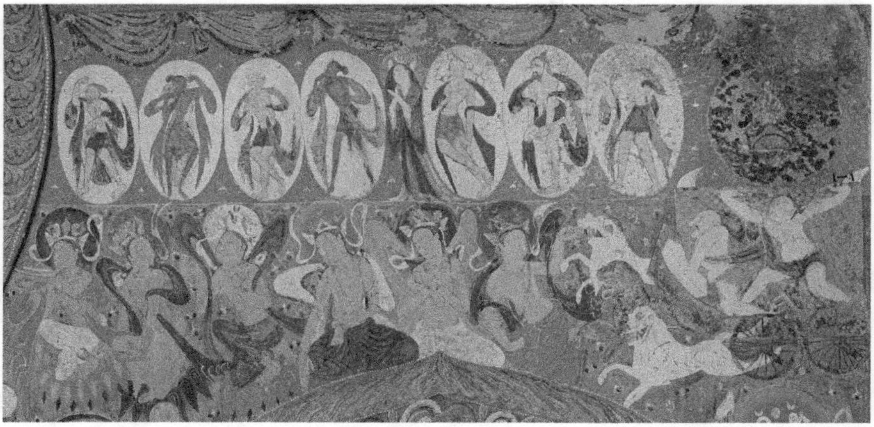

FIGURE 10.8 Candra with Saptarṣi and Arundhatī. Wall painting of the upper right section of the western wall of Mogao Cave 285, Dunhuang
© HERITAGE DIGITIZATION INSTITUTE, THE DUNHUANG ACADEMY 敦煌研究院文物数字化研究所

section of the western wall. This strip of imagery clearly represents the celestial sphere, as indicated by the blue background that contrasts with the orange of the main section of the wall and by the depiction of draped curtains, a motif for the celestial sphere that is attested in Central Asia.[27] This observation is further reinforced by the presence of Sūrya and Candra, the Indian sun god and moon god, at either end of the strip.[28] It is important to note that sets of the sun and moon are represented four times in this cave, including the disks held by Fuxi and Nüwa on the ceiling, two pairs held separately by Śiva and Viṣṇu on the western wall, and here again personified as Sūrya and Candra. This fact alone indicates that a coherent, systematic visualization of astronomical phenomena was not at all of interest to those who arranged the imagery of this cave.

In the row between Sūrya and Candra are found fourteen intriguing figures, all surrounded by white aureoles except for one female. Six Bodhisattva-like

27 See the wall paintings in the niche of the colossal Western Buddha of the Bamiyan Caves (Miyaji, *Nehan to Miroku*, 558). The German team has radiocarbon dated the organic materials from this Buddha statue to around 600 CE. See Miyaji, "Art-Historical Study on Bāmiyān," 139.

28 For the iconographical features of Sūrya and Candra in this cave see Hiyama, "Tonkō Bakkōkutsu," 15, 18–19. Their closest iconographical parallels can be found in the wall paintings of the Bamiyan Caves 111 (dated to around the middle of the sixth century) and 330 (dated to the late half of the seventh century to the first half of the eighth century). See Miyaji, "Parinirvāṇa Scenes," 28–29, figs. 23–25.

male figures with nimbuses behind their heads sit on the viewer's left (fig. 10.7), and seven Brahmins accompanied by one female figure stand on the right (fig. 10.8). While their placement secures their identification as personifications of astral bodies, there is a lack of comparable examples or iconographic prototypes, allowing for a wide range of interpretations.

The most widely accepted interpretation of these astral figures was proposed by He Shizhe, who regarded them as personifications of the Southern Dipper (Nandou liuxing 南斗六星) and the Northern Dipper (Beidou qixing 北斗七星).[29] The Beidou/Nandou pair is one of the most important constellation sets in China, believed in the Taoist tradition to be responsible for life and death from an early period to the present day.[30] Generally, the Southern Dipper, in charge of life and longevity, is visualized as a young male, while the Northern Dipper, in charge of death, is shown as an aged male—though visual evidence for these forms can be found only at a much later date.

The interpretation by He is supported by both the numbers and ages of the figures on each side, six younger figures on the left and seven aged figures on the right. Furthermore, the fact that only the former are represented with nimbuses can be explained by the stronger luminosity of the Southern Dipper as compared to the Northern Dipper, emphasized by the double-use of the nimbus and aureole for the six figures on the left. The largest problem of this interpretation, however, is that there is no iconographic convention in East Asia to depict the Southern Dipper as six young male figures or the Northern Dipper as seven Brahmins accompanied by a female figure.[31]

This weakness of He's interpretation was partly resolved by Ritsuko Sasaki with reference to the Indian interpretation of the Northern Dipper as the Saptarṣi (Seven Sages).[32] My recent paper further reinforced this identification by referring to the *Bṛhatsaṃhitā* of Varāhamihira.[33] The thirteenth chapter of the *Bṛhatsaṃhitā*, "Course of the Seven Sages," describes the circular movements of

29 He, "Dunhuang Mogaoku di 285 ku xibi," 380n13. Sasaki supported this interpretation ("Tonkō Bakkōkutsu," 126–127).

30 See Li Gang, "Wudoumai."

31 For the typology of the iconography of the Northern Dipper in East Asian Buddhist art, see Takeda, *Hoshi Mandala no Kenkyū*, 53–55; Ariga, "Hoshi Mandala to Myōken Bosatsu," 36–40. For the various forms of visual representation of the Northern Dipper in Taoist cults, see Mollier, *Buddhism and Taoism*, 134–173; Huang, *Picturing the True Form*, 38–52.

32 Sasaki, "Tonkō Bakkōkutsu," 127. The personification of the Northern Dipper as the seven sages can be traced back to the *Śatapatha Brāhmaṇa* (XIV. 5. 2. 5, 6; II. 1. 2. 4). Cf. Mani, *Purāṇic Encyclopaedia*, 56, 691, 835; Shastri, *India as Seen*, 168–169.

33 Hiyama, "Tonkō Bakkōkutsu," 15–17; "New Identification."

the Saptarṣi as if they are dancing under the instruction of the Pole Star.[34] This can provide a satisfactory explanation for the depiction of the seven Brahmins, who show various dynamic postures. The female standing between the fourth and fifth Brahmins (counting from the center of the wall), in a pose of devotion of the Brahmin in front of her, can be interpreted as the virtuous lady Arundhatī, the personification of Alcor and the devout wife of Vasiṣṭha, the greatest among the seven sages. Her lack of an aureole illustrates the weaker luminosity of Alcor in comparison to the other seven stars.[35] The fourth Brahmin from the left, to whom Arundhatī folds her hands, is thus most likely the great sage Vasiṣṭha, Arundhatī's husband. The *Bṛhatsaṃhitā*, however, contains neither descriptions on the Southern Dipper nor any reference associable with the six Bodhisattva-like figures.

As a solution for elucidating both these six astral figures and the western wall painting as a whole, I suggest reference to the *Sūryagarbha-parivarta* (*Rizang fen* 日藏分), one of sixty chapters of an extant Chinese recension of the *Mahāsaṃnipāta-sūtra* (*Dafangdeng daji jing* 大方等大集經, T 397).[36] The main composition of the western wall matches a spectacle vividly described in the seventh chapter of the *Sūryagarbha-parivarta*, which is called "The Chapter of the Buddha's Miraculous Performance" (Fo xian shentong pin 佛現神通品).[37] This begins with a scene in which Bodhisattvas worshipped and made donations to Śākyamuni Buddha. The Bodhisattvas then appeared in various forms, such as those of Pratyeka-Buddhas, Arhats, Brahma, Indra, the Four Heavenly Kings, Nārāyaṇa (Viṣṇu), Maheśvara (Śiva), astral deities, Asura, Cakravartin,

---

34 "I shall now expound, according to the view of the Sage Vṛddha (Senior) Garga, the movements of the Seven Seers (the Great Bear), through whom the northern region shines as through bedecked with a pearl necklace, like a maiden with a smiling countenance wearing a garland of white lotuses, by whom she appears to have a lord (or husband), **and by whose circular movements the northern quarter seems verily dancing at the instruction of the leader, viz., the Pole-Star.** [...] Among the Sages the revered Marīci is situated in the east; to his west is Vasiṣṭha, to his west is Aṅgiras; to his west is situated Atri; and close to him is Pulastya. Next to him are in order Pulaha and Kratu. **Arundhatī, the paragon of virtue, is close to the great Sage Vasiṣṭha among them.**" (Chap. XIII, 1–2; 4–6). Bhat, *Varāhamihira's Bṛhat Saṃhitā*, 161, 163.

35 For the localized elements as seen in the representation of Arundhatī see Hiyama, "Tonkō Bakkōkutsu," 17; "New Identification."

36 Hiyama, "Tonkō Bakkōkutsu," 17–19; "New Identification." The *Sūryagarbha-parivarta* was first translated into Chinese by Narendrayaśas in 585 and was incorporated into the *Mahāsaṃnipāta-sūtra* by Sengjiu 僧就 in 586; previously, the Sanskrit version of this text had existed as an independent text that was available in East-Turkestan. See Mak, "Transmission," 64–65.

37 T 397: 270a8–b1. A related passage can also be found in the *Baochuang fen* 寶幢分 (*Ratnaketu-parivarta*, T 397: 138b15–24) of the *Mahāsaṃnipāta-sūtra*.

and so forth, in temples in various locations. Since only those appearing as Arhats became subjects of worship by sentient beings, the other Bodhisattvas gave themselves the appearance of Arhats, too. The sentient beings then found the one emitting the greatest radiance, Śākyamuni, and they became delighted and fully satisfied.

This chapter enables an interpretation of the large Buddha statue set in the middle of the western wall, surrounded by a multilayered nimbus and mandorla, as Śākyamuni showing his great radiance, while the two statues of monks wearing patchwork robes and seated in meditation can be understood as the Arhats into which Bodhisattvas transformed themselves. The various deities surrounding them, including six-armed Śiva, Nārāyaṇa, Indra, Brahma, the Four Heavenly Kings, and the astral deities, can also be interpreted as the temporary forms of the Bodhisattvas.[38]

The "Chapter of Nakṣatras" (Xingsu pin 星宿品) of the *Sūryagarbha-parivarta*, in turn, refers to a star cluster that can be associated with the visual image of the six male figures: the Kṛttikā (Maosu 昴宿, the Pleiades). When the sage Kharoṣṭhī asked the heavenly beings whether they agreed to place the Kṛttikā as the first among all the *nakṣatras*, they answered: "We went through the lunar mansions and knew that the Kṛttikā is the nephew of the most exalted Mahātejas. They have six sons who move in the sky. This is the reason why the Kṛttikā can be placed as the first among them."[39] In another section of the same chapter, a similar conversation between the sage and the heavenly beings is repeated, where the sun god Sūrya explains the superior role of the Kṛttikā among the lunar mansions, steadily moving through the sky over the four continents and always doing good deeds and bringing benefits.[40]

These passages provide a good reason for the Kṛttikā to be placed right next to Sūrya in the wall painting. Because the star cluster Kṛttikā has strong luminosity, the visual emphasis on its brightness through overlapping nimbuses

---

38 It has to be noted, though, that not all the Indian deities surrounding the central Buddha niche are mentioned in this text, especially Gaṇeśa and Skanda Kārttikēya, who are seated below Śiva. In fact, this is the earliest evidence indicating the father-son relationship of Śiva and Gaṇeśa known to date, being earlier than its first textual description in the *Skanda-purāṇa*. See Törzsök, "Three Chapters," 22n20. I would like to express my deep gratitude to Prof. Yuko Yokochi for providing me with this information.

39 T 397: 276a21–23 (日藏分中星宿品第八之二): "我等經歷星宿。知昴最尊大威德天之外甥也。其有六子運行虛空。是故昴星可爲先首" (translated by the author). The author expresses her deep gratitude to the late Prof. Seishi Karashima (Soka University, Tokyo) for his advice on translating *Da-weide-tian* (大威德天) as Mahātejas, given in a personal correspondence by email (April 17, 2018).

40 T 397: 274c12–16 (日藏分中星宿品第八之一).

and aureoles can also be reasonably explained. As for the pairing of the Kṛttikā with the Saptarṣi, one reason might be provided in a reference to the heavenly sages (*tianxian* 天仙) who are said to have arranged all the heavenly bodies in the sky to protect the country and aid sentient beings, as described in the *Candragarbha-parivarta* (*Yuezang fen* 月藏分), another chapter of the *Mahāsaṃnipāta-sūtra* that is closely related to the *Sūryagarbha-parivarta*.[41] This would be an unparalleled visual representation of Kṛttikā based on the *Sūryagarbha-parivarta*, lacking any supporting comparative materials elsewhere—although this holds true for the representation of the Saptarṣi as well.

In previous papers, I was inclined to reject the Southern and Northern Dipper theory, not only because of the lack of comparable visual material in early medieval China but also because the pairing of the Southern Dipper with the Northern Dipper is unknown in the Indian astral tradition, on which the iconography of the Saptarṣi apparently relied; nor is any star cluster considered to be a counterpart of the Northern Dipper in India. Examining the syncretistic images that allow double-reading in the ceilings of Caves 249 and 285, however, I have since considered the possibility that these astral figures might also have borne overlapping meanings. Both groups of astral figures are unique and unparalleled. The Northern Dipper in this cave is its earliest-known personified representation, both in Indian and Chinese cultural spheres. Comparative visual representations in India can only be found in later Pāla art from the eleventh through twelfth centuries,[42] while the earliest textual reference that could mention a personified image of the Southern and Northern Dippers is a record in the *Xuanhe huapu* 宣和畫譜 (compiled in the twelfth century) about Zhuyao 朱繇, a late Tang era painter, who reportedly painted "the true form of the Southern and Northern Dippers."[43]

Considering the novelty of this pair of astral images, which does not seem to follow any fixed iconographic prototype, they could have been perceived as the Southern and Northern Dippers by those familiar with the Chinese Nandou/Beidou cult but as the Kṛttikā and the Saptarṣi by those who were instead familiar with the Indian tradition or knew the account of the *Sūryagarbha-parivarta*. In both cases, these astral deities could have been considered power-

41 This information is repeatedly mentioned in this text, for example in T 397: 373b25–c12. See Hiyama, "Tonkō Bakkōkutsu," 18.

42 See Bautze-Picron, "Unpublished or Little Known," pl. 4.19; Hiyama, "Tonkō Bakkōkutsu," 16–17, fig. 6; "New Identification."

43 Cf. Hayashi, *Myōken Bosatsu*, 30–34.

ful protectors of the Buddha, his disciples, and even the devout visitors of this cave, regardless of their religious affiliation. This way of reading enables us to consider Cave 285 as an extraordinary creative enterprise, providing through its decoration an overlapping cosmological system in which astral and cosmological ideas with completely different origins coexisted and were merged into one visual system without conflict.

6 Overlapping Cosmologies in Mogao Cave 285

The unique syncretistic visual space of Mogao Cave 285 bears witness to a dynamic Sino-Buddhist-Indian encounter of astral notions. It was triggered by highly political factors, such as the long-distance migration of the local ruler and possibly his intention to strengthen diplomatic relationships with the multicultural residents of the region through this large-scale cultural project—a strategic act for maintaining and strengthening the safety and stability of the multiethnic oasis town of Dunhuang in the middle of, and dependent on, the long-distance trade of the Silk Road. The visual syncretism must have been a smart strategy to involve powerful local donors with different religious backgrounds in the commission of this syncretistic worship place embedded in a Buddhist framework. Previous studies have often interpreted the wall paintings of this cave to be underpinned by the philosophy of Three-in-One, which considers Confucian, Taoist, and Buddhist systems to form a harmonious aggregate and each be explicable in terms of the other.[44] This argument may hold some truth, since the philosophy was fashionable in Luoyang at that time, but the rather jumbled arrangement of cosmological representations—with the sun and moon in different shapes appearing multiple times within the same cave—does not seem to reflect an intention to present a single coherent worldview in which different cosmological systems are incontrovertibly harmonized. The visual bricolage of the Sino-Buddhist-Indian heavens in this cave allowed for a unique aesthetic-cognitive experience, through which widely varied bodies of cosmological notions could have been constantly reinterpreted by each individual who visited this temple.

44 He, "Mogaoku di 285 ku kuding," 7–11; Li Zhengyu, "Xina xiaohua huabi weiwo."

Acknowledgments

I had opportunities to present the progress of this research in various international conferences and academic events from 2017 to 2020. My paper presented in the panel "Buddhist Cosmology and Astral Science" at the seventeenth Congress of the International Association of Buddhist Studies (IABS) in Toronto formed the basis of the present study. I would like to give a special thanks to Dr. Bill M. Mak for encouraging me to present a paper in his panel and generously helping to develop my research on this topic. Along with a great number of international colleagues who have kindly given helpful feedback for this research, I also wish to express my gratitude to the editors and reviewers of this volume for their critical and constructive suggestions, and to David Hendrix for improving my English text. This research was supported by JSPS KAKENHI No. 16J02828 "Study on the transregional context of the Indo-Iranian style paintings in the Buddhist rock monasteries alongside the Northern Silk Road Research Project."

Bibliography

Alram, Michael. *Das Antlitz des Fremden: Die Münzprägung der Hunnen und Westtürken in Zentralasien und Indien*. Vienna: Verlag der Österreichischen Akademie der Wissenschaften, 2016.

Ariga Takumi 有賀匠. "Hoshi Mandala to Myōken Bosatsu no zuzōgaku teki kenkyū" 星曼荼羅と妙見菩薩の図像学的研究. *Mikkyō Bunka* 密教文化 204 (2000): 25–63.

Bautze-Picron, Claudine. "Unpublished or Little Known Visnu Visvarupa Images from Eastern India." *Journal of Bengal Art* 25 (2020): 53–80.

Bhat, M. Ramakrishna. *Varāhamihira's Bṛhat Saṃhitā with English Translation, Exhaustive Notes and Literary Comments, Parts I–II*. Delhi: Motilal Banarsidass, 1981–1982.

De la Vaissière, Étienne. *Sogdian Traders*. Leiden: Brill, 2005.

Duan Wenjie 段文傑, *Dunhuang Shiku Yishu: Mogaoku di erbawu ku (Xi Wei)* 敦煌石窟藝術: 莫高窟第二八五窟（西魏）. Nanjing: Jiangsu Meishu Chubanshe, 1995.

Dunhuang Grottoes (*Chūgoku Sekkutsu: Tonkō Bakkōkutsu* 中国石窟：敦煌莫高窟), edited by Tonkō Bunbutsu Kenkyūjo 敦煌文物研究所. Vols. 1–5. Tokyo: Heibonsha, 1980–1982.

Enoki Kazuo 榎一雄. "Katsukoku ni kansuru Ryō-Shokko-zu no kiji ni tsuite" 滑国に關する梁職貢図の記事について. *Tohogaku* 東方学 27 (1964): 12–21.

Grenet, Franz. "Regional Interaction in Central Asia and Northwest India in the Kidarite and Hephthalite Periods." In *Indo-Iranian Languages and Peoples*, edited by Nicholas Sims-Williams, 203–224. Proceedings of the British Academy 116. Oxford: The British Academy and Oxford University Press, 2002.

Hayashi On 林温. *Myōken Bosatsu to Hoshi Mandara (Nihon no Bijutsu 377)* 妙見菩薩と星曼荼羅（日本の美術377）. Tokyo: Shibunsha, 1997.

He Shizhe 賀世哲. Dunhuang Mogaoku di 249 ku kuding xipo bihua neirong kaoshi 敦煌莫高窟第二四九窟窟頂西破壁画内容考釈. *Dunhuangxue Jikan* 敦煌学輯刊 3 (1982): 28–32.

He Shizhe 賀世哲. "Dunhuang Mogaoku di 285 ku xibi neirong kaoshi" 敦煌莫高窟第285窟西壁内容考释. In *Dunhuang Shiku yanjiu guoji taolunhui wenji: Shiku kaogu* 敦煌石窟研究国際討論会文集：石窟考古, 350–382. Shenyang: Liaoning Meishu Chubanshe, 1990.

He Shizhe 賀世哲. "Guanyu 285 ku zhi Baoyingsheng Pusa yu Baojixiang Pusa" 関于二八五窟之宝応声菩薩与宝吉祥菩薩. *Dunhuang Yanjiu* 敦煌研究 3 (1985): 37–40.

He Shizhe 賀世哲. "Mogaoku di 285 ku kuding tianxiangtu kaolun" 莫高窟第285窟窟頂天象圖考論. *Dunhuang Yanjiu* 敦煌研究 2 (1987): 1–13.

Hiyama Satomi 檜山智美. "New Identification of Astral Deities Represented on the Western Wall of Mogao Cave 285, Dunhuang." *Dunhuang and Cultural Contact along the Silk Road*, edited by Imre Galambos. Budapest: Hungarian Academy of Sciences, 2021 (forthcoming).

Hiyama Satomi 檜山智美. "Tonkō Bakkōkutsu dai 285 kutsu seiheki hekiga ni mirareru seishuku zuzō to sekkutsu zentai no kōsō ni tsuite" 敦煌莫高窟第二八五窟西壁壁画に見られる星宿図像と石窟全体の構想について. *Bukkyo Geijutsu* 仏教芸術 5 (2020): 11–29.

Hiyama Satomi. "Transmission of the "World": Sumeru Cosmology as Seen in Central Asian Buddhist Paintings Around 500 AD." *NTM Zeitschrift für Geschichte der Wissenschaften, Technik und Medizin* 28 (2020): 411–429.

Huang Shih-shan Susan. *Picturing the True Form: Daoist Visual Culture in Traditional China*. Cambridge: Harvard University Press, 2012.

Inamoto Yasuo 稲本泰生. "Tonkō dai 249, 285 kutsu ni okeru kamigami no zuzō no igi" 敦煌第二四九・二八五窟における神々の図像の意義. In *Bijutsushi Rekisan* 美術史歴参, 361–378. Tokyo: Chūō Koron Bijutsu Shuppan, 2013.

Ishimatsu Hinako 石松日奈子. "Tonkō Bakkōkutsu dai 285 kutsu hokuheki no kuyōshazō to kuyōsha daiki" 敦煌莫高窟第二八五窟北壁の供養者像と供養者題記. *Ryūkoku Shidan* 龍谷史壇 131 (2010): 43–87.

Jiang Boqin 姜伯勤. " 'Tian' de tuxiang yu jieshi—Yi Dunhuang Mogaoku di 285 ku kuding tuxiang wei zhongxin" 「天」的图像与解释—以敦煌莫高窟第二八五窟窟頂图像為中心. In *Dunhuang Yishu Zongjiao yu Lile Wenming: Dunhuang Xinshi Sanlun* 敦煌芸術宗教与礼梁文明：敦煌心史散論, 55–76. Beijing: Zhongguo Shehui Kexue Chubanshe, 1996.

Kuwayama Shoshin 桑山正進. *Kāpishī: Gandāra kenkyū* カーピシー: ガンダーラ史研究. Kyoto: Kyoto Daigaku Jinbun Kagaku Kenkyujo, 1990.

Li Gang 李剛. "Wudoumai, Beidou, Nandou." In *Daojiao yu xingdou xinyao* 道教與星

斗信仰, vol. 1, edited by Pan Chongxian 潘崇贤 and Liang Fa 梁发, 269–303. Jinan: Qilu Shushe, 2014.

Li Zhengyu 李正宇. "Xina xiaohua huabi weiwo: Tan Mogaoku Beiciao dongku 'Shenhua, Daojiao ticai' de shuxing" 吸納消化 化彼為我—談莫高窟北朝洞窟 "神話, 道教題材" 的属性. *Dunhuang Yanjiu* 敦煌研究 139 (2013): 47–50.

Lin Sheng-chih 林聖智. "Hokuchō jidai ni okeru sōgu no zuzō to kinō—Sekikanyuka ihei no boshushozō to kōshidenzu wo rei to shite" 北朝時代における葬具の図像と機能—石棺床囲屏の墓主肖像と孝子伝図を例として. *Bijutushi* 美術史 52, no. 2 (2003): 207–226.

Mak, Bill M. "The Transmission of Buddhist Astral Science from India to East Asia: The Central Asian Connection." *Historia Scientiarum* 24, no. 2 (2015): 59–75.

Mani, Vettam. *Purāṇic Encyclopaedia: A Comprehensive Work with Special Reference to the Epic and Puranic Literature*. Delhi: Motilal Banarsidass, 1964.

Miyaji Akira. "The Art-Historical Study on Bāmiyān and the Radiocarbon Dating." In *Recent Cultural Heritage Issues in Afghanistan vol. 2: Radiocarbon Dating of the Bamiyan Mural Paintings*, edited by National Research Institute for Cultural Properties, Japan et al., 131–143. Tokyo: Akashi Shoten, 2006.

Miyaji Akira 宮治昭. *Nehan to Miroku no zuzōgaku* 涅槃と弥勒の図像学. Tokyo: Yoshikawa Kobunkan, 1992.

Miyaji Akira 宮治昭, "Parinirvāṇa Scenes in Bāmiyān Murals: Transformation of Gandhāran Art in Central Asia." In *Mural Art: Studies on Paintings in Asia*, edited by Munier-Gaillard, Cristophe, 18–41. Bangkok: River Books, 2018.

Mollier, Christine, 2009, *Buddhism and Taoism Fact to Face: Scripture, Ritual, and Iconographic Exchange in Medieval China*. Honolulu: University of Hawai'i Press.

Neelis, Jason. *Early Buddhist Transmission and Trade Networks: Mobility and Exchange within and beyond the Northwestern Borderlands of South Asia*. Leiden: Brill, 2011.

Ning Qiang 寧強. "Shangshi dengxian tu yu Weimojie jingbian—Mogaoku di 249 ku Kuding bihua Zaitan" 上士登仙図与維摩詰経変 莫高窟第二四九窟窟頂壁画再探. *Dunhuang Yanjiu* 敦煌研究 1 (1990): 30–37.

Rong Xinjiang. *Eighteen Lectures on Dunhuang*. Leiden: Brill, 2013.

Saito Rieko 齋藤理恵子. "Tonkō dai 249 kutsu tenjō ni okeru Chūgoku teki zuzō no juyō keitai" 敦煌第249窟天井における中国的図像の受容形態. *Ars Buddhica* 佛教藝術 218 (1995): 39–56.

Sasaki Ritsuko 佐々木律子. "Tonkō Bakkōkutsu dai 285 kutsu seiheki naiyō kaishaku shiron" 敦煌莫高窟第二八五窟西壁内容解釈試論. *Ars Buddhica* 佛教藝術 142 (1997): 121–138.

Shastri, Ajay Mitra. *India as Seen in the Bṛhatsaṃhitā of Varāhamihira*. Delhi: Motilal Banarsidass, 1969.

Su Bai 宿白. "Canguan Dunhuang Mogaoku di 285 haoku zhaji" 参観敦煌莫高窟第285號窟札記. In *Wenwu Cankao Ziliao* 文物参孜資料 2 (1956), 206–213. Beijing: Cultural Relics Publishing House, 1996.

Su Bai 宿白. "Dongyangwang yu Jianpinggong (ergao)" 東陽王與建平公（二稿）. *Dunhuang Tulufan Wenxian Yanjiu Lunji* 煌吐魯番文献研究論集 4 (1987), 244–259. Beijing: Cultural Relics Publishing House, 1996.

Sudō Hirotoshi 須藤弘敏. "Zenjō biku zuzō to Tonkō dai 285 kutsu" 禅定比丘図像と敦煌第285窟. *Ars Buddhica* 佛教藝術 183 (1989): 11–27.

Tabayashi Kei 田林啓. "Tonkō Bakkōkutsu dai 285 kutsu kaisaku no haikei ni kanshite" 敦煌莫高窟第二八五窟開鑿の背景に関して. In *Bijutsushi Rekisan* 美術史歴参, 501–518. Tokyo: Chūō Kōron Bijutsu Shuppan, 2013.

Tabayashi Kei 田林啓. "Tonkō Bakkōkutsu dai 285 kutsu no Bukkyō sekai ni tsuite—Tenjō hekiga wo chushin to shite" 敦煌莫高窟第二八五窟の仏教世界について—天井壁画を中心として. *Ars Buddhica* 佛教藝術 170 (2011): 229–245.

Taguchi Eiichi 田口榮一. "Hekiga ni tsuite" 壁画について. In *Tonkō sekkutsu gakujutsu chōsa (dai 1 ji) hōkokusyo* 敦煌石窟学術調査 (第1次) 報告書, 29–57. Tokyo: Tokyo Geijutsu Daigaku Bijutsu Gakubu, 1985.

Takeda Kazuaki 武田和昭. *Hoshi Mandala no kenkyū* 星曼荼羅の研究. Kyoto: Hozokan, 1995.

Törzsök, Judit. "Three Chapters of Śaiva Material Added to the Earliest Known Recension of the Skandapurāṇa." In *Origin and Growth of the Purāṇic Text Corpus, with Special Reference to the Skandapurāṇa*, edited by Hans T. Bakker. *Papers of the 12th World Sanskrit Conference*, vol. 3.2, 17–39. Delhi: Motilal Banarsidass, 2004.

Vondrovec, Klaus. *Coinage of the Iranian Huns and their Successors from Bactria to Gandhara (4th to 8th century* CE), vols. 1–2. Vienna: Verlag der österreichiscnen Akademie der Wissenschaften, 2014.

Waldschmidt, Ernst. "Über den Stil der Wandgemälde." In *Die buddhistische Spätantike in Mittelasien* VII, *Neue Bildwerke* 3, edited by Albert von Le Coq and Ernst Waldschmidt, 24–31. Berlin: Reimer and Vohsen, 1933.

Wang Huimin 王惠民. *Dunhuang Fojiao yu shiku yingjian* 敦煌佛教与石窟営建. Lanzhou: Gansu Jiaoyu Chubanshe, 2017.

Wen Mengxia 文夢霞. "Zailun Dongyangwang Yuan Rong ling Guazhou cishi de shijian" 再論東陽王元榮領瓜州刺史的時間. *Dunhuang Yanjiu* 敦煌研究 2 (2006): 101–105.

Xiang Da 向達. "Mogao, Yulin erku zakao" 莫高, 楡林二窟雑考. *Wenwu Cankao Ziliao* 文物参考資料 2, no. 5 (1951): 76–95.

Yagi Haruo 八木春生. *Chūgoku Bukkyō bijutsu to Kanminzoku-ka* 中国仏教美術と漢民族化. Kyoto: Hōzōkan, 2004.

Yamabe Nobuyoshi 山部能宣. "'Zenjō-kutsu' saiko—Indo, Chūō-Ajia kara Tonkō ni itaru '*vihāra*' kutsu no tenkai" 「禅定窟」再考—インド・中央アジアから敦煌にいたる「ヴィハーラ」窟の展開. In *Asia Bukkyō Bijutsu Ronshu, Chūō Asia* I, *Gandhāra—Tōzai Turkestan* アジア仏教美術論集 中央アジアI ガンダーラ〜東西トルキスタン, edited by Miyaji Akira, 473–498. Tokyo: Chūō Kōron Bijutsu Shuppan, 2017.

Yin Guangming 殷光明. “Tonkō Seishinbo shutsudo no bokusho daiki gazō tenbō wo meguru kōsatsu” 敦煌西晋墓出土の墨書題記画像磚墓をめぐる考察. Translated by Kitamura Haruka 北村永. *Ars Buddhica* 佛教藝術 285 (2006): 45–72.

Yoshida Yutaka 吉田豊. “Sogudojin to Sogudo no rekishi” ソグド人とソグドの歴史. In *Sogudojin no bijutsu to gengo* ソグド人の美術と言語, edited by Hiroshi Sofugawa and Yutaka Yoshida, 7–78. Kyoto: Rinsen Shoten, 2011.

Yoshida Yutaka 吉田豊. “Some Reflections about the Origin of Čamuk.” In *Chūō Asia shutsudo bunbutsu ronsō* 中央アジア出土文物論叢, edited by Takao Moriyasu, 129–135. Kyoto: Hōyu shoten 2004.

Zhang Fan. “Chinese-Buddhist Encounter: Synthesis of Fuxi-Nüwa and *Cintamani* in Early Medieval Chinese Art.” *Asian Studies* 7 (23), no. 2 (2019): 87–111.

Zhang Jianyu 張建宇. “Dunhuang Xiwei huafeng xinquan—yi Mogaoku di 285 ku gongjiang ji fenben wenti wei hexin” 敦煌西魏画風新詮－以莫高窟第285窟工匠及粉本問題為核心. *Dunhuang Yanjiu* 敦煌研究 150 (2015): 4–14.

Zhao Wanli 趙万里. *Han Wei Nanbeichao muzhi jishi* 漢魏南北朝墓誌集释. Beijing: Science Press, 1956.

Zhao Wanli 趙万里. “Wei zongshi Dongyangwang Rong yu Dunhuang xiejing” 魏宗室東陽王榮与敦煌写経. *Zhongde Zazhi* 中徳雑誌 5, no. 3 (1943): 7–12.

Index

www.ingramcontent.com/pod-product-compliance
Lightning Source LLC
LaVergne TN
LVHW012053011026
849162LV00018B/227
* 9 7 8 9 0 0 4 5 1 1 4 1 5 *